Frontiers in Civil Engineering

(Volume 2)

(Water Savings in Buildings)

Editor

Enedir Ghisi
Department of Civil Engineering, Federal University of Santa Catarina, Florianópolis, Brazil

Frontiers in Civil Engineering

Volume # 2

Water Savings in Buildings

Editor: Enedir Ghisi

eISSN (Online): 2468-4708

ISSN (Print): 2468-4694

eISBN (Online): 978-1-68108-483-1

ISBN (Print): 978-1-68108-484-8

Bentham Science Publishers Ltd.
Executive Suite Y - 2
PO Box 7917, Saif Zone
Sharjah, U.A.E.
Email: subscriptions@benthamscience.org

BENTHAM
SCIENCE

CONTENTS

PREFACE

Water savings in buildings has been a matter of concern all over the world. Alternative water sources as well as water saving appliances have been studied by many researchers in order to try to promote water savings in buildings. Rainwater tank sizing and modelling, wastewater treatment and reuse, relationship between user behaviour and water savings, health issues related to water savings and environmental analysis of rainwater and grey water use in buildings are subjects related to water savings in buildings. Thus, the objective of the eBook is to put together some of these aspects by means of seven chapters written by renowned researchers.

Chapter 1, written by Andrea Teston, Barbara Müller Colasio and Enedir Ghisi, of the Federal University of Santa Catarina, Brazil, presents the state of the art on water savings in buildings in Brazil. The authors conclude that there is a high potential for potable water savings in buildings by using rainwater for non-potable purposes in Brazil.

Chapter 2, written by M. Ashiqur Rahman, Md Mahmudul Haque, Amir Ahmed and Ataur Rahman, of the Western Sydney University, Australia, focus on rainwater harvesting to reduce potable water demand in buildings in Australia.

Ilaria Gnecco and Anna Palla (of the University of Genova), and Alberto Campisano and Carlo Modica (of the University of Catania), Italy, wrote Chapter 3. It assesses methodologies for designing rainwater harvesting systems. The impact of European precipitation regimes on the management of rainwater tanks as well as on the influence of water demands on the performance of the system was also analysed.

Chapter 4 was written by Ghazaleh Vaseghi, Ilke Celik and Defne Apul (of the University of Toledo) and Steven Burian (of University of Utah), USA. It contains an approach for using multi-criteria decision analysis to study the tradeoffs of rainwater harvesting system designs.

Chapter 5, written by Asher Kiperstok and Alice Costa Kiperstok, of the Federal University of Bahia, Brazil, discusses the implementation of water saving programmes in institutional and university buildings over seventeen years in Bahia.

Chapter 6 was written by Cristina Matos Silva and Vitor Sousa, of the University of Lisbon, and Inês Meireles, of the University of Aveiro, Portugal. The performance of rainwater harvesting systems in residential buildings, a shopping centre and a university building in Portugal, covering different water use patterns and geographical locations, is shown in this chapter.

And last but not least, Cristina Santos (of the University of Porto), Cristina Matos (of the University of Beira Interior and University of Trás-os-Montes and Alto Douro) and Armando Silva-Afonso (of the University of Aveiro), Portugal, are the authors of Chapter 7. This chapter assesses health issues related to the application of water saving systems in buildings.

It is possible to observe that most of the locations presented in this eBook demonstrate that there is a great potential for potable water saving through the use of rainwater in buildings. Rainwater harvesting can increase water availability and, as a consequence, improve the population's life quality even in semi-arid regions, such as the Brazilian Northeast. However, the rainwater harvesting systems should be used in combination with other forms of water conservation for efficient water management in cities, such as water saving appliances and greywater reuse.

The more developed the country, the greater the water consumption in the buildings. For this reason, European standards recommend the conservation of water. Potable water savings is an ally for rainwater harvesting systems due to the retention time which is inversely proportional to the water quality in the reservoir. Asher Kiperstok and Alice Costa Kiperstok pointed out that waste of water is constantly driven by weak maintenance schemes and due to the fact that water is free to users of public facilities.

In Australia, according to M. Ashiqur Rahman, Md Mahmudul Haque, Amir Ahmed and Ataur Rahman, favourable regulations and public awareness have made use of rainwater as a significant alternative source of water in buildings. However, there is still room for improvement in systems by designing the project for greater efficiency. In Brazil, for example, there is still the need for more investments in national support programmes for water savings in buildings. But one of the most important programmes in the country is the programme called one million cisterns. This programme aims to provide a rainwater harvesting system for families living in areas where water is scarce or where the water utilities do not supply water.

As for methodologies for designing rainwater harvesting systems, Ilaria Gnecco, Anna Palla, Alberto Campisano and Carlo Modica explain the critical correlation between the rainfall event characteristic, antecedent dry weather period, and the system performance: the water saving efficiency is in inverse proportion while the water quality degradation (expressed as detention time) is directly proportional to the antecedent dry weather period.

The Utah case study presented by Ghazaleh Vaseghi, Ilke Celik, Defne Apul and Steven Burian in Chapter 4 shows three scenarios of reservoir positioning: (1) two reservoirs, one on the ground floor and one on the roof; (2) a reservoir on the ground floor; and (3) a reservoir on the roof. The conclusion suggests that using only one reservoir for rainwater collection and distribution is the most viable option, although they have not evaluated the impact of the need for greater structural support to withstand reservoir weight.

One of the problems that is still considered as an obstacle to the efficient use of water is health issues. That is why it is important to clearly state how to install a system and keep it safe and feasible, explaining what treatments are needed for rainwater and greywater to prevent human contamination.

Although the obstacles still met for the use of rainwater harvesting systems, such as population awareness, water waste and issues related to water quality, Cristina Matos Silva, Vitor Sousa and Inês Meireles have shown that rainwater is a relevant alternative for water conservation in Mediterranean countries, where water savings can reach up to 60% in most of the buildings studied.

I think this eBook will be a valuable resource to researchers, post-graduate students, undergraduate students, engineers, water utilities, environment agencies, industries, and many others interested in water savings in buildings.

I would like to thank all the authors who have contributed to this ebook, and the editorial team for their valuable work and completion of this eBook.

The views and opinions expressed in each chapter of this eBook are those of the authors.

Prof. Enedir Ghisi
Department of Civil Engineering,
Federal University of Santa Catarina,
Florianópolis,
Brazil.

List of Contributors

Alberto Campisano	Department of Civil Engineering and Architecture, University of Catania, Catania, Italy
Alice Costa Kiperstok	Teclim, Clean Technology Network, Department of Environmental Engineering, Polytechnics School, Federal University of Bahia, Salvador, State of Bahia, Brazil
Amir Ahmed	School of Computing, Engineering and Mathematics, Western Sydney University, Sydney, NSW 2751, Australia
Andrea Teston	Department of Civil Engineering, Laboratory of Energy Efficiency in Buildings, Federal University of Santa Catarina, Florianópolis, SC, 88040-900, Brazil
Anna Palla	Department of Civil, Chemical and Environmental Engineering, University of Genova, Genova, Italy
Armando Silva-Afonso	Department of Civil Engineering, University of Aveiro; the Board of ANQIP (National Association for Quality in Building Services), Aveiro, Portugal
Asher Kiperstok	Teclim, Clean Technology Network, Department of Environmental Engineering, Polytechnics School, Federal University of Bahia, Salvador, State of Bahia, Brazil
Ataur Rahman	School of Computing, Engineering and Mathematics, Western Sydney University, Building XB2.48, Locked Bag 1797, Penrith, NSW 2751, Australia
Barbara Müller Colasio	Department of Civil Engineering, Laboratory of Energy Efficiency in Buildings, Federal University of Santa Catarina, Florianópolis, SC, 88040-900, Brazil
Carlo Modica	Department of Civil Engineering and Architecture, University of Catania, Catania, Italy
Cristina Matos	Departamento de Engenharias, Escola de Ciências e Tecnologia, Universidade de Trás-os-Montes e Alto Douro, C-MADE- Centre of Materials and Building Technologies; University of Beira Interior, Vila Real, Portugal
Cristina Matos Silva	CERIS, Department of Civil Engineering, Architecture and Georesources, Instituto Superior Técnico, Universidade de Lisboa, Lisbon, Portugal
Cristina Santos	Department of Civil Engineering, Faculdade de Engenharia da Universidade do Porto, Porto, Portugal
Defne Apul	Civil Engineering Department, University of Toledo, Ohio, USA
Enedir Ghisi	Department of Civil Engineering, Laboratory of Energy Efficiency in Buildings, Federal University of Santa Catarina, Florianópolis, SC, 88040-900, Brazil
Ghazaleh Vaseghi	Chemical Engineering Department, University of Toledo, Ohio, USA
Ilaria Gnecco	Department of Civil, Chemical and Environmental Engineering, University of Genova, Genova, Italy
Ilke Celik	Civil Engineering Department, University of Toledo, Ohio, USA
Inês Meireles	RISCO, Department of Civil Engineering, Universidade de Aveiro, Aveiro, Portugal
M. Ashiqur Rahman	School of Computing, Engineering and Mathematics, Western Sydney University, Sydney, NSW 2751, Australia

Md Mahmudul Haque School of Computing, Engineering and Mathematics, Western Sydney University, Sydney, NSW 2751, Australia

Steven Burian Civil Engineering Department, University of Utah, Utah, USA

Vitor Sousa CERIS, Department of Civil Engineering, Architecture and Georesources, Instituto Superior Técnico, Universidade de Lisboa, Lisbon, Portugal

CHAPTER 1

State of the Art on Water Savings in Buildings in Brazil: A Literature Review

Andrea Teston, Barbara Müller Colasio and **Enedir Ghisi**[*]

Department of Civil Engineering, Laboratory of Energy Efficiency in Buildings, Federal University of Santa Catarina, Florianópolis, SC, 88040-900, Brazil

Abstract: Water savings in buildings has become a matter of concern due to water demand increase in cities and to problems with water quality and availability. This chapter presents the state of the art on water savings in buildings in Brazil. Journal and conference papers, reports and theses published from 2003 to 2016 were reviewed. First, in order to show the main shortage problems faced by Brazilian people, the national panorama of water resources is shown. Then, papers about the use of rainwater, water quality, potential for potable water savings, investment feasibility analysis, greywater reuse and water-efficient appliances are also addressed. Water end-uses in buildings, environmental education, and national water savings support programmes are also presented. It was observed that there is a high potential for potable water savings in buildings by using rainwater for non-potable purposes in Brazil.

Keywords: Alternative water sources, Brazil, Environmental education, Greywater, National programmes, Payback, Rainwater, Sizing, Social acceptability, Water end-uses, Water-efficient appliances, Water savings, Water quality.

INTRODUCTION

In Brazil, the National Water Agency (ANA – *Agência Nacional de Águas*) conducts studies and plans that aim to support and guide the implementation of national policies and management on water resources.

In addition to being the country with the most voluminous river in the world (Amazonas), Brazil still shares the largest underground water source in the world with three countries (Uruguay, Paraguay, Argentina): the Guaraní Aquifer, which has an area of 1.2 km^2. The country is among the nine who owns 60.0% of

[*] **Corresponding author Enedir Ghisi:** Department of Civil Engineering, Laboratory of Energy Efficiency in Buildings, Federal University of Santa Catarina, Florianópolis, SC 88040-900, Brazil; Tel: +55 48 3721 2115; E-mail: enedirghisi@gmail.com

renewable freshwater sources in the world. Considering only the American continent, Brazil has 34.9% of surface freshwater (Fig. **1**) [1].

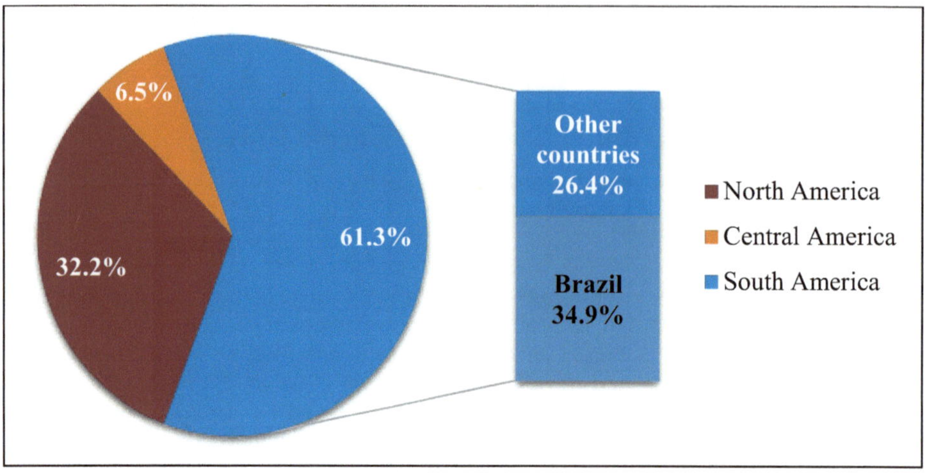

Fig. (1). Distribution of surface freshwater in the Americas [1].

However, although it has one of the largest water reservoirs in the world, Brazil faces shortages. Geographical and historical features justified this contradiction. The Brazilian urban expansion happened quickly, disorderly and recently; today 84.0% of Brazilians live in cities and most of them are located in regions where the supply of water is poor [2]. While the river basin district of greater scarcity shows availability of less than 100 m^3/s, in the Amazon River basin the availability reaches the flow of 74,000 m^3/s [1]. That is, the Amazon region, which has 69.0% of the water resources in Brazil, contains only 8.0% of the population, while the Northeast has 28.0% of the population and only 3.0% of the water resources [3].

In the Brazilian north-eastern semi-arid, water availability is an obstacle for the social and economic development of the rural population. The water availability is characterized by dry and rainy periods, and some regions have an annual rainfall of less than 200 mm. Most of the population live in this region collecting water from alternative sources (generally lakes and ponds) because they have no access to public supply. These sources present a serious risk to the health of the population, contributing to the maintenance of endemic cycles, with high infant mortality [4].

According to the historical series, the largest amount of dry records happens in the Northeast, followed by the southern region, where the most affected states are Bahia and Rio Grande do Sul. Fig. (**2**) shows a map of Brazil with the municipalities that have enacted emergency due to drought in 2014.

Fig. (2). Brazilian municipalities with dry critical events in 2014 [5].

The river system also suffers variations throughout the year due to rainfall [5]. So that, it is necessary to create water reservoirs able to supply the demand during dry periods. Also, the water supply often becomes impaired because of the climate change and extreme events. The year 2014 was characterized by both: excess and lack of rainfall. The southeast region suffered due to shortage, while the south and part of the north and Midwest regions suffered due to high rainfall. Fig. (**3**) shows the extreme events that happened in the country in 2014 and their return periods [5].

The lack of rainfall in the southeast should be highlighted on the map shown in Fig. (**3**), whose drought return periods were greater than 100 years in several places. This becomes even more problematic because it is the most populous region of the country, owning 42.1% of the population [6]. It was found that the level of rainfall was among the three worst ever recorded in 25.0% of weather stations with more than 50 years of data. Only in the state of São Paulo, 30.0% of the stations recorded the driest event since they have data about rainfall [7].

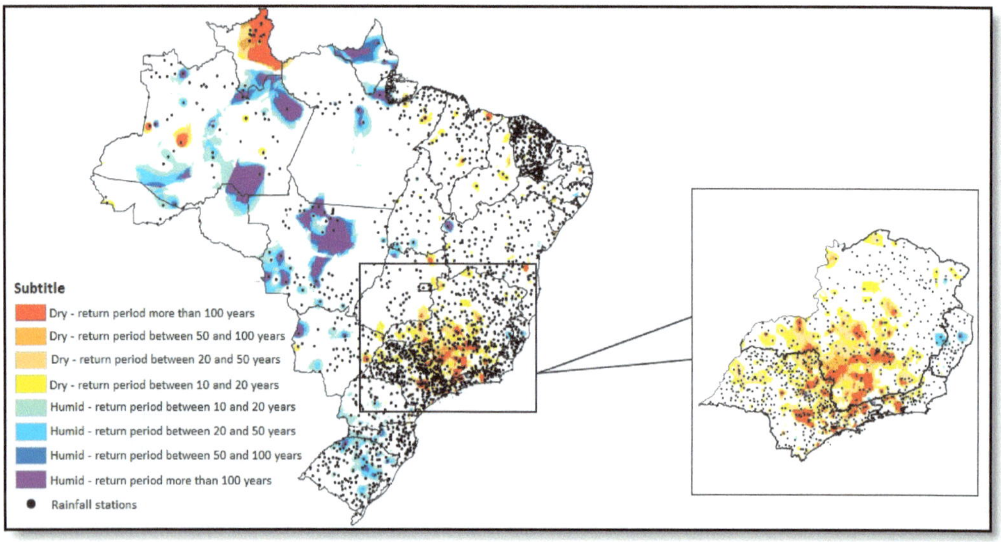

Fig. (3). Return periods of the annual rainfall in Brazil for the hydrological year 2014, especially the southeast [5].

Due to the drought, which began in 2013 and was intensified in 2014, the *Cantareira* system that supplies about 9 million people in São Paulo, was impaired. From May 2014 it was necessary to use the technical reserve or dead volume, which is the volume of water stored below the reservoir levels, requiring pumping to use. This volume corresponds to 486 billion litres of water (33.0% of total storage capacity). The use of this dead volume occurred in two stages, and after the second stage, there were only 77 billion litres of water left [7].

Another problem is the water quality of rivers that supply the cities. ANA made a compilation of water quality parameters collected by the state water systems in 2013, and presented the Water Quality Index (IQA – *Índice de Qualidade da Água*) to 1,683 of 2,800 points monitored in rivers [5]. The results showed that 7.0% of points were classified as bad or very bad and considering only the points located in urban areas this rate increases to 21.0% (Fig. **4**).

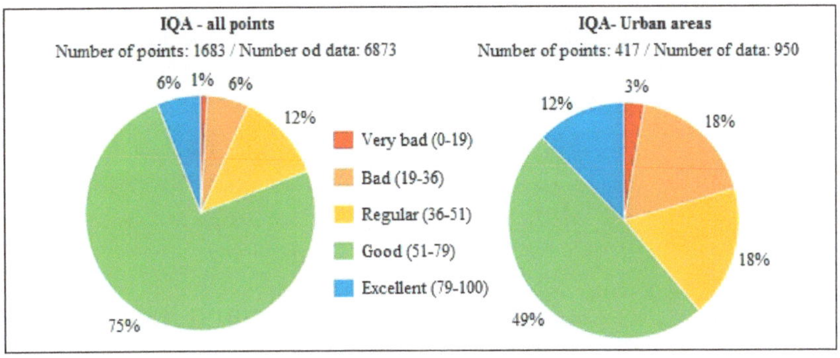

Fig. (4). Water Quality Index of rivers in Brazil in 2013 [5].

This shows that the quality of water in the cities is worrying. Several factors are responsible for this urban rivers degradation, from the uncontrolled growth and increasing soil sealing to the lack of basic sanitation. One of the greatest problems is the sewage, especially in areas of high density and where there are rivers with a low assimilative capacity of pollutant loads. Thus, improving the sanitary sewer system should be paramount to the urban centers to ensure health to the population. Fig. (**5**) shows the distribution of municipalities with and without public sewage collection and treatment systems.

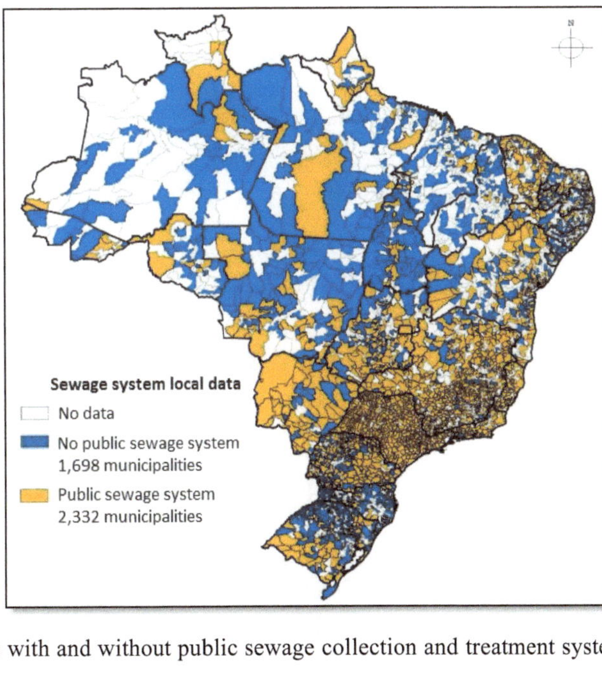

Fig. (5). Municipalities with and without public sewage collection and treatment systems in Brazil in 2014 [8].

Moreover, there is still a lot of wastewater in cities, because the availability of water in Brazil is still satisfactory, excluding extreme events. This wastage starts with the urban water supply, where the average water loss was 36.7% of the collected water to supply in 2014 [8]. Also, there is much waste of water in the end-uses. According to the Rational Use of Water Programme at the University of São Paulo, 1/3 of all drinking water consumed in Brazilian residential buildings could be saved by using an alternative source of water, such as the use of rainwater, for example [9].

It is necessary to know the water consumption pattern in order to improve the water management in buildings and thus improving supply through alternative sources and reducing demand. For that, several studies have been conducted about the water end-uses in Brazil, with the goal to show in which activities more or less water is consumed.

OBJECTIVE

The objective of this chapter is to present the state of the art on water savings in buildings in Brazil.

WATER END-USES IN BUILDINGS

For the implementation of a Water Conservation Programme it is necessary to know the distribution of water consumption that varies according to the type of building, the specificity of hydraulic systems and the users [10]. Similarly, to determine an alternative system of water supply it is important to know the demand and the end-uses, so that the system can be correctly sized and meet the needs of users.

Approximately 8.0% of the total water flow is consumed for urban human supply, which represents about 100 m^3/s [5]. Residential consumption (single-family and multi-family buildings) makes up more than half of the total water consumption in urban areas [11]. But, each building typology shows a different pattern of water end-uses. Thus, the water end-uses in buildings in Brazil (residential, public and commercial) are presented below.

Water End-Uses in Residential Buildings

Water end-uses were assessed for multi-family residential building composed of three blocks in Florianópolis, Santa Catarina [12]; a multi-family building in São José, Santa Catarina [13]; a single-family house in Panambi, Rio Grande do Sul [14]; six houses in Blumenau, Santa Catarina [15]; households located in the city of São Paulo, São Paulo [16]; low-income households in Florianópolis, Santa

Catarina [17]; 48 households in Florianópolis, Santa Catarina [18]; a single-family house in São José, Santa Catarina [19]; another multi-family residential building composed of three blocks in Florianópolis, Santa Catarina [20]. The results are shown in Table **1**.

Table 1. Water end-uses in residential buildings in Brazil.

End-use	Water End-uses (%)								
	[12]	[13]	[14]	[15]	[16]	[17]	[18]	[19]	[20]
Showers	22.6	46.5	32.3	10.8	13.9	33.3	32.7	13.0	21.4
Toilet flushing	33.2	18.3	20.8	18.5	5.5	19.3	19.4	55.0	9.1
Kitchen tap	20.7	13.6	31.6	12.6	12.0	17.0	18.0	4.0	22.8
Lavatories	16.0	3.2	5.4	1.7	4.2	4.7	5.2	2.0	31.5
Cleaning	2.9	1.9	-	-	-	-	-	-	-
Washing machine	4.7	-	5.7	23.5	10.9	12.7	10.7	12.0	13.7
Others	-	0.1	3.3	32.1	30.6	8.3	7.9	8.0	-
Laundry tub	-	-	0.9	1.4	14.6	4.3	6.1	6.0	1.4
Laundry tub with washing machine	-	16.4	-	-	8.3	-	-	-	-

"Others" in Table **1** include: garden irrigation [13 - 15, 18, 19]; garage tap and laundry tap [19]; car washing [17 - 19]; and sidewalk washing [19]. Barreto [16] did not state what is included in "others".

The variation of some water end-uses is significant, *i.e.*, it ranges from 5.5% [16] to 55.0% [19] for toilet flushing. Another example is the kitchen tap, which ranged from 4.0% [19] to 31.6% [14]. In this case, water consumption is related to the lifestyle of the residents, where there is a considerable difference between the families that prepare their meals at home and the families that have meals at restaurants [17].

Based on the data shown in Table **1**, the average water end-uses were estimated, as shown in Fig. (**6**), which also shows the maximum and minimum values for each water end-use. Washing machine, laundry tub with washing machine, and laundry tub, were joined together to avoid repetition.

Thus, it can be observed that on average showers are responsible for the largest water consumption in households (25.2%), followed by toilet flushing (22.1%), and washing clothes (17.0%). Considering only the largest end-uses, 39.1% of the water is used for non-potable purposes (toilet flushing and washing clothes). Fig. (**7**) shows the potable and non-potable water end-uses based on the given references.

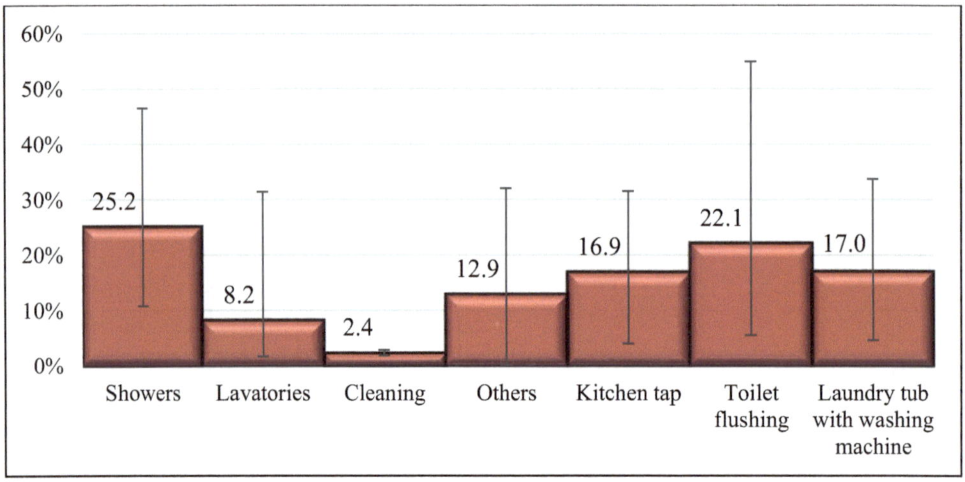

Fig. (6). Average, maximum and minimum water end-uses in residential buildings in Brazil.

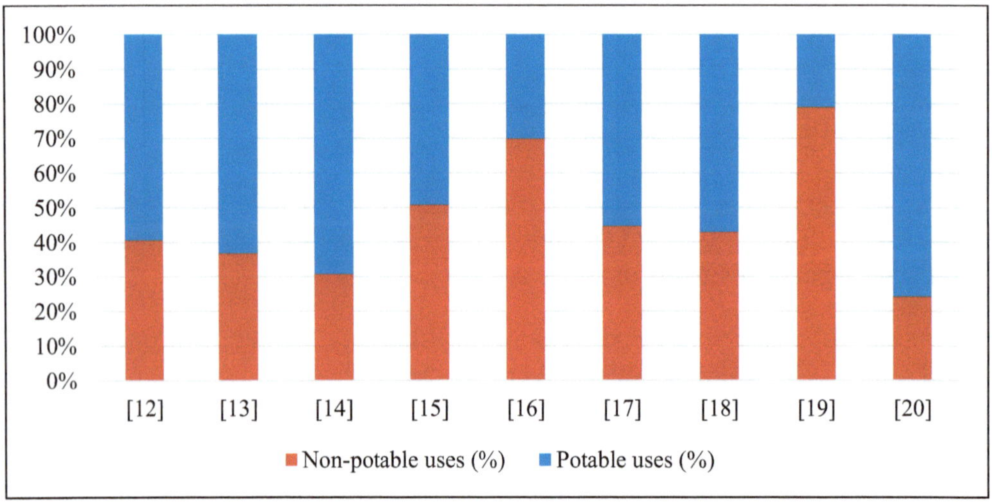

Fig. (7). Comparison between potable and non-potable water end-uses in residential buildings in Brazil.

Thus, the potential for potable water savings by using alternative water sources varies from 24.0% [20] to 79.0% [19], considering toilet flushing, washing clothes, and "other" uses when they are non-potable.

Nagali *et al.* [21] studied the influence of children, pets, housekeepers, and the genre of the residents on the water consumption in a residential condominium in Curitiba, Paraná. They obtained that the family composition influences the water consumption, and households with pets had water consumption on average 23.0% higher than those without pets. Children consume about 22.0% less water than adults. Single women or couples consume up to 25.0% more water than single men. And domestic workers do not significantly affect the water consumption (1.0% to 3.0%).

Water End-Uses in Public Buildings

The papers related to water end-uses in buildings of public category refer to ten public buildings [22], a public university [23], a Technology Center in Automation and Information [24] and two schools [25] all located in Florianópolis, Santa Catarina. The results are shown in Table **2**.

Table 2. Water end-uses in public buildings in Brazil.

End-use	Water End-uses (%)			
	[22]	[23]	[24]	[25]
Toilet flushing	44.5	62.3	45.0	24.7
Urinal	29.2	9.0	11.1	19.0
Bathroom tap	14.2	25.3	18.7	3.5
Kitchen	17.9	-	15.2	46.5
Washing cars	2.0	-	2.0	-
Cooling tower	22.7	-	-	-
Others	6.5	-	6.7	5.4
Drinking fountain	-	3.7	1.2	1.0

"Others" in Table **2** include: cleaning [22, 24, 25]; garden irrigation [22, 24]; laboratories tanks and showers [24]. Also, Botelho [23] did not consider the existing restaurant at the university to estimate the water end-uses and made an estimate of the consumption of the university taking the survey conducted in only two classroom blocks as a reference. Another important fact is that among the two schools evaluated by Fasola *et al.* [25], meals are cooked to students in only one of them.

Fig. (**8**) shows the average, maximum and minimum water end-uses in the public buildings considering the data shown in Table **2**.

The highest water consumption in public buildings, as shown in Fig. (**8**), is due to toilet flushing (44.1%); followed by kitchen (26.5%) and cooling towers for air-conditioning (22.7%). The urinal is the fourth largest water consumption. Fig. (**9**) shows a comparison between potable and non-potable end-uses in public buildings, whereas the percentages related to toilet flushing, urinal, washing cars and "others" were considered as non-potable uses.

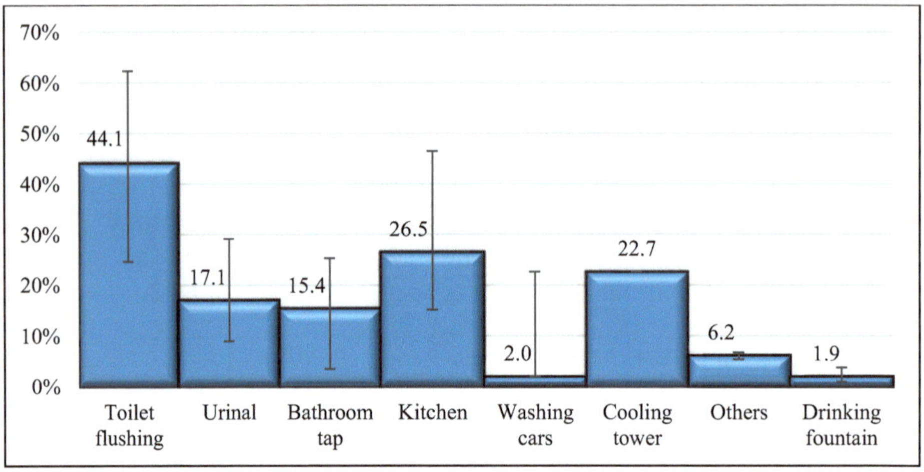

Fig. (8). Average, maximum and minimum water end-uses in public buildings in Brazil.

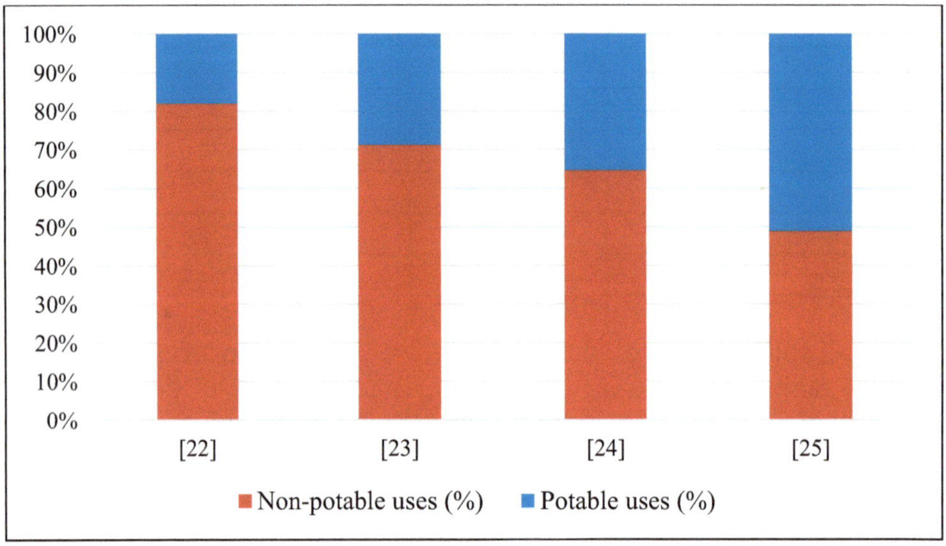

Fig. (9). Comparison between potable and non-potable water end-uses in public buildings in Brazil.

The lowest non-potable end-uses was approximately 49.1% [25], while the highest was 82.1% [22]; the average non-potable end-use was 66.8%. This shows that there is a high potential for potable water savings in public buildings.

Water End-Uses in Commercial Buildings

There are few papers that approach the water end-uses in commercial buildings in Brazil. The first study was conducted by Proença e Ghisi [26], who estimated the water end-uses for ten office buildings located in the city centre of Florianópolis,

Santa Catarina. Thus, the results of the water end-uses are shown in Table **3**.

Toilet flushing corresponds to the highest water consumption, with 67.6%.The second highest water consumption corresponds to the taps, with 25.4%. Moreover, it is noted that these buildings also show a great potential for potable water savings.

Table 3. Water end-uses in ten commercial buildings in Florianópolis [26].

Building	**Water End-uses (%)**			
	Taps	**Toilet flushing**	**Cleaning**	**Others**
Aliança	8.4	54.0	2.6	35.0
Exaldo Moritz	6.8	84.6	1.6	7.0
Granemann	29.3	64.6	2.1	6.0
Ilha de Santorini	23.9	79.1	1.5	4.9
Ilha dos Ventos	7.4	81.0	2.1	3.5
Manhattan	22.1	71.5	4.3	4.8
Olmiro Faraco	24.5	78.5	1.8	2.6
Pedro Xavier	62.4	54.8	2.1	18.7
Trajanus	56.6	55.8	1.2	4.5
Via Venneto	12.4	52.0	3.9	29.7
Average	**25.4**	**67.6**	**2.3**	**11.7**

Nascimento e Sant'ana [27] conducted a study in which the water end-uses of two hotels in Brasília, the capital city of Brazil, were obtained. The first building was conceptualized as typology A, as a three-story building. The second, as height building, typology B, which showed a higher number of employees, apartments, guests and also has shops, swimming pools, laundry, event rooms, and restaurant, among other facilities. Table **4** shows the water end-uses for the hotel A and the hotel B.

Table 4. Water end-uses in two hotels in Brasília [27].

End-uses	**Water End-uses (%)**		
	Hotel A	**Hotel B**	**Average**
Shower	71.3	38.7	55.0
Toilet	4.0	4.3	4.2
Apartments taps	3.5	4.4	4.0
Hygienic shower	0.2	0.1	0.2

(Table 4) contd.....

End-uses	Water End-uses (%)		
	Hotel A	**Hotel B**	**Average**
Bathtub	-	1.5	0.8
Employees	8.0	26.0	17.0
Kitchen	7.0	5.0	6.0
Cleaning	4.0	12.0	8.0
Garden irrigation	2.0	1.0	1.5
Laundry	-	7.0	3.5

Among the end-uses presented in Table **4**, the consumption regarding the apartments resulted in 79.0% to hotel A and 49.0% to hotel B, considering a shower, toilet, taps, hygienic shower and bathtub. For "Employees" were considered the water consumptions of the toilet (1.6% hotel A and 7.8% hotel B), lavatory (1.6% hotel A and 2.3% hotel B), shower (4.8% hotel A and 14.3% hotel B) and urinal (1.6% hotel B).

As observed for residential buildings, the shower was responsible for the largest water consumption in the two hotels in Brasília, *i.e.*, 55.0% on average. The activities performed by the employees corresponded to the second highest consumption (17.0%). Based on the data presented, this study [27] presented the lowest potential for potable water savings, due to the low water consumption for non-potable uses. A comparison between potable and non-potable end-uses for commercial buildings is shown in Fig. (**10**).

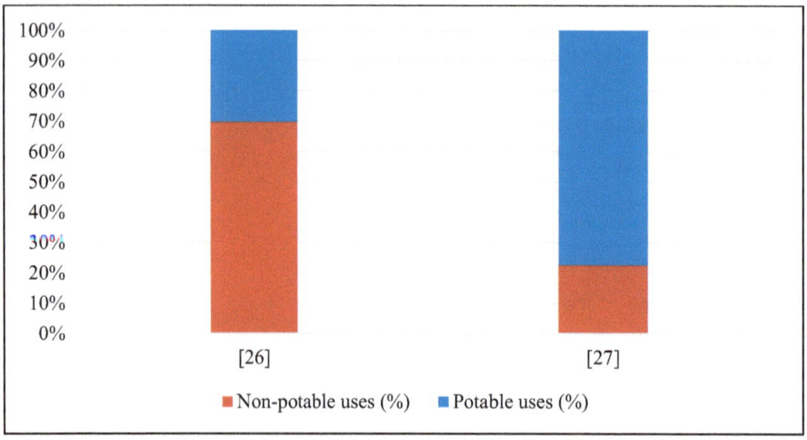

Fig. (10). Comparison between potable and non-potable water end-uses in commercial buildings in Brazil.

The non-potable uses corresponded to approximately 69.0% [26] e 23.0% [27]. This difference may be related to the fact that in the first study, 67.6% of the total water consumption in the buildings was due to toilet flushing, while the highest consumption in the second study was due to shower, *i.e.*, 55.0%.

WATER SAVINGS IN BLUINDINGS

Water conservation in buildings requires both water demand management and water supply management. It includes reducing the consumption of potable water and offer water through alternative sources. Thus, less noble uses can be supplied by lower quality water. The appropriate action to minimize the consumption of potable water is the replacement of common sanitary appliances for water-efficient appliances. They decrease the flow, and thus consumption. On the supply side, it is possible to make use of rainwater and also the greywater reuse for non-potable purposes such as flushing toilets, garden irrigation and washing cars and floors.

The rainwater utilization is related to the use of water seeping from impervious surfaces, usually the rooftops. The greywater of a building equivalent to the effluent from sinks, showers, bathtubs, washing machine and laundry tub. These effluents are treated and can be used for non-potable uses.

In Brazil, rainwater harvesting is the most investigated issue due to high rainfall in the country, to the simple system maintainability and to the quality of water, which usually does not require advanced treatments. However, depending on the collection area, there is not sufficient availability to supply demand in buildings. In some cases, such as in residential apartment buildings, rainwater collection area is small compared to the number of inhabitants. Thus, it becomes attractive to study the reuse of greywater.

Kammers [13], for example, to measure the rainwater utilization system for a multi-family building in São José, has not reached the potential of desirable potable water economy. The project was developed to meet a rainwater demand of 36.7% of the potable water demand, based on the non-potable uses. However, considering a population of 130 inhabitants and a daily demand of 200 litres per person, the potential savings achieved was 4.21%. This is due a very large population in a small rainwater collection area.

As for water quality, there is a difference between the research conducted in the northeast and those in the rest of the country. It is because in the north-eastern rainwater collected on roofs; it is often the only available source, or of better quality. So this is not an alternative source. In other Brazilian regions research also address the use of rainwater from other impermeable surfaces such as urban roads and parking yards.

The greywater reuse and rainwater harvesting should follow four criteria, according to May [28]: hygienic safety, aesthetics, environmental protection and technical and financial feasibility. Both systems address the water conservation premises, as an alternative source. Besides contributing to the potable water saving, reducing the load withdrawal of water sources and water treatment plants, the rainwater harvesting still has to feature the potential to decrease the runoff, and the greywater reuse minimizes the burden of residential wastewater, commercial and industrial. According to Ribeiro [18] by using greywater, there is a decrease in the volume of effluent directed to the public network, the demand for potable water in toilets and, consequently, the energy consumption of water and sewage. The studies on alternative sources primarily address the potable water savings, the design of systems and water quality.

Potable Water Savings

The potential for potable water savings through rainwater utilization varies widely, depending on demand, the collection area, and the local rainfall. Through research conducted by Ghisi [3] it was possible to verify that Brazil has high potential savings due to high rainfall. The southeast region had less potential than the others due to the demand for potable water, which is higher (Table **5**).

Table 5. Potential water savings, demand, and rainfall from each Brazilian region [3].

Region	Rainfall (mm/year)	Potable Water Demand (Litres per Capita/Day)	Potential for Potable Water Savings (%)
North	2182	88	100.0
Northeast	1146	97	61.0
Southeast	1362	158	48.0
South	1615	117	82.0
Central–West	1540	120	74.0

There are papers on different types of buildings: residential, public and commercial. Residential are those with the highest similarity and therefore can be compared. In public and commercial buildings there are singularities that affect the results, as the demand for non-potable water, which varies according to the type of building.

Public buildings can be office buildings, universities, and schools. In office buildings, the main water consumption occurs in toilets, but in schools and universities, there is a large consumption in restaurants. This consumption difference affects the system design and the potential savings. The commercial buildings still exhibit more variations and can be represented by the office, markets, vehicle washing stations, restaurants, etc. As a result, the water end-uses and potential savings may vary. However, what can be observed in the vast majority of research is that the higher the daily potable water demand, the lower the potential savings; and the greater the rainfall, the greater the potential savings.

Potential for Potable Water Saving in Residential Buildings

In the residential buildings, there are studies being developed for both: single-family buildings and multi-family buildings. Scientific papers usually look for display the results from the variation of some parameters such as the rainfall, the collection area and the demand for potable water and rainwater. Table **6** shows the maximum and minimum values, average and standard deviation of each parameter evaluated in all investigated research.

Table 6. Parameters that affect the potential savings [3, 12 - 14, 19, 20, 29 - 32].

Values	Daily Potable Water Demand (Litres/Day)	Rainwater Demand (% of Potable Water)	Roof Area (m^2)	Rainfall (mm/year)
Minimum	100.00	10.00%	38.00	580.00
Maximum	18,832.00	100.00%	470.00	3,409.00
Average	1,012.86	45.34%	122.54	1,843.94
Standard deviation	2,386.28	19.61%	76.14	1,013.82

It can be seen that the standard deviation of each variable is high. It is due to the wide range of variation between the values of each search, because some of them are about houses and others are about multifamily buildings. That is, although research has a wide range of values, the intervals between them are too far away. Vill [31] evaluated the potential for potable water savings considering figured values of daily potable water demand. He varied daily potable water demand from 100 litres per day to 1000 litres per day, at intervals of 100 litres per day. Further studies assessed only for the actual demand value studied of the building, such as search Dalsenter [20] evaluated three blocks of buildings to demands of 16,220 litres per day, 16,006 litres per day and 18,832 litres per day. It would be interesting to reduce the standard deviation of values to evaluate which parameter influences on the potential savings.

The daily potable water demand also interferes in the demand for rainwater, because it is a percentage of the demand for drinking water to be replaced. Although some studies have evaluated rainwater demand ranging from 10% to 100% of the potable water demand, the majority evaluated the rainwater demand according to the non-potable end-uses. From Fig. (**11**), it can be seen that most research is being developed with a rainwater demand from around 30.0% to

60.0% of the potable water demand.

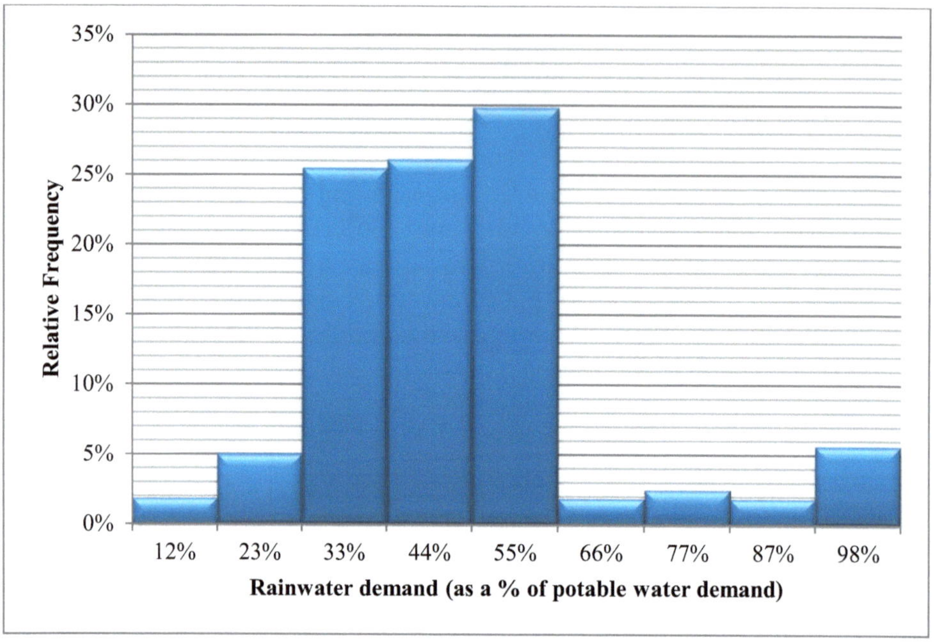

Fig. (11). Relative frequency of rainwater demand considered in residential buildings [3, 12 - 14, 19, 20, 29 - 32].

The rainwater demand ranging from 30.0 to 60.0% of potable water was adopted approximately 80.0% of the cases investigated. It is noted in the graph that, also, about 5.0% of cases considered the possibility that rainwater demand could be 100.0% of potable water demand. This data refers to studies that have estimated the potential for potable water savings in a city, state or region if all the residential buildings do use rainwater.

Ghisi *et al.* [29], for example, evaluated the potential for water savings through rainwater harvesting in residential buildings in Santa Catarina state. It required knowing the local rainfall, consumption of potable water, the number of inhabitants and housing in each town of Santa Catarina. As it was not possible to obtain this information in all cities in the state, data from 62 municipalities, representing 33.0% of the state area and 41.0% of the population, were used. The average found for the potential for potable water savings through rainwater use in Santa Catarina was 69.0%.

Similar research was developed by Ghisi *et al.* [30], which evaluated a sample of 195 cities in south-eastern Brazil. The main conclusion indicated by the authors showed that the average potential savings to the southeast were 41.0%. When they analyze the four states separately, the potential savings was 42.0% for Minas Gerais and São Paulo, 35.0% for Espítito Santo and 29.0% for Rio de Janeiro. The results were lower than those found in Santa Catarina due to lower rainfall and increased demand in the Southeast.

The roof area and rainfall were also varied in some research. Some authors varied the city to cover cities with different rainfall rates as Ghisi [3], Ghisi *et al.* [30] and Vill [31]. Through this, it was concluded that the higher annual rainfall, the greater the potential savings. The same proportion relationship was found between potential savings and roof area, regarded as the area of collecting rainwater [29, 32].

The average potential saving all addressed research was 34.5%. But, as already mentioned, this result depends on the variables used in each case. From Fig. (**11**), it can be observed that most used by researchers range were 50.0% to 60.0% substitution of potable water rainwater (adopted in 30.0% of cases). The potential savings water average was below this range because the trend is that the potential saving water becomes constant, even if the rainwater demand continues to grow up. Through Ghisi *et al.* [30], ranging rainwater demand of 10.0% to 100.0%, it can be seen that trend (Fig. **12**). Data presented are for a part of the research conducted by authors. This part is related to the potential for potable water savings to three cities in the state of São Paulo: Itaquaquecetuba, Jales, and Santos. These cities have different rainfall, demands of potable water and roof areas (Table **7**).

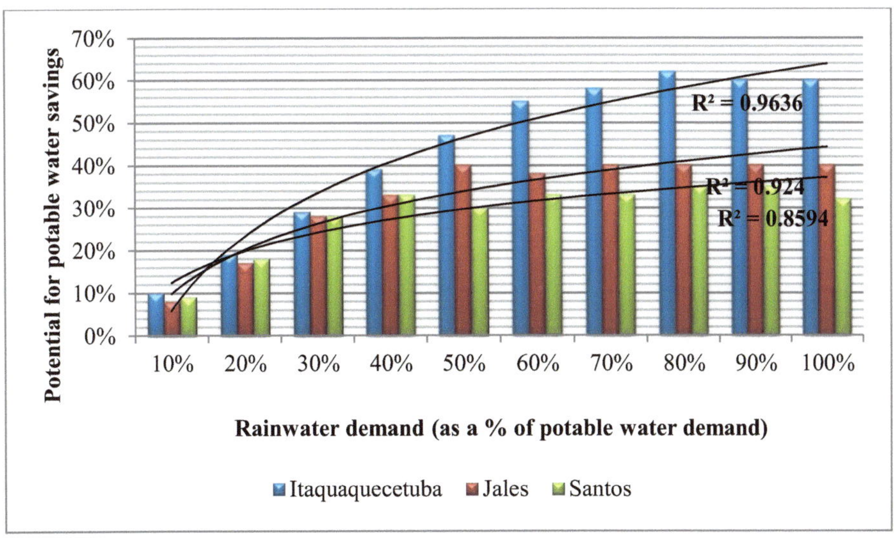

Fig. (12). Potential for potable water savings as a function of rainwater demand [30].

Table 7. Daily potable water demand, roof area and rainfall from the cities of Itaquaquecetuba, Jales and Santos [30].

Cities	Daily potable water demand (litres/day)	Roof area (m^2)	Rainfall (mm/year)
Itaquaquecetuba	373.18	82.00	1,340
Jales	534.52	84.00	1,166
Santos	813.45	38.00	3,395

Also about this research, it can be confirmed that the higher the roof area and lower demand, the greater the potential of water saving. Even with a smaller amount of rainfall than a half of Santos rainfall, Itaquaquecetuba showed higher potential savings due to low demand and greater collection area than Santos.

Finally, it is also important to present the effectiveness of surveyed systems, which is the ratio of the potential savings and the established demand. In the addressed research, it could be seen that this ratio is high. That is, the demands adopted for the use of rainwater are being met in most cases, as shown in Fig. (**13**) which shows the histogram of the effectiveness of the studies surveyed systems.

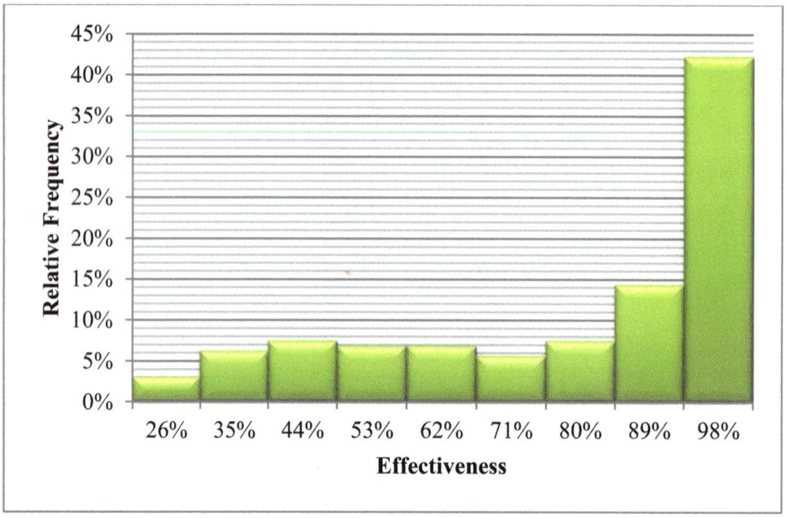

Fig. (13). Effectiveness of cases investigated [3, 12 - 14, 19, 20, 29 - 32].

Around 60.0% of the cases addressed in the research presented effectiveness above 80.0%. Most of the research presented considered input data consistent with single-family homes. It contributed to this high frequency of cases with effective systems. The studies that were developed for multi-family buildings showed efficacy ranging from 30.0 to 40.0%. That is, in these cases, the best solution to improve the supply of water is a combination of rainwater utilization with the greywater reuse.

However, there is not a significant amount of work that addresses the potential water savings through the greywater reuse. Among the collected works, to consider greywater demand to supply the toilets in 48 homes in Florianópolis, Ribeiro [18] found the supply of water, coming from shower waste, sink, washing machine and laundry tub, was 54.7%, and the greywater demand average was 19.4%.

By studying low-income households also in Florianópolis, Marinoski *et al.* [17] concluded that 42.0% to 45.0% of total water consumption were intended for non-potable uses. Moreover, the demand for non-potable water in discharges and external taps could be supplied through the greywater reuse, which would correspond from 23.0% to 32.0% of total water demand in homes.

Paes *et al.* [33] obtained an average percentage of 7.9% to 26.4% reuse when installing a water reuse system in six residences Cuiabá, MT.

Still, some papers conduct a comparison between the potential saving of rainwater utilization systems, greywater reuse, and systems that include the two alternatives. Ghisi and Ferreira [12] obtained the potential savings when considering the application of each one of the three systems in a three blocks residential condominium in Florianópolis. Ghisi and Oliveira [34] conduct the same analysis type in two residences Palhoça, SC. Through the results obtained by the authors, it was possible to achieve a potential of water savings average, as shown in Fig. (**14**).

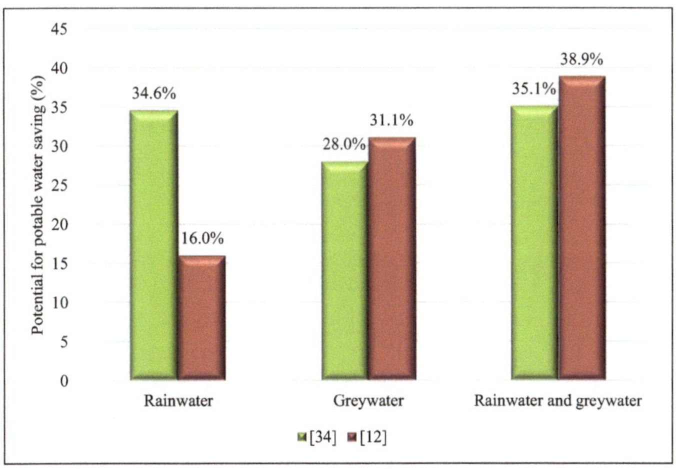

Fig. (14). Comparison between alternative water systems in the residential category.

From Fig. (**14**), it can be seen that in both cases, when considering the rainwater harvesting and greywater reuse, they obtained the greatest potential for potable water savings with 35.1% [34] and 38.9% [12], with a difference of 3.8% between studies. Considering only the greywater reuse the potential obtained were 28.0% [34] and 31.1% [12], that is, the difference was 3.1%. However, when considering only the rainwater harvesting, it is observed that the biggest difference between the studies, equivalent to 18.6%, then being 34.6% [34] and 16.0% [12] was presented.

Potential for Potable Water Saving in Public Buildings

Office and education buildings characterize public buildings surveyed on papers. The demand for non-potable water of this type of buildings is higher than residential demand. For this reason, there is a greater frequency of cases studied using as input rainwater demand rate ranging from around 45.0% to 90.0% (Fig. **15**).

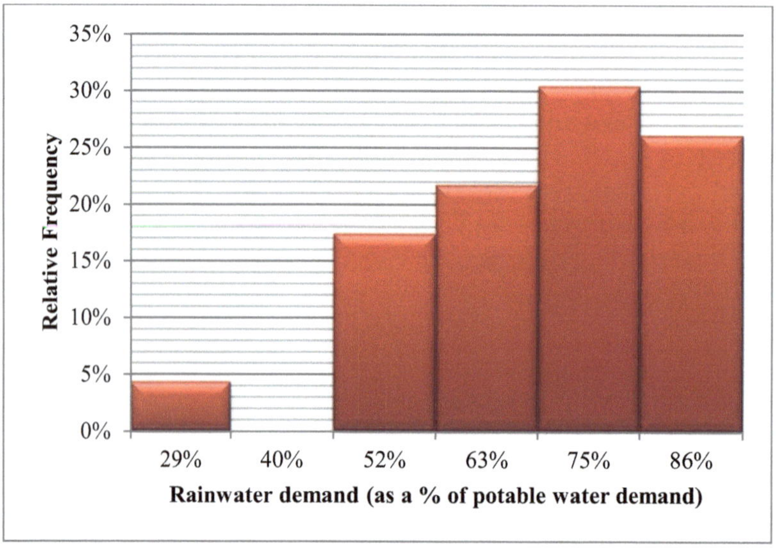

Fig. (15). Relative frequency of rainwater demand for public buildings [24, 25, 35, 36].

As can be seen in Fig. (**15**), only one addressed research presented a case with rainwater demand below 45.0%. This case was a public school located in the city of Florianópolis and the replacement percentage of potable water by rainwater was set at 26.1% after analysis the water end-uses in the institution. This analysis found that the largest school use of water was in the food preparation, therefore consumption for non-potable purposes represented a lower percentage.

In the study of Marinoski and Ghisi [24], developed at the Centre for Technology in Automation and Informatics in Florianópolis, the potential for potable water savings was estimated through rainwater harvesting for non-potable end-uses, accounting for 63.6% of the total water consumed in the building. After the sizing of the reservoir, the potential for potable water saving achieved was 42.6%.

Paes [36] developed a rainwater utilization project for building the technology development centre and training people. Through this project, the author has estimated the potential savings and the economic feasibility of implementation. Based on literature review the author varied data per capita demand for drinking water in amounts of 15 litres/day, 20 litres/day, 25 litres/day to 50 litres/day. For each of these values, 50.0%, 60.0%, 70.0% and 80.0% of the consumed potable water was replaced by rainwater. Within this high variation demand and combinations of them, there was also a wide variation of potential savings, getting between 26.7% and 64.7%. However, excluding the value of 50 litres/day, as the author pointed out as the least realistic, the potential savings was between 39.7% and 64.7%. The greatest potential for savings was observed for the combination of lower daily demand for potable water (15 litres/day) with the greatest demand for rainwater (80.0%).

Ghisi *et al.* [37] raised the potential for potable water savings through rainwater harvesting systems, greywater reuse and the combination of both in a school in Florianópolis, Santa Catarina. The results are shown in Fig. (**16**).

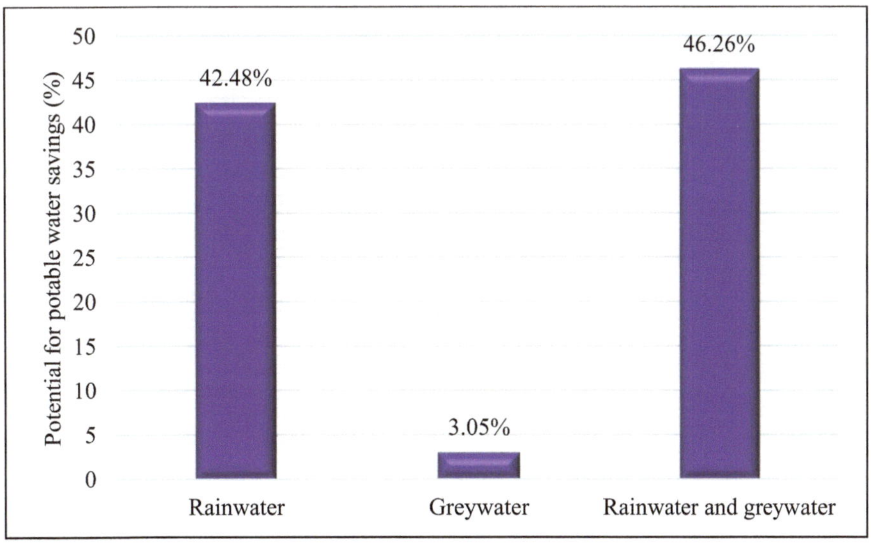

Fig. (16). Comparison of potential of water savings for each alternative system of water in the public category [37].

Through Fig. (**16**), it can be observed that while the rainwater harvesting system has a great potential for potable water savings, with 42.5%, the greywater reuse system resulted in low potential, only 3.1%. However, the combination of the two systems had a potential of potable water saving 46.2%, with a difference of 3.8% to rainwater harvesting system.

Potential for Potable Water Saving in Commercial Buildings

In commercial buildings there is a greater variety of water end-uses. Therefore it is difficult to make an analogy results. Usually, surveys are developed in commercial buildings which are characterized by a high consumption of non-potable water. Ghisi *et al.* [38] evaluated the potential for potable water savings by using rainwater for washing vehicles at fuel stations in Brasília, through *Netuno*. The results showed a variation of 9.2% to 57.2% of water saved. For this study, different areas of collection, storage capacity, the number of washes and demands of potable water and rainwater were considered. Daily rainfall data were obtained from two rainfall stations. Through the results, the authors concluded that:

- the higher the daily water demand, the lower the potential for potable water savings;
- the larger the area of the roof, the greater the potential for potable water savings;
- the higher the level of precipitation, the greater the potential for potable water savings.

Lage [9] developed a similar survey in car shops. The study was conducted in Belo Horizonte, and the effectiveness of the system was higher, mainly due to the collection areas, ranging from 2,833 m^2 to 4,800 m^2. The average daily demand of rainwater was also lower than that used in previous research. Ghisi *et al.* [38] obtained an average effectiveness of 31.5% while Lage [9] obtained 81.0%. This demand has been met satisfactorily, because of the collection areas were higher than that and demand was lower than that.

Depending on the type of commercial building, where there is the great availability of water for reuse, it becomes more attractive the combination of greywater reuse and rainwater harvesting. Fasola *et al.* [25] concluded that the four Florianópolis commercial buildings studied by them showed a great potential for reducing consumption of drinking water because more than half of the water was used for non-potable purposes. When considering the greywater reuse or the rainwater harvesting, the potential for potable water savings ranged between 55.9% and 80.6%.

The association between the greywater reuse and rainwater harvesting was also analysed by Carvalho *et al.* [39] in Galeão Airport, where these water alternatives were used in the air conditioning system. Thus, the airport recorded a reduction in water consumption of 150,000 m^3 in 2001 to 99,200 m^3 in 2010 (a savings of 33.9%), while there was an increase of 106.0% in the number of passengers.

Sizing the Rainwater Tank

The rainwater tank sizing is one of the critical points in the system according to Ghisi *et al.* [38] and is constantly being investigated by Brazilian scientific papers. It is because it is one of the most expensive items in the system implementation and influence on its effectiveness. Therefore, the design of the tank is important to avoid unnecessary expenses and demand to be met satisfactorily.

Brazilian standard NBR 15527 [40] shows six tank sizing methods: Rippl, Simulation, Australian, Azevedo Neto, Practical English, and Practical German. Besides these, it also presents the calculation recommended for the volume of rainwater that could be harvested. Equations (1) - (7) show the six standard methods suggested, and equation 9 shows the calculation of the volume of rainwater that could be harvested [40, 41].

a. Method Rippl:

$$S_{(t)} = D_{(t)} - Q_{(t)} \quad (1)$$

Where:

$Q_{(t)} = C \times \text{rainfall}_{(t)} \times \text{roof area}$

$V = \sum S_{(t)}$, only values $S_{(t)} > 0$,

since: $\sum D_{(t)} < \sum Q_{(t)}$

Where $S_{(t)}$ is the volume of rainwater in the tank at time t, $Q_{(t)}$ is the volume of rainwater that could be harvested at time t, $D_{(t)}$ is the demand or consumption at time t, V is the reservoir volume, C is the runoff coefficient.

b. Method Azevedo Neto:

$$V = 0.042 \times P \times A \times T \quad (2)$$

Where P is the average of annual rainfall data (mm), A is the roof area (m^2), T is the number of months of low rainfall or drought, and V is the volume of rainwater that could be harvested (litres).

c. Method Practical English:

$$V = 0.05 \times P \times A \qquad (3)$$

Where P is the average of annual rainfall data (mm), A is the roof area (m^2), and V is the volume of rainwater that could be harvested (litres).

d. Method Practical German:

It is an empirical method, where the volume of the tank is the lowest value of 6.0% of the annual volume of consumption or 6.0% of the annual volume of usable precipitation.

$$V_{adotado} = \text{mín. } (V,D) \times 0{,}06 \qquad (4)$$

Where V is the volume of rainwater that could be harvested (litres) in one year, D is the consumption or demand f potable water (litres) in a year, and e $V_{adotado}$ is the volume of the tank (litres).

e. Simulation Method:

$$V_{adotado} = S_{(t)} = Q_{(t)} + S_{(t-1)} - D_{(t)} \qquad (5)$$

Where:

$Q_{(t)} = C \times \text{rainfall}_{(t)} \times \text{roof area}$

Since: $0 \leq St \leq V$

Where $S_{(t)}$ is the volume of rainwater in the tank at time t, $S_{(t-1)}$ is the volume of rainwater in the tank at time t-1, $Q_{(t)}$ is the volume of rainwater that could be harvested at time t, $D_{(t)}$ is the demand or consumption at time t, V is the fixed volume, and C is the runoff coefficient.

f. Australian method:

$$Q = A \times C \times (P - I) \qquad (6)$$

Where Q is the monthly production volume of rainwater, A is the roof area, C is the runoff coefficient (usually 0.8), P is the average monthly rainfall data (mm), and I is the interception of the water wets surfaces and loss by evaporation (usually 2 mm).

The calculation of the tank volume is done by attempts to reach optimum values of system reliability:

$$V_t = V_{(t-1)} + Q_{(t)} - D_{(t)} \quad \textbf{(7)}$$

Where $Q_{(t)}$ is the monthly volume produced by rain in the month t, V_t is the volume of water in the tank at the end of month t, $V_{(t-1)}$ is the volume of water in the tank at the start of the month (t) and D is the monthly demand. In the first month is considered to be an empty shell.

The effectiveness of the system is given by Eq. (8):

$$P_r = N_r / N \quad \textbf{(8)}$$

Where P_r is the failure, N_r is the number of months in which the reservoir met the demand ($V_{(t)} = 0$), and N is the number of months.

g. Volume of rainwater that could be harvested is presented on Eq. (9):

$$V_{apr.} = P \times A \times C \times n \quad \textbf{(9)}$$

Where $V_{apr.}$ is the annual volume, monthly or daily Volume of rainwater that could be harvested, P is the average annual rainfall, monthly or daily rainfall, A is the roof area, C is the runoff coefficient and n is the capture factor that depends on system efficiency, taking into account the solids disposal device and the diversion of the first flush, if the latter is used.

According to May [28], losses are estimated from the reserved volume ranging from 10.0% to 33.0% of the precipitate volume. For Bezerra *et al.* [41] the capture factor is established 0.85 in practice.

According to Dornelles [42], practical methods cited by the standard are those with major problems because they adopt simple proportionality relations between rainfall and the volume of the reservoir. So that, they do not assess the seasonal precipitation. However, these are the easiest methods and are probably the most used.

Beyond the six methods proposed by the standard, The Probable Period of Days without Rain Method is also very widespread. This method has limitations on the need for daily rainfall data for a sufficient time to obtain statistical representativeness. The simulation method is the most suitable for the design of the reservoir, it provides performance indices such as the level of service demand

and percentage of extravasation, even though the more work [42].

Bezerra *et al.* [41] performed a comparative analysis of the methods proposed by the NBR 15527 in commercial buildings located in Curitiba / Paraná. They pointed out as one of the problems that the proposed methods have naming differences in an indication of variables, thus compromising the understanding of formulas.

As a result of his research, Bezerra *et al.* [41] obtained very discrepant tank volumes in each method, and it was not possible to determine what is the most appropriate method. Ghisi *et al.* [38] also developed a survey to compare the methods of NBR 15527 in three cities with different rainfall regimes. The authors also analysed the method proposed by the Netuno programme in this study. In the precipitation rate, the factors were varied: the roof area, demand for potable water, percentage substitution of potable water rainwater and therefore rainwater demand. 27 different cases were analysed.

The analysis pointed to some problems in the methods of the Brazilian standard. In the methods Azevedo Neto and Practical English, for example, dimensioning constants resulted in reservoir sizes for each city analysed and independent of rainwater demand. That is the higher the precipitation, the larger the volume of the reservoir. By Rippl methods, simulation and practical Australian was only possible to measure the reservoir when demand exceeds the volume of rainwater, making it impossible to use in cities with high rainfall. The Practical German method could be calculated for any case without restrictions [38].

An important consideration of the Practical German method is that, as adopts the minimum between the volume of usable rain or demand, it is not oversized, because when there is a demand greater than the rainfall can meet the volume is scaled by the rainfall index. When there are high rainfall and little demand, the volume is then calculated by demand. However, the factor of 6.0% from the formula was empirically set to some precipitation and demand data, not varying depending on the local rainfall, for example. Furthermore, the risk of using this type of design does not know the effectiveness of the system and potential savings.

Through Netuno it is possible to obtain the potential of rainwater saving for several volumes of reservoirs, in a practical way. One factor that ends up preventing their use by professionals is that the programme requires daily data of rain, often difficult to access or unavailable. However, it is the best way to scale the capacity of the reservoir, among the presented research, since it allows the simulation volume, potential savings, and financial returns.

Netuno allows two forms of system design: with lower and upper tank, and only with the lower tank. In the simulation, it calculates the volume of rainwater that could be harvested, the available volume in the tank before consumption; the volume consumed in a day (given by the minimum between demand and available volume in the reservoir), and the volume of the tank after the daily demand. Thus, it is possible to obtain the volume of overflowed rain and the potential savings, given by dividing the volume consumed per day and the daily demand.

The programme allows to enter the desired value for the volume of the upper tank or that this is equal to the average daily demand of rainwater. To the lower tank, it is possible to choose a fixed value or range of values. In this case, it is necessary to specify values for the maximum volume and the range of volumes. To indicate the ideal volume of the tank, it required the user to indicate the difference between the percentages of potable water savings through the desired rainwater utilization (in % per m^3). For example, when the simulation is made to an intervals volume of 1000 litres and the difference adopted is 0.5% per m^3, it indicate that the optimal capacity of the tank is taken as one in which the potential drinking water economy grows up by 0.5% when the tank to increase its capacity to 1000 litres [37, 38, 43 - 45].

Ghisi *et al.* [30] also analysed the reservoir capacity for three cities located in the state of São Paulo: Itaquecetuba, Jales, and Santos. Rainfall is lower in Itaquecetuba, a little higher in Jales and Santos has the highest rainfall index. The results confirm that the higher the rainfall index, the lower the volume of the reservoir needed to meet demand, and the higher the potential savings, the greater the volume of the reservoir.

Amorim and Pereira [46] performed the design of a rainwater tank for a building of a university in São Carlos, São Paulo, using seven methods. They are Rippl method - analytical and graphical mode; Probable Period of Days without Rain Method; Simulation Method; Volume of Rainwater that could be Harvested Method (NBR 15527); Method Practical German (NBR 15527); Method Practical English (NBR 15527); and Australian Method (NBR 15527).

The authors concluded that the Rippl method (analytical model) shows a wide variation of the final volume, mainly due to the use of rainfall data, and achieved an effectiveness of the system to 100.0%. The Rippl method (graphical model) presents a greater simplicity that the analytical model and results in volumes of similar values (with 86.0% efficiency). Then it becomes interesting to the reservoir sizing for family homes. Furthermore, the use of Rippl Method capturing area is inversely proportional to the volume of the reservoir for the same water demand [46].

The Probable Period of Days without Rain Method showed lower volumes and system effectiveness (73.2%), being interesting for situations aimed at reducing costs and have other sources to meet demand. The simulation method provides a detailed analysis of the variation of the reservoir volume over the years since the system efficiency is adopted. Regarding the methods of NBR 15527, the Volume of Rainwater that could be Harvested Method and Practical English resulted in high volumes, while the Method Practical German and Australian Practical provided more conservative values [46].

Ghisi [43] analysed different data rainfall, water demand, the number of people in the house and roof area. Found interesting results, relating these variables to the graphic function. The cities chosen by the author were Itaquaquecetuba with average annual rainfall of 1380 mm/year; Espírito Santo do Pinhal with average 1537 mm/year and Santos with an average of 3321 mm/year. As the distribution of rains proved similar to the cities of Espírito Santo do Pinhal and Itaquaquecetuba, was only made comparison Itaquaquecetuba with the city of Santos.

The author has some interesting comparisons to obtain conclusive results on the volume of the reservoir. By linking the tank to the coverage area, in general, to the coverage area larger than 100 m^2, the tank capacity tends to be greater as the roof area increases. However, for small areas of the roof and lower levels of precipitation, as in Itaquaquecetuba, the tank capacity decreases as the demand for drinking water and rainwater increase. It demonstrates that increasing the volume of the reservoir when the catchment area is small and demand is high does not cause the system has good results in efficiency and savings potential [43].

To scale a reservoir for rainwater utilization for a multifamily building in Santa Maria, Pozzebon and Gastaldini [47] used the Simulation Method and Method Practical English presented by NBR 15527. They obtained a small difference between the volumes of the tank. The resulting of Method Practical English is 5.05% higher than the Simulation Method. However, the authors show that the Practical English method considers only the average annual rainfall and the area of collection, therefore effective only for the calculation of large volumes, such as for irrigation on an annual basis. Thus they judged the Simulation Method most suitable for the design of tanks for residential purposes because this method considered seasonal variations in the analysed period.

Therefore, the design of the tank must be prepared as the final interests, targeting the characteristics of the region where the system will be deployed. So, it tries to avoid, for example, oversizing a system located in high rainfall areas, while in low rainfall regions, it is necessary to store water to supply the dry season.

Water Quality

Rainwater Harvesting

Andrade *et al.* [48] in some situations rainwater can mean the most viable source or better among the available sources, as to semi-arid regions and small islands. Elsewhere, where rainfall is favorable, rainwater can be used as an alternative water resource for non-potable purposes.

In Brazil, research on rainwater harvesting quality is being developed for both: as potable purposes, as for non-potable purposes. There is not, in a single Brazilian standard determination of all required water quality parameters for each of the end-uses presented in Water end-uses in buildings. The Ministry of Health Ordinance number 2914/2011 addresses the water quality standards for potable water and the Brazilian NBR 15527/2007 has quality rainwater parameters collected from roofs for non-potable purposes. However, this standard has few parameters, so there is the need to complement them so that the studies can be compared.

The National Environmental Council (CONAMA – *Conselho Nacional do Meio Ambiente*) presents the maximum allowable values for some water quality parameters in rivers, depending on your environment. As a result of the classification of rivers, Resolution 357/2005 determines the permitted uses for each class (divided into special class, 1, 2, 3 and 4). Class 2 and class 3 are intended for primary and secondary contact activities and will be presented for comparison activities that do not require potability, as a complement to NBR 15527/2007. Table **8** provides a summary of some parameters set by these documents [49 - 51].

About water quality, it can realize that the potability parameters has been the subject of studies conducted in the Brazilian semi-arid northeast. This region is characterized by the scarcity of water resources, and the use of rainwater use of tanks is often the only source for human consumption. However, there are still caring for the maintenance and management of water systems installed. It causes recurrent health problems outbreaks in the population.

Unlike the traditional potable water system, which is sealed to contaminants, tanks are an open system in the north-eastern, in which users collect water. Often this water is collected improperly using ropes and buckets, and the maintenance of quality is the result of practical knowledge of the residents.

Another common practice in the semiarid region is the supply of tanks by tank trucks to complement the system when rainwater is not enough to meet the population. This practice becomes more a potential source of contamination due to vulnerability during transport and the conditions of hygiene and cleanliness of trucks [52].

Table 8. Water quality parameters set by the documents.

Parameter	Unit	Ordinance 2.914/2011	NBR 15427	CONAMA 357	
				Class 2	Class 3
Dissolved aluminum	mg/L	0.200	-	0.100	0.200
Ammonia	mg/L	1.500	-	-	-
Lead	mg/L	0.010	-	0.010	0.033
Chlorides	mg/L	250.000	-	250.000	250.000
Free residual chlorine	mg/L	2.000	0.5 a 3	*	*
Dissolved copper	mg/L	2.000	-	0.009	0.013
Chrome	mg/L	0.050	-	0.050	0.050
Iron	mg/L	0.300	-	0.300	5.000
Total phosphorus	mg/L	-	-	0.030	0.050
Phosphate	mg/L	-	-	-	-
Nitrate	mg/L	10.000	-	10.000	10.000
Nitrite	mg/L	1.000	-	1.000	1.000
Selenium	mg/L	0.010	-	0.010	0.050
Zinc	mg/L	5.000	-	0.180	5.000
BOD5[1]	mg/L	-	-	5.000	10.000
Dissolved oxygen	mg/L	-	-	> 5	> 4
Total hardness	mg/L	500.000	-	-	-
pH	-	6 a 9.5	6 a 8	*	*
Turbidity	uT[2]	5.000	< 5	*	*
Apparent colour	uH[3]	15.000	15.000	*	*
Total coliforms	NMP/100mL	0.000	0.000	*	*
Escherichia coli	NMP/100mL	0.000	0.000	*	*
Total heterotrophic bacteria	UFC/mL	500.000	-	-	-
[1] Biochemical oxygen demands five days - 20°C					
[2] Turbidity Unit					
[3] Hazen unit					
* Respect the NBR 15527 standards					

It is recommended to make the disposal of the first millimetres of rain seeping through the roof to improve storm water quality collected from the roofs. It allows the impurities present on the surface are carried by the water and does not contaminate the water stored in the tanks. According to Souza *et al.* [53], incorrect disposal of this water and the wrong management of water in the tank are the main components of the deterioration of the reserved water quality. The NBR 15527/2007 [49] recommends that these devices are automatic and dimensioned by the designer disposal of 2 mm from the initial precipitation.

Souza *et al.* [53] conducted a survey which the aim was to evaluate the applicability of the automatic discharge devices such as sanitary barriers in the north-eastern hinterland. The results indicated that the devices were effective for reduction of turbidity, total coliforms, and total heterotrophic bacteria, reducing 79.0%, 95.5%, and 94.4% respectively. Although the water in the tanks does not meet the standard of microbiological quality established by Ordinance 2914/11 the Ministry of Health, these reductions help the treatment, reducing the consumption of chemicals for disinfection such as chlorination, for example.

Several studies point to the presence of *Escherichia coli* in rainwater tanks [54, 55]. A search result for Souza *et al.* [53] was that in one of the systems analysed by the investigators was not found contamination by *Escherichia coli*. This system consisted of the pump for removal of water from the cistern. Thus, the assumption of the authors is that the presence of this bacterium in other studies may be related to improper handling by users of the water tanks.

Silva *et al.* [56] did a field survey of 1,866 families in the Brazilian semi-arid rural. The results indicated that in most tanks of households (approximately 79.15%) occurred lack of Total Coliforms and 94.08% of Escherichia coli. They also allowed quantifying, in 22.65% of the cases the water was not potable and the drinkability recommended by the WHO (World Health Organization) was found at 76.83% of the samples.

Taking into account the problems encountered on the disposal of the first-flush and financial conditions of the north-eastern population, a low-cost automatic discharge device of the first-flush was developed at the Federal University of Pernambuco (Fig. **17**). It is rooted in the physical principles of communicating vessels and water closure, where water accumulates in standpipes and only after these are filled it is directed to the water tank. After each rainfall event, it´s essential that the device is emptied through the discharge opening. It has to be closed again after emptying, allowing operation of the device for the automatic shift in the next rain. The device constituent materials are completely PVC which allows ease of assembly and disassembly and ensure tightness. Thus it is not necessary specialized hand labour and is adjustable to the required volume of discharge depending on the catchment area [57].

According to Lima [57], there was an average reduction of 93.2% for apparent colour and 98.0% for turbidity after the offset of the first millimetres of rain in residence in the Pernambuco hinterland using the device. About the bacteriological parameters, he was also observed the removal of coliforms and *Escherichia coli*, reaching levels of 98.0% and 100.0%, respectively.

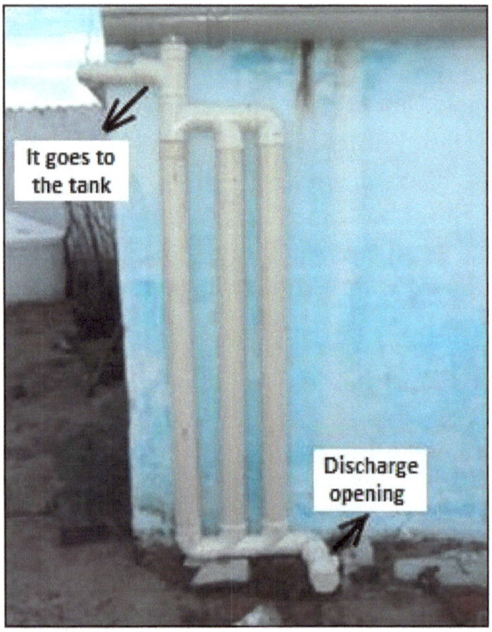

Fig. (17). Discharge device of the first-flush [58].

About the research found on the quality of water in the tank where there is consumption for drinking purposes can be concluded that all parameters are by the standards established by Ordinance 2914/2011, except *Escherichia coli* and total coliforms. So it´s necessary disinfection of water to be consumed [52, 59 - 61].

Some studies have been developed to check the quality of water during precipitation, to quantify the required volume of water that must be discarded for quality improvement. Jaques [62] evaluated the quality parameters through samples collected at 0, 10, 30 and 60 minutes of rain. Most physicochemical parameters showed a decrease of concentration *versus* time. However, this study did not quantify the amount of rain that should be discarded, only analysed the results in function of time for different rainfall.

Annecchini [63] made a qualitative assessment of rainwater considering some steps. First evaluated the rain quality parameters in the atmosphere, checking the water quality in the first, second and third millimetre. In the second step assessed the quality of water collected from the roof without any prior treatment or disposal sheets, the first rain nor disposal. In the other steps, she evaluated the quality of water in the tank when there are disposed of 0.5 mm, 1.0 mm and 1.5 mm of rainfall (Table **9**).

Table 9. Annecchini search results for water quality in tanks with different treatments [63].

Parameter	Unit	Tank - Without Treatment	Tank 0.5 mm	Tank 1.0 mm	Tank 1.5 mm
Ammonia	(mg/L)	0.48	0.36	0.06	0.13
Chlorides	(mg/L)	14.50	-	-	-
Total phosphorus	(mg/L)	0.15	0.02	0.01	0.02
Nitrate	(mg/L)	0.34	0.14	0.14	0.04
Nitrite	(mg/L)	0.04	0.01	0.01	0.00
BOD5	(mg/L)	6.40	2.00	1.20	1.30
Total hardness	(mg/L)	19.80	-	-	-
pH	-	6.84	6.39	6,.3	6.21
Turbidity	(uT)	10.40	0.40	0.50	0.20
Apparent colour	(uH)	-	2.50	2.10	0.70
Total coliforms	(NMP / 100mL)	146.00	360.00	150.00	47.00

The results showed that there was a significant increase in the hardness parameters, chloride, sulphate, chemical oxygen demand and total solids when the water from the atmosphere comes into contact with the roof. With the elimination of the first rain, there was a significant improvement in quality parameters: eliminating only the first 0.5 mm is possible for Turbidity reduction of 15.9 NTU to 0.4 NTU. There was no great difference between the eliminations of 0.5 mm, 1.0 mm or 1.5 mm for biochemical demand parameters of oxygen, chemical oxygen demand, chloride, and sulphate. However, for total suspended solids and total dissolved solids, there was a significant reduction from the first millimetre of rain eliminated. The author concluded that promoting the simplified treatment of rainwater, by removing the first millimetre, the quality is compatible for non-potable purposes, in compliance with Resolution 357/05 of CONAMA [63].

Some filter types are also being investigated. Silva *et al.* [64] evaluated the treatment of rainwater through the ascending rapid filtration trilayer and ultraviolet radiation. It was installed a first discharge device with a volume of

41.40 litres, to improve the water quality before reaching the filter. Conductivity, pH, temperature, apparent colour, dissolved oxygen, turbidity, ammonia, nitrite, nitrate, phosphate, sulphate, solids, chemical oxygen demand and coliforms were analysed to characterize the water before and after the treatment. The mean values and standard deviation obtained for the parameters pH, colour, dissolved oxygen, turbidity, nitrite, nitrate, ammonia and sulphate for rainwater samples collected at post treatment (tank 2) meet the standards set by CONAMA (2005) and Ministry of Health (2012). The efficiency of treatment for colour removal was 42.0%, 67.0% Turbidity, chemical oxygen demands 73.0%, 16.7% nitrite, nitrate 48.0%, 60.0% ammonia, phosphate and sulphate 96.8% and 50.0%. According to Resolution 357/05 (CONAMA), rainwater treated fits in Class 1 classification of rivers.

Vieira *et al.* [65] developed a prototype filter with PVC pipes and fittings for drainage of rainwater. They evaluated three geotextile filter devices used for existing drainages in Brazil: Bidim RT31, CASFIL 4100TMC, and CASFIL 5689TMC. The concept for water filtration rain developed in this study employed up-flow filtration and backwash down-flow. System Tests have proven that the filtration systems that use this concept could operate efficiently with the correct selection of the filter medium. Using a polypropylene filter fabric with 75-micron pore size and thickness 1.07 mm, the prototype system operated treatment with 68.0% efficiency and 100.0% backwash efficiency.

Other studies conducted in Brazil evaluated the influence of the type of surface on which the water is collected. Budel *et al.* [66] made a comparison of rainwater seeping quality by a roof with concrete tiles and a green roof. From the results, the parameters that had major differences were pH and turbidity (Fig. **18**).

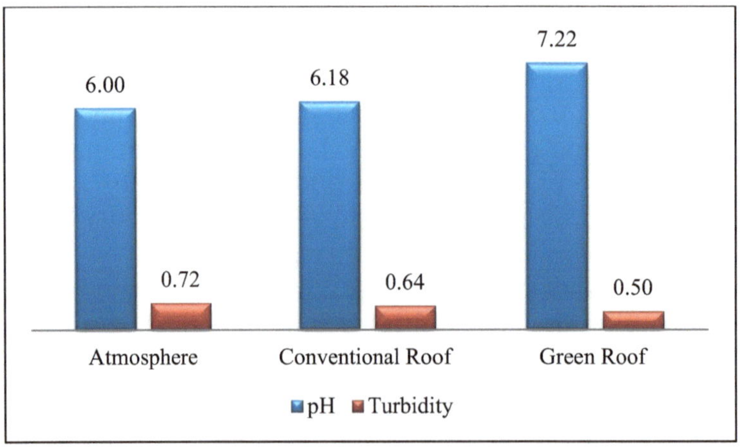

Fig. (18). Comparison between the conventional roof and a green roof [66].

There are studies being carried out to evaluate the capacity and infiltration of rainwater filtration through porous pavements [67, 68]. Following this same line of research, Antunes *et al.* [69] estimated the potential for potable water savings through rainwater utilization draining pavements in the city of Florianópolis. Brecht [68] made a qualitative analysis of runoff water that seeps into the porous pavement. Table **10** shows the results for Brecht [68]. It is noticed that there is an improvement in almost all presented parameters, to infiltrate the water collected from the street (runoff) in the systems A and B analysed by the author.

Table 10. Brecht quality results for runoff water that infiltrates into different porous pavement systems [68].

Parameter	Unit	Runoff	System A	System B
Dissolved aluminium	mg/L	0.06	0.03	0.03
Ammonia	mg/L	0.25	0.10	0.10
Dissolved copper	mg/L	0.95	0.00	0.49
Chrome	mg/L	0.06	0.02	0.04
Iron	mg/L	0.93	0.77	0.40
Total phosphorus	mg/L	1.22	0.60	0.36
Nitrite	mg/L	0.04	0.02	0.01
Zinc	mg/L	0.16	0.06	0.13
Dissolved oxygen	mg/L	8.86	9.00	8.86
pH	-	6.64	6.57	6.93

There are no standard parameters that are being analysed in rainwater utilization systems in raised studies. It makes difficult the comparison between them and the comparison between studies results and standards that dictate the patterns to be achieved.

Greywater Reuse

Most research focuses on the treatment of the effluent when it comes to the reuse water quality. Among the treatments, the most usual are composed of a sand filtration step followed by disinfection [70, 71]. However, there are studies that seek different treatment alternatives, such as a system composed of an Anaerobic Reactor Compartmented a Biological Aerated Filter Submerged, and Tertiary Filter [72]; Contact biological treatment system (bio-discs) [28]. However, it should be noted that the choice of treatment system varies depending on the quality of the greywater and the end-uses assigned.

The standard NBR 13969/97 [73] describes the water quality parameters for treated domestic wastewater reuse for irrigation of gardens, washing floors and automotive vehicles, discharge toilets, landscape maintenance of lakes and canals, and irrigation according to classes and purposes for which they are intended (Table **11**).

The standard gives the classes:

- Class 1 - Carwash and other uses that require direct user contact with the water, with possible inhalation of aerosols by the operator, including fountains;
- Class 2 - Washes floors, sidewalks, and irrigation of gardens, maintenance of lakes and channels for landscaping purposes, except fountains;
- Class 3 - Reuse in discharges of toilets;
- Class 4 - Reuse in orchards, cereals, forage, pasture for cattle and other crops through runoff or by the timely irrigation system.

Table 11. Water quality standard for reuse by NBR 13969/97.

Parameter	Unit	NBR 13969/97			
		Class 1	Class 2	Class 3	Class 4
pH	-	6 a 8	-	-	-
Turbidity	uT	<5	<5	<10	-
Fecal coliform	NMP/100ml	<200	<500	<500	<5000
Total dissolved solid	mg/L	<200	-	-	-
Residual chlorine	mg/L	0.5 a 1.5	>0.5	-	-
Dissolved oxygen	mg/L	-	-	-	>2

The results raised regarding the qualitative analysis of greywater are presented below, aiming qualitative parameters shown in Table **11**. The results refer to an experimental system of residences in São Paulo, SP [28], 26 residential buildings of Curitiba, Paraná [71], a building of a university located in Vitoria, Espírito Santo [72], residential buildings of Passo Fundo, Rio Grande do Sul [74], and a sports complex of a university also in Passo Fundo, Rio Grande do Sul [75]. Thus, pH analysis results are shown in Fig. (**19**), the turbidity in Fig. (**20**), the total dissolved solids in Fig. (**21**), and the oxygen dissolved in Fig. (**22**).

It can be seen from Fig. (**19**), there is a higher concentration of pH results between 6.5 and 8.5, but contained a minimum value close to 5, which corresponds to a sample of the acid character analysed greywater [72]. In Fig. (**20**), Fig. (**21**) and Fig. (**22**), it can be seen a wide range of values between the results obtained by the authors, and the higher turbidity values were observed for Fiori *et al.* [74], total dissolved solids by Borges [71] and dissolved oxygen by Bazzarella [72]. The results of the fecal coliform parameter are shown in Table **12**.

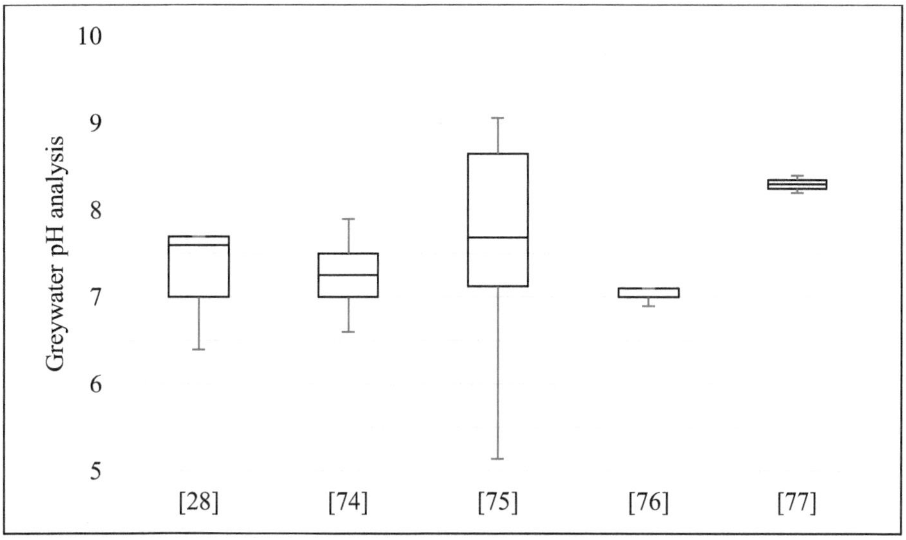

Fig. (19). Results of greywater pH analysis.

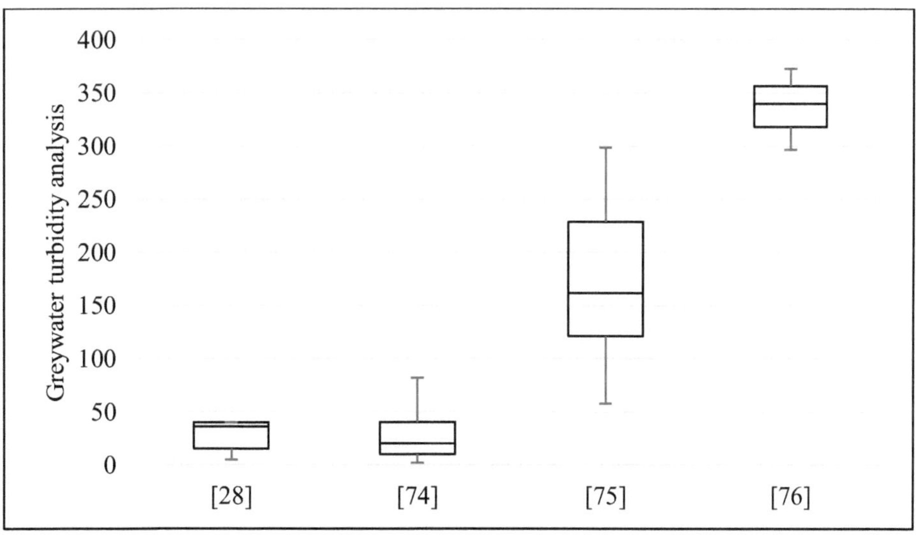

Fig. (20). Results of greywater turbidity analysis.

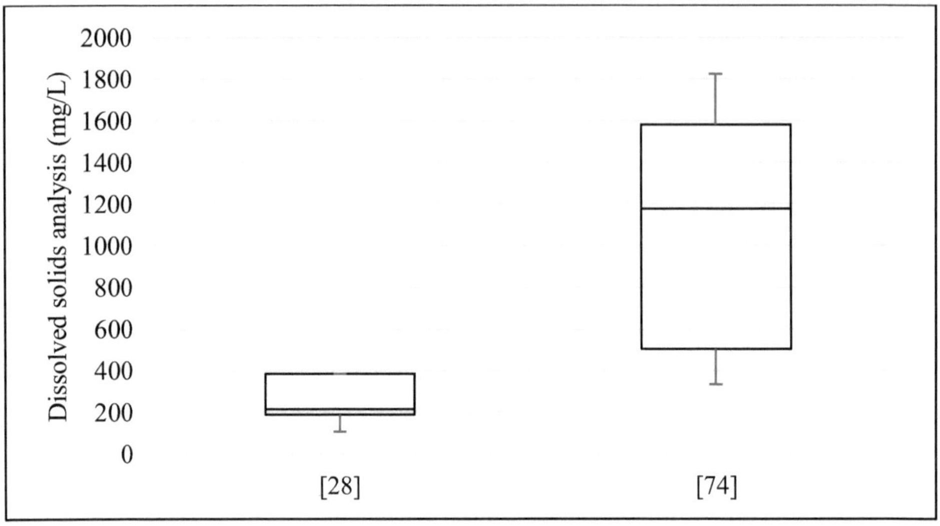

Fig. (21). Results of greywater total dissolved solids analysis.

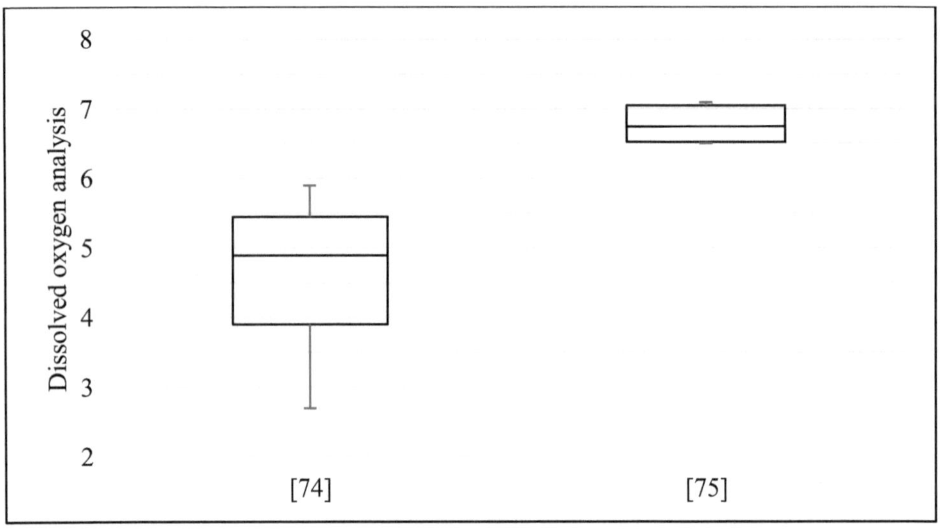

Fig. (22). Results of greywater dissolved oxygen analysis.

In Table **12**, it is observed meaningful results in some analysis for fecal coliforms, confirming the greywater pollution by these microorganisms in other residential water appliances, not only in toilets, which many believe is the unique coliform contamination. This fact, together with the other qualitative analysis of greywater, reinforces that it is necessary to perform the treatment of this water for greater security who will handle it, depending on the reuse the greywater will be used.

Table 12. Results of the qualitative analysis of fecal coliform greywater.

Reference	Samples	Fecal Coliforms (NMP/100 ml)		
		Minimum	Medium	Maximum
May [28]	Washing	1.6×10^3	7.9×10^3	1.6×10^4
	1st rinse	2.0	8.5	30.0
	2nd rinse	2.0	2.3	3.0
	Lavatory and shower	1.1×10^3	4.5×10^3	1.6×10^4
	Mixture	1.1×10^3	3.0×10^3	1.6×10^4
Borges [74]	Shower, lavatory, bidet and bathtub	2.0	4.0×10^2	1.6×10^7
Bazzarella [75]	Lavatory	1.0	1.01×10^1	9.0×10^1
	Shower	2.42×10^3	2.65×10^4	1.98×10^5
	Tank	1.0	2.87×10^1	2.14×10^3
	Washing machine	1.0	2.73×10^1	2.61×10^4
	Kitchen	1.0	6.47×10^2	1.9×10^5
	Mixture	1.01×10^4	3.25×10^4	1.3×10^5

Water-Efficient Appliances

More important than dealing with the management of supply, is dealing with the management of water demand in buildings. The higher the demand, the greater the need to supply it, either by conventional sources or alternative sources. Thus, the costs and energy consumption also increase, causing a greater environmental impact. Therefore, a way of managing the demand is the reduction of water consumption by replacing common water appliances for water-efficient appliances.

Botelho [76] conducted a survey of water-efficient appliances that could be deployed on the campus of the Federal University of Santa Catarina in Florianópolis. The water-efficient appliances proved applicable were: hydra-taps; automatic taps; aerators for internal taps, for showers and external taps; urinals with hydro valves, electronic valves, and waterless urinals; toilets with low flush volume; among others. The author noted that the Socio-Economic Center of the Federal University of Santa Catarina already had some water-efficient appliances installed in its structure, such as urinal with electronic valve and hydromechanical tap. These devices were not observed in the Technology Center, in which the corresponding devices showed higher flow than Socio-Economic Center. Also, he drafted a campaign to raise awareness about the rational use of water.

Santos *et al.* [77] established a hierarchy of water conservation actions that could be applied in buildings, based on a residential building in Curitiba, Paraná. Thus it was obtained the most preferred conservation measure is the use of greywater in toilets. The other steps were the detection and correction of leaks, the replacement of conventional toilets, and finally, the use of rainwater for irrigation of green areas, cleaning sidewalks and parking.

For reducing potable water consumption in the survey conducted by Fasola *et al.* [25], the option with the highest potential for savings was the installation of water-efficient appliances, with a reduction of 15.3% for the municipal school and 55.9% in the state school. Already the combination that best performance was the installation saving equipment followed by rainwater utilization, with a 27.8% reduction potential for the municipal school and 72.7% in the state school.

Ghisi *et al.* [37] evaluated the implementation of water-efficient appliances at a school in Florianópolis, considering the replacement of common taps for sensor taps, common valves for infrared valves, and common toilets for dual-flush toilets. Thus, they obtained a potential for potable water savings equivalent to 55.9%.

So, it can be seen that the implementation of water-efficient appliances provides a rational use of water by reducing consumption independent of user action. However, it is important to increase the conscious use of water with the installation of water-efficient appliances.

BENEFITS OF THE RATIONAL WATER USE IN BUILDINGS

Ghisi *et al.* [29] demonstrated that through rainwater harvesting can improve water availability over time in a certain region. In the study conducted in Santa Catarina, the results pointed to the prediction that, by rainwater harvesting with a potential saving of 69.0%, water availability in the year 2100 would be 5,000 m^3 per capita per year, against to 2,000 m^3 per capita per year if it was not made rainwater harvesting from the roofs. In addition to the direct benefit of potable water saving, the rational water use in buildings and the use of alternative sources can bring other benefits, such as savings in energy consumption in a city and the financial return to the user.

Through the evaluation of the life cycle is possible to assess environmental aspects of a product and the effects caused by this product throughout its life: from manufacturing to final disposal or recycling. Thus, it is a method that assists in making decisions in the comparison between systems or products. Kalbush e Ghisi [78, 79] evaluated the environmental and economic impacts resulting from the replacement of ordinary taps for water saving taps through life cycle analysis at a university in southern Brazil. It was considered as categories for analysis of the global warming potential: the destruction of the ozone layer, the human toxicity, the acidification, the water consumption and the energy consumption. They obtained that the replacement of taps influences in the global warming potential, in the destruction of the ozone layer, in the water consumption and the energy consumption, and the water saving taps showed superior environmental performance in all categories of environmental impacts. That is, there would be a reduction in the water consumption of 26.2%, in the energy consumption of 13.6%, in the human toxicity of 4.6%, in the acidification of 0.2%, of the global warming potential of 14.8% and the destruction of the ozone layer of 15.8%. However, this replacement action is not economically viable, since the costs of life cycles for ordinary and saving water taps are R$ 245.99 and R$ 316.20, respectively, wherein are necessary public incentives.

By evaluating the life cycle of water supply scenarios assessed by Ribeiro [18], with a scenario considering the integration of rainwater utilization system and greywater and other scenario considering only public potable water supply, it is concluded that to improve the environmental performance is essential to promote energy saving in the operation phase. An alternative is through the configuration of the integrated systems, seeking to design an elevated reservoir system that works by gravity, which would save energy consumption in repression and water distribution. Also, it is advisable to associate the rainwater harvesting, the greywater reuse, the installation of water-efficient appliances and promoting educational campaigns to raise awareness of the users.

To assess the potential for energy savings in water and sewage services in low-income households in southern Brazil, Vieira e Ghisi [80, 81] found the potential for saving water and energy for rainwater harvesting and greywater reuse for later comparison with the water and centralized sewage services. On average, for the total water consumption of households, the potential water savings were equivalent to 24.0% for greywater reuse and 43.0% for rainwater harvesting. The rainwater harvesting was the most intensive strategy (0.86 kWh/m^3), concerning energy saving, followed by centralized systems (0.84 kWh/m^3) and finally the greywater reuse, which it was the most efficient strategy for energy (0.54 kWh/m^3). The results suggest that the alternative water and sewage services can promote energy saving compared with the centralized mechanisms only when a concomitant reduction in the wastewater production is achieved.

Proença *et al.* [82] estimated the potential electrical energy savings in water and sewage utility, reducing the consumption of potable water in residential, commercial and public sectors in the city of Florianópolis, Southern Brazil. These three sectors represent 98.9% of total water consumption in the city. It was possible to estimate the potential of energy saving, reducing the consumption of potable water and the wastewater treatment, using data related to the energy consumption and to the costs that apply to the local water company for water and sewage treatment, and also the potential for potable water savings over the three sectors. There were considered three options: installation of dual-flush toilets, greywater reuse, and rainwater harvesting. The average potential for potable water savings was 30.0%, 53.4% and 60.3%, respectively, for the residential, commercial and public sectors. Thus, the average amount of potable water saving was about 10,153,835 m^3/year, and the electricity savings was 4.4 GWh/year, which would be enough to supply 1,217 homes and apartments in Florianópolis, with an average energy consumption of 300 kWh/month.

The economic benefits of rainwater harvesting are given by the reduction in water consumption rate, so it is easily measured. Nevertheless, the economic viability depends on each location, sizing and system characteristics, becoming little valid the generalization of results [42]. However, as a basis, it is important to know some results already obtained.

In a survey carried out in the metropolitan region of São Paulo, for an industry with a catchment area of 16,960 m^2, a fixed demand of 60 m^3/day and reservation volume of 100 m^3 and 500 m^3, the financial return for the two reservoirs was 1.81 years and 2.37 years, respectively. The aspects that significantly influenced to make feasible the rainwater harvesting were the water storage capacity and the demand to be attended [42].

Gomes *et al.* [83] presented a method for the volume setting of rainwater harvesting reservoir through a financial criterion. The basic idea was to define the volume of the reservoir by comparing two alternatives: one concerning the building without rainwater harvesting and the other with the collection, storage and use of rainwater system. So, the financial analysis was made comparing the costs of the two alternatives, where the first was represented by the cost of treated water consumption billed and the second concerning the implementation, maintenance and operation of the system coupled with the cost of treated water consumption. Among the conclusions reached can be highlighted:

- greater or equal demands to the volume raised generate reduced spending on the water volume of concessionary, reducing the time for the financial return. However, the periods of total exhaustion of the rainwater reserves captured are more frequent;
- while that larger reservoirs decrease the consumption of water billed, the cost of reservoir construction grows with its size, requiring a search for the optimum point, for a given financial return time desirable;

- harnessing systems with a large catchment area tend to satisfactorily attend the demand, reducing the return time of investment. In this type of system, the reservoir is almost always full, causing frequent extravasation.

Research about financial viability is being developed both in the residential sector also in the commercial and the public. Most of the residential sector research is carried out on the feasibility of implementation of rainwater harvesting systems in single-family residences. In the commercial sector, the objective of the research is to evaluate the financial viability of implementation of rainwater harvesting systems in buildings whose activities require the use of non-potable water, such as vehicle washing stations, for example. In the public sector, in addition to the harnessing research, there are being conducted assessments of replacement of hydraulic equipment for more efficient and automatic parts.

Financial Viability in Residential Buildings

The biggest number of research is being conducted in the residential sector, but it is not possible yet to compare them, once the results are very discrepant from each other. Table **13** presents a summary of the payback time (months) reached in some research, through the maximum value found, minimum, average and also the standard deviation.

Table 13. Payback time of the researches (in months).

Values	[32]	[31]	[20]	[19]	[12]
Minimum	160.08	32.00	12.00	11.00	28.80
Maximun	252.24	226.00	13.00	16.00	60.00
Average	211.92	103.40	12.33	13.50	44.40
Standard deviation	28.49	52.95	0.58	3.54	22.06

Berwanger e Ghisi [32] developed a study on the investment viability for the use of rainwater harvesting systems in households in the city of Itapiranga, in southern Brazil. The analysis was performed using the computer system Netuno and the daily precipitation data. By combining some variables, were obtained 144 cases evaluated. There were varied: catchment areas, consumption of potable water per day, the number of inhabitants, rainwater harvesting coefficient, and demand. Of these cases, 45 showed economic viability, those who had the highest potable water consumption and the largest roof area to the catchment. According to the authors, there isn't financial viability for residences with water consumption lower than 10 m^3 per month. It is due to the pricing of Brazilian utilities, who work with a flat rate per track consumption, *i.e.* up to 10 m^3 consumed, the amount charged the same, regardless of whether the consumption was 1 m^3 or 9 m^3 [32]. The same was observed in Florianópolis by Marinoski *et al.* [17], where only 33.0% of households studied showed monthly water consumption of up to 10 m^3, which shows that it hasn't an incentive for these families to reduce their consumption or use alternative sources of water.

In estimating the potential for potable water savings through rainwater harvesting and analyze their economic impacts on residences of São Miguel do Oeste, Santa Catarina, Martini [84] reached a result of saving up to R$ 93.84 per month. Moreover, of the 108 cases studied, the payback period of rainwater harvesting system ranged between 3.2 years and 21.5 years. However, it was considered feasible only the cases which reached, a maximum, six years, and equivalent to only 11 cases.

Dalsenter [20] obtained the financial investment and the payback time of a rainwater harvesting system in a condominium composed of three blocks in Florianópolis. For the first block, Baía Norte, with 110 residents, the percentage of potable water which could be replaced by rainwater was 2.0%, giving a lower reservoir volume of 15,000 litres, and the potential for saving water resulted in 7.21%. For the second block, Baía Sul, with 101 residents, the percentage of potable water which could be replaced by rainwater was 26.6%, with a lower reservoir volume of 15,000 litres, and the potential water savings resulting from 8.42%. Finally, for the third block, Ilha dos Guarás with 106 residents, the percentage of potable water which could be replaced by rainwater was 22.1%, obtaining a lower reservoir volume of 15,000 litres, and the potential water savings resulted in 8.21%. The investment for the system deployment would be R$ 10,260.58, and the payback period estimated for the first and second block was 12 months and for the third block was 13 months.

When assessing the implementation of a rainwater harvesting system in a family residence of São José, Santa Catarina, Meincheim [19] presented two possible scenarios. The first scenario provides for the replacement of potable water by rainwater in the devices: tank tap, two external taps and washing machine. In the second scenario, the evaluated devices are equivalent to the first scenario along with two toilets. With this, it was found that the cost of the system would be R$ 15,557.25 and the payback period of 16 months, for the first scenario. For the second scenario, the cost of the system would be R$ 25,641.77 and the payback period of 11 months.

Ghisi e Ferreira [12] concluded that, by using as an alternative source the rainwater harvesting or greywater reuse separately, the payback period of the investment would be less than five years. By using both sources jointly, the payback period would increase, but would still profitable. Thus, assessing the economic returns for buildings similar to the study, it would be possible to apply one of the three alternative sources systems analysed since the consumption of potable water of each apartment exceeds 10 m^3.

Júnior *et al.* [85] conducted a study of the economic viability of rainwater harvesting in residences of João Pessoa, Paraíba. The authors considered three typologies socioeconomic of residences: the popular standard (4 inhabitants, consumption per capita of 130.0 litres per capita/day and catchment area of 60 m^2), the average standard (5 inhabitants, 162.0 litres per capita/day and 120 m^2), and the high standard (6 inhabitants, 192.8 litres per capita/day and 300 m^2). Also, they contemplated that only the kitchen tap would receive potable water from the concessionary. They obtained that, for the rates applicable at the time of the studies, the rainwater harvesting system was not economically feasible for the residences of popular and average standards. In contrast, to the high standard residences, in which the tariff and non-potable water consumption are high, the return periods found ranged from 8.2 to 10.2 years, conceptualizing as economically viable.

Financial Viability in Commercial Buildings

Using the same reservoir sizing programme Netuno, Ghisi *et al.* [38] evaluated the feasibility of investment for replacement of potable water by rainwater for washing vehicles at fuel stations in Brasília. For each case were obtained the cost of materials, installation, and maintenance through local shops and the Syndicate of Industry of Civil Construction of the Federal District (Sinduscon-DF – *Sindicato da Indústria da Construção Civil do Distrito Federal*). An interest rate of 1% per month and a useful life of 15 years were also taken into account. Regarding the payment rates for water consumption and the use of the collection and treatment sewage system, it was adopted two scenarios: (1) considering the payment of sanitary sewer rate for rainwater used and (2) not considering. The results indicate that, for the scenario (1), the three cases were viable, with a return time of 43, 31 and 62 months, respectively. The case 3 had the highest return time because it is the one with lower demand. In the scenario (2), when considering the sewage rate for the rainwater used, the return time increased to 165 (case 1) and 94 (case 2). The case 3 was not evaluated in the scenario (2).

Lage [9] conducted a study of the potential for potable water savings through rainwater harvesting in a car wash in six vehicle dealerships Belo Horizonte, Minas Gerais. Through this study, the author was able to perform a financial analysis of the proposed system. With that, the author got that investment ranging from R$ 7,194.02 and R$ 43,268.33, which the dealership with the lowest cost of deployment had an average of 406 vehicles washed per month and the dealership with the higher cost had an average of 920. However, the dealership with the largest investment had the lowest return time, 75 months, while the dealership with the lowest investment showed the largest return time, equivalent to 143 months.

Financial Viability in Public Buildings

The case study carried out in Florianópolis by Paes [36], analysed the financial feasibility of implementation of rainwater harvesting system in the integrated centre of human resources and technological development through of Netuno. They were considered the equipment costs, workforce to installation, maintenance, inflation and a useful life of 20 years. In higher demand scenario (50 litres per capita/day and 80.0%), the payback period was three years. In the most unfavorable case scenario, the time increased to 14 years.

Fernandes *et al.* [86] conducted a viability study of a rainwater harvesting system in a building of a University of Natal, Rio Grande do Norte. Through the calculation of the net present value considering the provision of public water and the installation of a rainwater harvesting system, the authors obtained that is more economically viable to opt for the system installation than keep the water supply only with public water distribution.

When analyzing the cost of cisterns construction for a rainwater harvesting system in a public university building in the state of São Paulo, Maeda e Brandão [87] obtained values between R$ 387.50 (10 m^3) and R$ 6,612.00 (50 m^3). Thus, they concluded that it is financially advantageous build more than one tank with a smaller volume than a single tank with a greater volume.

NATIONAL PROGRAMMES, ENVIRONMENTAL EDUCATION AND ACCEPTANCE

As described in section 1, Brazil is a rich country in water resources, but the water distribution in the territory is very uneven, which generates differentiated needs throughout the country. In the Brazilian semiarid, the concern is related to environmental education. Without the primordial knowledge of installation and maintenance of rainwater harvesting systems, it can impair the benefits that could be provided. In other regions, the research focus is the acceptance of the use of alternative sources. Despite knowing the potential for potable water savings through the use of alternative sources, there are still doubts regarding the acceptance of these systems, according to Marinoski *et al.* [88].

In the semi-arid, the need for water resources management improvement resulted in the creation of a programme for the installation of rainwater harvesting cisterns.

Started in 2003, the One Million Cisterns Programme (P1MC – *Programa Um Milhão de Cisternas*) aims to meet the need for potable water of the population in rural areas of the Brazilian semiarid region. This rainwater harvesting system is recognized and legitimized by the National Council for Food and Nutritional Security (Consea – *Conselho Nacional de Segurança Alimentar e Nutricional*) as an element of water and food [89]. The programme recommends the involvement and empowerment of rural families, providing courses before the cisterns construction, called "Capacity building in the management of water resources", aiming at user awareness regarding the vulnerability of rainwater contamination [90].

Until 2014 approximately 470,000 cisterns had been installed, and the goal of the federal government is to reach 1.5 million cisterns. The government recommendations for the operation and maintenance of the cisterns include a manual bypass of the first rain; use only pumps to remove the water from cisterns, keep the reservoirs capped, and store only rainwater [60]. However, the adoption of appropriate practices involved in handling the different components of the programme depends on a greater awareness among the beneficiaries. It represents one of main challenge for the programme's effectiveness [91]. Thus, among the papers, there are highlights the related with integrated actions seeking improvement in water quality [4, 92 - 96]. According to Santos *et al.* [93], there is the need for inclusion of environmental education in the subjects of the school's curriculum to expand the understanding of the local reality.

Santos e Dias [94] made an assessment of the training course used as part of the programme. As a result of their research, they pointed to the importance of two tools that support learning: the language adopted by the instructor, which should be clear and objective always leaving open to dialogue and interaction; and educational material that should facilitate the training. The authors showed that the most impacting result of the programme is the significant improvement in the health of the families, although it requires more representation and coverage and less punctuality of the programme. According to Gomes *et al.* [91], 70.0% of programme beneficiaries in Minas Gerais did not receive information about the water treatment.

About knowledge about water quality, it is perceived that in the northeastern hinterlands, even where the water is available for local utilities, the population uses the cistern water for drinking and the concessionaire's water for less noble purposes, considering that the rainwater has better quality [97]. It demonstrates the need to improve the coverage of capacity building courses.

Still in northeastern Brazil, Júnior *et al.* [85] interviewed some residents of João Pessoa, Paraíba, through the application of forms. Among the interviewees, 66.1% claim to have some knowledge about rainwater harvesting in residences, and 54.4% of them use rainwater, even eventually. Most of who use rainwater (99.6%) don't implement any treatment. The residents who use rainwater harvesting claimed: believe it's not necessary treatment (69.5%), didn't have knowledge (28.3%), and for not increase system cost (2.2%). Regarding the residents who have disclosed didn't have knowledge about rainwater harvesting (33.9%), 87.6% would accept use rainwater in their residences, and 9.8% would not use. As arguments for not using, mentioned: the difficulty of the catchment (47.1%), the lack of incentive (38.2%), the danger of contamination (11.8%), the lack of interest (8.8%), and other options (8.8%).

Almeida [98] conducted a study in the city of Feira de Santana, located in Bahia, in northeastern Brazil. The author performed interviews in 379 households to understand the behavior of people, the use, and customs related to water consumption. In the city of Feira de Santana, more than half of the population is low class, and the results of water consumption in buildings reflect that: 41.2% of the interviewed households consume from 0 m^3 to 10 m^3 of water per month, and 43.5% consume of 10 m^3 to 18 m^3 per month. Still, the family income in more than 70.0% of the investigated households is up to 5 minimum wages. Regarding acceptance issues of the use of alternative sources, 75.4% of interviewed said they would use the two sources: rainwater harvesting and greywater reuse. Furthermore, 10.3% would make use only rainwater, 7.1% would make use only greywater reuse, and 7.1% would not make use of any of these sources. About the knowledge of the use of alternative sources of supply, 65.0% said they already had some knowledge of the subject. On the importance, 97.0% reported be important to use alternative sources nowadays, and the main factor presented was the water scarcity in the region.

In the other regions of Brazil, where water availability is not critical, the concern of researchers is related to the acceptance of the use of alternative sources for improvement of water resources in cities. Thus, Marinoski *et al.* [88] interviewed low-income families, totalling 44 residences in southern Brazil. The results showed that 27.0% to 59.0% of householders would use rainwater for potable purposes, 84.0% to 96.0% would use for non-potable purposes and 80.0% to 90.0% would use greywater for non-potable purposes. Through the correlation between the results, it was found that the higher the age, the lower the acceptance of greywater reuse as an alternative source. The higher the education level, the lower the acceptance of rainwater. And the greater the number of occupants of the residence, the lower the acceptance of rainwater for washing dishes.

Also in southern Brazil, among the residents interviewed by Ribeiro [18] in Florianópolis, 84.0% to 95.0% would use rainwater for non-potable purposes and 80.0% to 91.0% would use greywater treated for non-potable purposes, demonstrating a high degree of social acceptability of alternative sources. It is worth noting that, about the rainwater harvesting, the maximum percentage of 95.0% corresponds to the use of this water for the discharge of toilets.

Peixoto *et al.* [99] conducted a study with teachers, staff and students of a school in the city of Humaitá, Amazonas, aiming to introduce students to the importance of rainwater harvesting for non-potable uses. For this, they discussed environmental awareness through lectures and seminars, besides they distributed leaflets. As a positive result, the school showed interest in the implementation of a rainwater harvesting system in your building.

Although most research has shown that there is acceptance of the population for the use of alternative water sources, the initial cost of deployment of the systems remains an obstacle in the dissemination of these technologies. Given the need to improve the management of water resources, some Brazilian cities have programmes for the water saving, which makes it compulsory for new buildings the use of techniques that improve water management. The Municipality of Curitiba presents its regulation through the Conservation Programme and Rational Use of Water in Buildings (PURAE – *Programa de Conservação e Uso Racional da Água nas Edificações*) that makes demands according to the uses of the buildings, as can be seen in Table **14**.

Table 14. Legal requirements of PURAE for buildings typologies [100].

Uses of Buildings	Requirements *			
	1	2	3	4
Single-family habitation	x	x		
Housing of institutional use (hotel, student accommodation, asylum, convent, seminary, boarding school, and orphanage)	x	x		
Transitional housing (apart-hotel, hotel, pension, and motel)	x	x		
Building for community use (education, welfare, health, library, leisure, culture and religious cult)	x	x		
Collective housing (more than two grouped autonomous units vertically), with a total built area per unit **less** than 250 m^2	x	x		
Collective housing (more than two grouped autonomous units vertically), with a total built area per unit **equal to or greater** than 250 m^2	x	x	x	
Housing estate (more than twenty units and/or more than two buildings on the same ground), for total area built by apartment unit **smaller** than 250 m^2	x	x		

(Table 14) contd.....

Uses of Buildings	Requirements *			
	1	2	3	4
Housing estate (more than twenty units and/or more than two buildings on the same ground), for isolated residences and total area built by apartment unit **equal to or greater** than 250 m^2	x	x	x	
Single-family housing in series (between three and twenty isolated units, grouped horizontally)	x	x	x	
Popular houses in series (between three and twenty isolated units, grouped horizontally, whose total area divided by the number of beds is less than 10, as indicated in Decree 212/2007)	x	x	x	
Post of supply	x	x		
Commerce and service with area **less** than 5,000 m^2	x	x		
Commerce and service with area **greater** than 5,000 m^2	x	x		x
Office building, Administrative Headquarters, Public service, Commercial parking, Shopping Center, Mall, Super and Hypermarket, Washing fast, Clinic and Ambulatory with an area **less** than 5,000 m^2	x	x		
Office building, Administrative Headquarters, Public service, Commercial parking, Shopping Center, Mall, Super and Hypermarket, Washing fast, Clinic and Ambulatory with an area **greater** than 5,000 m^2	x	x		x
Industry with an area **less** than 5,000 m^2	x	x		
Industry with an area **greater** than 5,000 m^2	x	x		x
*1 = catchment and harvesting of rainwater from roofs; 2 = toilet of reduced discharge volume and taps equipped with aerators; 3 = hydrometers for individual metering of consumption; and 4 = collection and wastewater treatment system.				

This programme establishes measures to induce conservation, rational use of water and use of alternative sources of water abstraction for all new buildings. The operation of PURAE occurs from the request of the construction permit when the technician responsible for the work and the owner sign the Statement of Responsibility stating that they are aware of the requirements of the Municipal Decree N° 293/06 [101].

Another programme aimed the water conservation is the Water Rational Use Programme (PURA – *Programa de Uso Racional da Água*), stemming from the partnership between the Basic Sanitation Company of the State of São Paulo (SABESP – *Companhia de Saneamento Básico do Estado de São Paulo*), the University of São Paulo (USP – *Universidade de São Paulo*) and the Institute for Technological Research (IPT – *Instituto de Pesquisas Tecnológicas*). This programme is geared towards sanitary hydraulic building systems and seeks to encourage the rational water use. There are three basic objectives that need to be attended. One the creation of models that quantify in hydraulic building systems the efficiency in the use of water-efficient appliances in the required demand. Another is to develop and make available products that provide to the population the water use in a conservatively and rational way. And create technical and institutional documents substantiating the programme's actions.

At the national level, the National Water Resources Policy (PNRH – *Política Nacional de Recursos Hídricos*) was established by Federal Law 9.433/1997, which created the National System for Water Resources Management (SINGREH – *Sistema Nacional de Gerenciamento dos Recursos Hídricos*). Some agencies from the national and state systems for water resources. Brazil has the National Water Resources Council (CNRH – *Conselho Nacional de Recursos Hídricos*), wich is a superior organ deliberative and legislative. The National Water Agency (ANA) is an autarchy of special arrangements, under the Ministry of Environment (MMA – *Ministério do Meio Ambiente*), a member of SINGREH and with administrative and financial autonomy to ensure the implementation of PNRH. Still, there are the Councils of Water Resources of the States and the Federal District and the Committees of Watershed, which are a collegiate organ in which taken decisions about the watershed where they work. Others are the organs of public powers federal, state, Federal District and municipal, whose competencies relate to the management of water resources, and the Water Agencies, which, after the formation of the Basin Committees, can be created to act as executive secretariats of one or more Basin Committees. In general, the State Water Resources Policy (PERHs – *Políticas Estaduais de Recursos Hídricos*) enframe the realities of the states and follow the same principles, objectives, and guidelines of PNRH [90].

Among the national programmes, there is the National Programme against Water Wastage (PNCDA – *Programa Nacional de Combate ao Desperdício de Água*), which is administered by the Executive. The PNCDA determines actions to combat the waste of water in watersheds in public water supply systems and sanitary hydraulic building systems. There is also the Management Programme of Water Use in Buildings (PGUAE – *Programa de Gestão do Uso da Água em Edificações*), which includes actions such as characterization of water consumption; characterization of water conservation measures; analysis of the applicability and prioritization of water conservation measures; and the development of water use management plan in buildings. Due to these actions, in the PGAUE there is a prediction of qualitative and quantitative management of water [77].

One of the big difficulties in water resources management in the country is the lack of both institutional and political consensus to approach issues related to the rational water use and alternative sources. Even to integrate rainwater harvesting in water management, through the determination of responsible for the supplied water quality. Although there is a National Water Resources Policy (PNRH), public institutions don't work in an integrated and cohesive manner. According to Nóbrega *et al.* [90], the actors of water management in the country should generate new integrations that will result in a functional organization that identifies with the goal to ensure efficient management of alternative water sources and technologies for its implementation highlighting the rainwater harvesting systems by cisterns.

CONCLUSION

Based on the potential for potable water savings by using rainwater in buildings in Brazil, it can be observed that rainwater could be used as an alternative water source. It was observed in most papers that the higher the daily potable water demand, the lower the potential for water savings; and the greater the rainfall, the greater the potential for water savings. Besides the rainwater harvesting system provides immediate savings of potable water, which improves water availability for the future.

The results of the water end-uses also pointed to the great potential for potable water savings through the use of alternative water sources for non-potable uses; on average 39.1% of the water consumed in homes is used for non-potable purposes. Furthermore, there is the possibility of reusing greywater. The results of the researches presented herein showed that the combination of rainwater harvesting systems with greywater reuse systems has a higher potential for potable water savings than using just one system.

Regarding the water demand in the buildings, the investigation showed that the use of water-efficient appliances could also ensure savings, ranging from 15.3% to 72.7% potable water savings.

On the quality of water, there was a great difficulty of comparing research results due to the different parameters investigated in each one. It was also observed that more investments in research are necessary, either through funding by governments or private companies, because laboratory costs are fairly high. About this matter, it could be concluded that there is proper technology to allow the use rainwater for potable purposes. However, the survey indicated the presence of pathogens in the water used for potable purposes by households included in the programme 1 million cisterns, in northeast Brazil, because there is a lack of structure and education for the correct use of rainwater tanks.

Other benefits of using alternative water sources and decrease water demand were presented, such as energy savings and financial benefits. About the national support programmes for water savings in buildings, it could be concluded that there is still the need for more investments. One of the most important programmes being carried out is the programme 1 million cisterns, which aims to provide rainwater for families living in areas where there is no availability of water provided by the water utilities, or where the water is scarce. However, there are still flaws in the system, such as the responsibility for the quality of the water consumed by the population, which was passed on to users. In other regions, where there is availability of potable water, there is the need for greater incentives for the use of alternative water sources as a prevention for future shortages.

CONFLICT OF INTEREST

The authors confirm that they have no conflict of interest to declare for this publication.

ACKNOWLEDGEMENTS

The authors would like to thank CAPES, an agency of the Brazilian government for postgraduate education.

REFERENCES

[1] "Brazil. Agência Nacional de Águas", *Atlas Brasil: Abastecimento Urbano de Água (Panorama Nacional),* Brasília, . National Waters Agency. Atlas Brazil: Supply Urban Water (National Panorama) [in Portuguese].

[2] V. Andreu, "Água para as cidades brasileiras", In: Paulo São, Ed., *Mundo Sustentável.* 2nd Globolivros: SP, 2012. Water for Brazilian cities. Sustainable world [in Portuguese].

[3] E. Ghisi, "Potential for potable water savings by using rainwater in the residential sector of Brazil", *Build. Environ.,* vol. 41, no. 11, pp. 1544-1550, 2006. [http://dx.doi.org/10.1016/j.buildenv.2005.03.018]

[4] M. M. P. Silva, L. A. Oliveira, and C. R. Diniz, "Educação ambiental para o uso sustentável de cisternas em comunidades rurais da Paraíba", *Revista de Biologia e Ciências da Terra,* vol. 1, pp. 122-136, Jul 2006. Environmental education for sustainable use of tanks in rural communities of Paraiba. Journal of Biology and Earth Sciences.

[5] "Brazil. Agência Nacional de Água", *Conjuntura dos Recursos Hídricos no Brasil,* Brasília, . National Waters Agency. Conjuncture of Water Resources in Brazil [in Portuguese].

[6] "Brazil. Instituto Brasileiro de Geografia e Estatística", *Atlas do censo demográfico 2010,* Rio de Janeiro, . Brazilian Institute of Geography and Statistics. Atlas of the 2010 census [in Portuguese].

[7] "Brazil. Agência Nacional de Água", *Conjuntura dos Recursos Hídricos no Brasil: encarte especial sobre a crise hídrica,* Brasília, . National Waters Agency. Conjuncture of Water Resources in Brazil: special insert on the water crisis [in Portuguese].

[8] "Brazil. Sistema Nacional de Informações sobre Saneamento", *Diagnóstico dos Serviços de Água e Esgoto - 2014,* Brasília, . National Sanitation Information System. Diagnosis of Water and Sewer Services – 2014 [in Portuguese].

[9] E.S. Lage, Aproveitamento de água pluvial em concessionárias de veículos na cidade de belo horizonte: potencial de economia de água potável e estudo de viabilidade, Federal University of Santa Catarina: Brazil, 2010. Rainwater Harvesting in car dealerships in the city of Belo Horizonte: potential

for potable water savings and feasibility study [in Portuguese].

[10] "Brazil. Agência Nacional de Água", *Conservação e Reuso da Água em Edificações,* Brasília, . National Waters Agency. Water Conservation and Reuse in Buildings [in Portuguese].

[11] P. Tomaz, *Previsão de consumo de água.*. Navegar: São Paulo, SP, 2000. Prediction of water consumption [in Portuguese].

[12] E. Ghisi, and D.F. Ferreira, "Potential for potable water savings by using rainwater and greywater in a multi-storey residential building in southern Brazil", *Build. Environ.*, vol. 42, pp. 2512-2522, 2007. [http://dx.doi.org/10.1016/j.buildenv.2006.07.019]

[13] P.C. Kammers, *Projeto de instalações hidráulicas com aproveitamento de água pluvial em um edifício residencial multifamiliar localizado em Florianópolis.* Final Paper, Federal University of Santa Catarina: Brazil, 2007. Design of hydraulic systems with rainwater harvesting in a multi-family residential building in Florianópolis [in Portuguese].

[14] V. Senger, *Análise de viabilidade para aproveitamento de água de chuva e reúso de água em residência unifamiliar: estudo de caso em Panambi-RS.* Final Paper, Federal University of Santa Catarina: Brazil, 2007. Feasibility analysis for rainwater harvesting and water reuse in single-family residence: a case study in Panambi-RS [in Portuguese].

[15] G.M. Coelho, *Avaliação dos usos finais de água em residências unifamiliares localizadas em Blumenau-SC.* Undergraduate research report, Federal University of Santa Catarina: Brazil, 2008. Assessment of water end-uses in single-family homes located in Blumenau-SC [in Portuguese].

[16] D. Barreto, "Perfil do consumo residencial e usos finais da água", *Ambiente Construído,* vol. 8, no. 2, pp. 23-40, 2008. [Profile residential consumption and water end-uses]. [Built Environment].

[17] A.K. Marinoski, A.S. Vieira, A.S. Silva, and E. Ghisi, "Water End-Uses in Low-Income Houses in Southern Brazil", *Water,* vol. 6, pp. 1985-1999, 2014. [http://dx.doi.org/10.3390/w6071985]

[18] A.K. Ribeiro, Método para avaliação do impacto ambiental da implantação de sistemas integrados de aproveitamento de água pluvial e água cinza em residências unifamiliares a partir da análise do ciclo de vida, Federal University of Santa Catarina: Brazil, 2015. Method for assessing the environmental impact of the implementation of integrated rainwater harvesting and greywater systems in single-family homes from the life cycle analysis [in Portuguese].

[19] D.L. Meincheim, *Potencial de economia de água potável através do aproveitamento de água pluvial em uma residência unifamiliar localizada em São José – SC.* Final Paper, Federal University of Santa Catarina: Brazil, 2015. Potential for potable water savings through rainwater harvesting in a single-family residence in São José – SC [in Portuguese].

[20] M.E. Dalsenter, *Estudo de potencial de economia de água potável por meio do aproveitamento de água pluvial em um condomínio residencial multifamiliar localizado em Florianópolis – SC.* Final Paper, Federal University of Santa Catarina: Brazil, 2016. Study of potential of potable water savings through rainwater harvesting in a multifamily residential condominium in Florianópolis – SC [in Portuguese].

[21] A. Nagalli, F. B. Freire, K. Q. Carvalho, F. H. Passig, and C. A. Teixeira, "Domestic water consumption on an upper middle class community in Curitiba, Brazil", *Advances in Water Resource and Protection,* vol. 1, no. 4, 2013.

[22] P. Kammers, and E. Ghisi, "Usos finais de água em edifícios públicos localizados em Florianópolis-SC", *Ambiente Construído,* vol. 6, no. 1, pp. 75-90, 2006. [Water end-uses in public buildings in Florianópolis, SC]. [Built Environment].

[23] A.N. Botelho, *Estimativa dos usos finais de água potável na Universidade Federal de Santa Catarina.* Undergraduate research report, Federal University of Santa Catarina: Brazil, 2008. Estimate of the potable water end-uses in the Federal University of Santa Catarina [in Portuguese].

[24] A.K. Marinoski, and E. Ghisi, "Aproveitamento de água pluvial para fins não potáveis em instituição de ensino: estudo de caso em Florianópolis – SC", *Ambiente Construído,* vol. 8, pp. 67-84, 2008. [Rainwater Harvesting for non-potable purposes in an educational institution: a case study in Florianópolis – SC]. [Built Environment].

[25] G.B. Fasola, E. Ghisi, A.K. Marinoski, and J.B. Borinelli, "Potencial de economia de água em duas escolas em Florianópolis, SC", *Ambiente Construído,* vol. 11, no. 4, pp. 65-78, 2011. [Potential of water savings in two schools in Florianópolis, SC]. [Built Environment]. [http://dx.doi.org/10.1590/S1678-86212011000400006]

[26] L.C. Proença, and E. Ghisi, "Assessment of Potable Water Savings in Office Buildings Considering Embodied Energy", *Water Resour. Manage.,* vol. 27, pp. 581-599, 2013. [http://dx.doi.org/10.1007/s11269-012-0203-1]

[27] E.A.A. Nascimento, and D. Sant'ana, "Caracterização dos usos-finais do consumo de água em edificações do Setor Hoteleiro de Brasília", In: *III Seminário nacional de construções sustentáveis.* Passo Fundo: Brazil, 2014. Characterization of the water end-uses consumption in the building Sector Hotelier Brasilia. III National sustainable buildings seminary [in Portuguese]. [http://dx.doi.org/10.18256/2318-1109/arqimed.v3n2p156-167]

[28] S. May, Caracterização, tratamento e reuso de águas cinzas e aproveitamento de águas pluviais em edificações, University of São Paulo. São Paulo: Brazil, 2009. Characterization, water treatment and greywater reuse and rainwater harvesting in buildings [in Portuguese].

[29] E. Ghisi, A. Montibeller, and R.W. Schmidt, "Potential for potable water savings by using rainwater: an analysis over 62 cities in southern Brazil", *Build. Environ.,* vol. 41, no. 2, pp. 204-210, 2006. [http://dx.doi.org/10.1016/j.buildenv.2005.01.014]

[30] E. Ghisi, D.L. Bressan, and M. Martini, "Rainwater tank capacity and potential for potable water savings by using rainwater in the residential sector of southeastern Brazil", *Build. Environ.,* vol. 42, pp. 1654-1666, 2007. [http://dx.doi.org/10.1016/j.buildenv.2006.02.007]

[31] M.R. Vill, *Aproveitamento de água pluvial para fins não potáveis: Viabilidade econômica para as cidades de Santana do Ipanema, Florianópolis e Santos.* Undergraduate research report, Federal University of Santa Catarina: Florianópolis, 2010. Rainwater Harvesting for non-potable purposes: Economic feasibility for the cities of Santana do Ipanema, Florianópolis and Santos [in Portuguese].

[32] H. Berwanger, and E. Ghisi, "Investment feasibility analysis of rainwater harvesting in the city of Itapiranga, Brazil", *International Journal of Sustainable Human Development,* vol. 2, pp. 104-114, 2014.

[33] R.P. Paes, G.C. Silva, J.C. Priante, E.B. Lima, and N.P. Filho, "Aplicação de tecnologias de conservação do uso da água através do reuso - Estudo de caso Cuiabá, MT", *Revista Brasileira de Recursos Hídricos,* vol. 15, no. 3, pp. 97-107, 2010. [Application of water conservation technologies through reuse - Case Study Cuiaba, MT]. [Brazilian Journal of Water Resources]. [http://dx.doi.org/10.21168/rbrh.v15n3.p97-107]

[34] E. Ghisi, and S.M. And, "Oliveira, "Potential for potable water savings by combining the use of rainwater and greywater in houses in southern Brazil", *Build. Environ.,* vol. 42, pp. 1731-1742, 2007. [http://dx.doi.org/10.1016/j.buildenv.2006.02.001]

[35] G.H. Poyer, *Análise do potencial de economia de água potável e da energia embutida na aplicação de sistema de aproveitamento de água da chuva em edifícios públicos de Florianópolis.* Final Paper, Federal University of Santa Catarina: Florianópolis, 2012. analysis of the potential for potable water savings and energy embodied in the application of rainwater utilization system in public buildings in Florianopolis[in Portuguese].

[36] R.F.W. Paes, *Elaboração de projeto de aproveitamento de água de chuva, para edificação de centro de desenvolvimento tecnológico e capacitação de pessoas, com estudo de água potável poupada e viabilidade econômica de implantação.* Final Paper, Federal University of Santa Catarina: Brazil,

2015. Preparation of rainwater harvesting project for the building of technology development center and training of people, with study of potential for potable water saving and economic feasibility of implementation [in Portuguese].

[37] E. Ghisi, R.F. Rupp, and Y. Triska, "Comparing indicators to rank strategies to save potable water in buildings", *Resour. Conserv. Recycling,* vol. 87, pp. 137-144, 2014. [http://dx.doi.org/10.1016/j.resconrec.2014.04.001]

[38] E. Ghisi, D.F. Tavares, and V.L. Rocha, "Rainwater harvesting in petrol stations in Brasília: potential for potable water savings and investment feasibility analysis", *Resour. Conserv. Recycling,* vol. 54, pp. 79-85, 2009. [http://dx.doi.org/10.1016/j.resconrec.2009.06.010]

[39] I.C. Carvalho, M.L. Calijuri, P.P. Assemany, M.D. Silva, R.F. Neto, A.F. Santiago, and M.H. Souza, "Sustainable airport environments: A review of water conservation practices in airports", *Resour. Conserv. Recycling,* no. 74, pp. 27-36, 2013. [http://dx.doi.org/10.1016/j.resconrec.2013.02.016]

[40] E. Ghisi, R.F. Rupp, and U. Munarim, "Comparação de métodos para dimensionamento de reservatórios de água pluvial", *Ambiente Construído,* vol. 11, pp. 47-64, 2011. [Comparison of methods for dimensioning of rainwater tanks]. [Built Environment]. [http://dx.doi.org/10.1590/S1678-86212011000400005]

[41] S. M. C. Bezerra, P. Christan, C. A. Teixeira, and K. Farahbakhsh, *Dimensionamento de reservatório para aproveitamento de água de chuva: comparação entre métodos da ABNT NBR 15527:2007 e Decreto Municipal 293/2006 de Curitiba, PR,* 2010. [http://dx.doi.org/10.1590/S1678-86212010000400015]

[42] F. Dornelles, Aproveitamento de água de chuva no meio urbano e seu efeito na drenagem pluvial, Federal University. of Rio Grande do Sul: Brazil, 2012. Rainwater use in the urban environment and its effect on rainwater drainage [in Portuguese].

[43] E. Ghisi, "Parameters Influencing the Sizing of Rainwater Tanks", *Water Resour. Manage.,* vol. 24, pp. 2381-2403, 2010. [http://dx.doi.org/10.1007/s11269-009-9557-4]

[44] M.M. Cordova, and E. Ghisi, "Analysis of Potable Water Savings Using Behavioural Models", *Water Conservation,* vol. 1, pp. 89-104, 2011.

[45] E. Ghisi, and M.M. Cordova, "Manual do usuário" labeee.ufsc,br", *[online]. Available: http://www.labeee.ufsc.br,* 2016.Accessed: 3th aug. 2016 Netuno 4: User Manual [in Portuguese].

[46] S. V. Amorim, and D. J. A. Pereira, "Estudo comparativo dos métodos de dimensionamento para reservatórios utilizados em aproveitamento de água pluvial", *Ambiente Construído,* vol. 8, no. 2, pp. 53-66, 2008. [Comparative study of design methods for tanks used in rainwater harvesting]. [Built Environment].

[47] P.H.B. Pozzebon, and M.C.C. Gastaldini, "Dimensionamento e simulação do comportamento de um reservatório para aproveitamento de água da chuva para uso não potável em Santa Maria – RS", *Simpósio brasileiro de recursos hídricos,* 2013pp. 1-8 Bento Gonçalves: Brazil Design and simulation of the behavior of a tank for rainwater utilization for non-potable use in Santa Maria – RS. Brazilian Symposium on water resources [in Portuguese].

[48] M. Andrade, A.K. Marinoski, and H.R. Becker, "Sistemas de aproveitamento de água de chuva", In: *Casa eficiente: uso racional da água.* UFSC: Florianópolis, SC, 2010. Rainwater harvesting systems. Efficient home: rational water use [in Portuguese].

[49] "Brazil. Associação Brasileira de Normas Técnicas", *NBR 15527 – Água da Chuva: Aproveitamento de coberturas em áreas urbanas para fins não potáveis – requisitos,* Rio de Janeiro, . Brazilian Association of Technical Standards. NBR 15527 - Rainwater: Harvesting covers in urban areas for non-potable purposes - requirements [in Portuguese].

[50] "Brazil. Ministério da Saúde", *Portaria 2.914, de 12 de dezembro de 2011,* Brasília, . Ministry of Health. Ordinance 2914 of December 12, 2011 [in Portuguese].

[51] "Brazil. Conselho Nacional do Meio Ambiente", *Resolução n° 357 de 17 de março de 2005,* Brasília, . National Environmental Council. Resolution no. 357 of March 17, 2005 [in Portuguese].

[52] M. C. C. Amorim, and E. R. Porto, "Considerações sobre controle e vigilância da qualidade de água de cisternas e seus tratamentos", *Simpósio Brasileiro de Captação e Manejo de Água de Chuva,* 2003 Juazeiro: Brazil Considerations about control and surveillance of cisterns water quality and its treatments. Brazilian Symposium on Capture and Rainwater Management [in Portuguese].

[53] E.H. Souza, S.M.G.L. Montenegro, S.M. Santos, and S.G.S. Pessoa, "Avaliação da qualidade da água e de barreiras sanitárias em sistemas de aproveitamento de águas de chuva", *Revista Brasileira de Recursos Hídricos,* vol. 16, pp. 81-93, 2011. [Water quality assessment and sanitary barriers in rainwater harvesting systems]. [Brazilian Journal of Water Resources]. [http://dx.doi.org/10.21168/rbrh.v16n3.p81-93]

[54] S. May, Estudo da viabilidade do aproveitamento de água de chuva para consumo não potável em edificações, University of São Paulo: Brazil, 2004. Study of the viability of rainwater harvesting for non-potable consumption in buildings [in Portuguese].

[55] V.C. Silva, and V.L. Pádua, "Qualidade microbiológica de água de chuva armazenada em cisternas de placas, construídas em comunidades rurais do município de Araçuaí – MG", *VI Brazilian Symposium on Capture and Rainwater Management,* 2007 Belo Horizonte: Brazil

[56] A.S. Silva, E.F. Fay, I.B. Sá, and L.T.L. Brito, "Perfil do semiárido rural brasileiro: avaliação microbiológica das águas de uso domiciliar", *IX Simpósio Brasileiro de Captação e Manejo de Água de Chuva,* 2014 Feira de Santana: Brazil Profile of the Brazilian semi-arid rural: microbiological evaluation of water of household use. IX Brazilian Symposium on Capture and Rainwater Management [in Portuguese].

[57] J.C.A.L. Lima, Avaliação do desempenho do dispositivo de desvio das primeiras águas da chuva utilizada em cisternas no semiárido pernambucano, Federal University of Pernambuco: Brazil, 2012. Performance evaluation of the bypass device of the first rainwater in cisterns used in the semiarid of Pernambuco [in Portuguese].

[58] M.L. Figueiras, *Avaliação da influência do descarte das primeiras águas de chuva sobre a qualidade bacteriológica da água captada em telhado.* Final Paper, Federal University of Pernambuco: Brazil, 2013. The evaluation of the influence of discarding of the first rainwater on the bacteriological quality of the water captured in roof [in Portuguese].

[59] L.B. Guilherme, Aproveitamento das águas de chuva para a cidade de Natal para fins potáveis, Federal University of Rio Grande do Norte: Brazil, 2006. Rainwater harvesting for the Natal city for potable purposes [in Portuguese].

[60] N.M.D. Silva, L.W. Perello, and L.R.S. Moraes, "Qualidade bacteriológica da água de chuva em cisternas e seus fatores intervenientes: estudo em Inhambupe-BA com a utilização de planejamento experimental fatorial", *IX Simpósio Brasileiro de Captação e Manejo de Água de Chuva,* 2014 Feira de Santana: Brazil Bacteriological quality of rainwater in cisterns and its intervening factors: a study in Inhambupe-BA with the use factorial experimental planning. IX Brazilian Symposium on Capture and Rainwater Management [in Portuguese].

[61] F. Alves, T. Köchling, J. Luz, S.M. Santos, and S. Gavazza, "Water quality and microbial diversity in cisterns from semiarid areas in Brazil", *J. Water Health,* vol. 12, no. 3, pp. 513-525, 2014. [http://dx.doi.org/10.2166/wh.2014.139] [PMID: 25252355]

[62] R.C. Jaques, Qualidade da água da chuva no município de Florianópolis e sua potencialidade para aproveitamento em edificações, Federal University of Santa Catarina: Brazil, 2005. Rainwater quality in the city of Florianópolis and its potential for use in buildings [in Portuguese].

[63] K.P.L. Annecchini, Aproveitamento de água de chuva para fins não potáveis na região metropolitana de Vitória (ES), Federal University of Espírito Santo: Brazil, 2005. Rainwater harvesting for non-potable purposes in the metropolitan region of Vitória (ES) [in Portuguese].

[64] F.A. Silva, K.Q. Carvalho, and C.A. Teixeira, "Tratamento da Água da Chuva Através de Filtração Rápida Ascendente em Tripla Camada e Radiação Ultravioleta", *28 Congresso Brasileiro de Engenharia Sanitária e Ambiental,* 2015 Rio de Janeiro: Brazil Treatment of Rainwater Through Ascending Quick Filtration in Triple Layer and Ultraviolet Radiation. 28 Brazilian Congress of Sanitary and Environmental Engineering [in Portuguese].

[65] A.S. Vieira, E. Ghisi, and M. Weeber, "Self-cleaning filtration: A novel concept for rainwater harvesting systems", *Resour. Conserv. Recycling,* vol. 78, pp. 67-73, 2013. [http://dx.doi.org/10.1016/j.resconrec.2013.06.008]

[66] M.A. Budel, S.M.C. Bezerra, M. Giacchini, K. Farahbakhsh, and C.A. Teixeira, "Estudo Comparativo da qualidade da água de chuva coletada em cobertura convencional e em telhado verde", In: *Prospectos Contemporâneos em Engenharia Civil.* 1st Pandion: Florianópolis, SC, 2013. Comparative study of rainwater quality collected in conventional roof and green roof. Contemporaries Prospects in Civil Engineering [in Portuguese].

[67] D.G. Brencht, and D.M. Pires, *Aproveitamento da água pluvial por meio de sua captação e filtragem através de sistemas de pavimentos drenantes.* Undergraduate research report, Federal University of Santa Catarina: Brazil, 2015. Rainwater harvesting through its catchment and filtering through draining flooring systems [in Portuguese].

[68] D.G. Brecht, *Sistemas de pavimento drenante: potencial de captação e filtragem da água pluvial para aproveitamento em edificações.* Undergraduate research report, Federal University of Santa Catarina: Brazil, 2015. Rainwater harvesting through its catchment and filtering through draining flooring systems [in Portuguese].

[69] L.N. Antunes, L.P. Thives, and E. Ghisi, "Potential for Potable Water Savings in Buildings by Using Stormwater Harvested from Porous Pavements", *Water,* vol. 8, pp. 110-128, 2016. [http://dx.doi.org/10.3390/w8040110]

[70] L.S. Philippi, P.H. Sezerino, M.R. Peters, and F.R. Lapolli, "Reuso combinado de águas cinzas (grey water) e água de chuva em uma unidade residencial", *23° Congresso Brasileiro de Engenharia Sanitária e Ambiental, ABES,* 2005pp. 1-7 Campo Grande: Brazil Combined reuse of greywater and rainwater in a residential unit. 23 Brazilian Congress of Sanitary and Environmental Engineering, ABES [in Portuguese].

[71] L.Z. Borges, Caracterização da água cinza para promoção da sustentabilidade dos recursos hídricos, Federal Unversity of Paraná: Brazil, 2003. Greywater characterization to promote the sustainability of water resources [in Portuguese].

[72] B.B. Bazzarella, Caracterização e aproveitamento de água cinza para uso não-potável em edificações, Federal University of Espírito Santo: Brazil, 2005. Characterization and harvesting of greywater for non-potable use in buildings [in Portuguese].

[73] "Brazil. Associação Brasileira de Normas Técnicas", *NBR 13969 - Tanques sépticos - Unidades de tratamento complementar e disposição final dos efluentes líquidos - Projeto, construção e operação,* Rio de Janeiro, . Brazilian Association of Technical Standards. NBR 13969 - Septic tanks - complementary treatment units and final disposal of liquid effluents - Project, construction and operation [in Portuguese].

[74] S. Fiori, V.M.C. Fernandes, and H.S. Piozzo, "Avaliação do potencial de reuso de águas cinzas em edificações", *I Conferência latino-americana de construção sustentável - X Encontro Nacional de Tecnologia do Ambiente Construído,* 2004 São Paulo: Brazil Assessment of greywater reuse potential in buildings. I Latin American conference of sustainable construction - X National Meeting of Built Environment Technology [in Portuguese].

[75] A. Fonini, V.M.C. Fernandes, and H.S. Pizzo, "Estudo para determinação da capacidade de aproveitamento das águas pluviais e das águas cinzas em um complexo esportivo universitário", *I Conferência latino-americana de construção sustentável - X Encontro Nacional de Tecnologia do Ambiente Construído,* 2004 São Paulo: Brazil Study to determine the capacity of rainwater harvesting and greywater in a university sports complex. I Latin American conference of sustainable construction - X National Meeting of Built Environment Technology [in Portuguese].

[76] A.N. Botelho, *Uso racional de água no campus da UFSC.* Undergraduate research report, Federal University of Santa Catarina: Florianópolis, 2006. Rational use of water on the campus of UFSC [in Portuguese].

[77] D.C. Santos, M.B. Lobato, N.M.P. Volpi, and L.Z. Borges, "Hierarquização de medidas de conservação de água em edificações residenciais com auxílio da análise multicritério", *Ambiente Construído,* vol. 6, no. 1, pp. 31-47, 2006. [Tiering water conservation measures in residential buildings with the aid of multi-criteria analysis]. [Built environment].

[78] A. Kalbusch, and E. Ghisi, "Método para quantificação do consumo energético no ciclo de vida de equipamentos hidrossanitários", *Ambiente Construído,* vol. 12, pp. 57-73, 2012. [Method for quantification of energy consumption in the life cycle of hydro-sanitary equipment]. [Built environment].
[http://dx.doi.org/10.1590/S1678-86212012000300005]

[79] A. Kalbusch, and E. Ghisi, "Comparative life-cycle assessment of ordinary and water-saving taps", *J. Clean. Prod.,* vol. 112, pp. 4585-4593, 2016.
[http://dx.doi.org/10.1016/j.jclepro.2015.06.075]

[80] A.S. Vieira, and E. Ghisi, "Water-energy nexus in houses in Brazil: comparing rainwater and gray water use with a centralized system", *Water Science and Technology: Water Supply (Print),* vol. 16, pp. 274-283, 2016.

[81] A.S. Vieira, and E. Ghisi, "Water-energy nexus in low-income houses in Brazil: the influence of integrated on-site water and sewage management strategies on the energy consumption of water and sewerage services", *J. Clean. Prod.,* vol. 133, pp. 145-162, 2016.
[http://dx.doi.org/10.1016/j.jclepro.2016.05.104]

[82] L.C. Proença, E. Ghisi, D.F. Tavares, and G.M. Coelho, "Potential for electricity savings by reducing potable water consumption in a city scale", *Resour. Conserv. Recycling,* vol. 55, pp. 960-965, 2011.
[http://dx.doi.org/10.1016/j.resconrec.2011.05.003]

[83] J. Gomes, D. C. Weber, and M. D. Delong, "Dimensionamento do reservatório de águas pluviais, usando um critério financeiro", *Revista Brasileira de Recursos hídricos,* vol. 15, no. 1, pp. 89-100, 2010. Dimensioning of rainwater reservoir, using financial criteria. Brazilian Journal of Water Resources [in Portuguese].

[84] F. Martini, *Potencial de economia de água potável por meio do uso de água de chuva em São Miguel do Oeste – SC.* Final Paper, Federal University of Santa Catarina: Brazil, 2009. Potential for potable water savings through rainwater harvesting in São Miguel do Oeste – SC. [in Portuguese]

[85] G.B.A. Júnior, I.C.S. Dias, and C.L.M. Gadelha, "Viabilidade econômica e aceitação social do aproveitamento de águas pluviais em residências na cidade de João Pessoa", *Ambiente Construído,* vol. 8, no. 2, pp. 85-98, 2008. [Economic viability and social acceptance of the rainwater harvesting in residences in João Pessoa city]. [Built environment].

[86] D.R.M. Fernandes, V.B.M. Neto, and K.M.C. Mattos, "Viabilidade econômica do uso da água da chuva: um estudo de caso da implantação de cisterna na UFRN/RN", *27 National Production Engineering Meeting,* 2007pp. 1-9 Foz do Iguaçu: Brazil

[87] L.K. Maeda, and J.L.B. Brandão, "Avaliação técnica e econômica de um sistema de captação de água pluvial", *XVIII Brazilian Symposium on Water Resources,* 2009pp. 1-20 Campo Grande: Brazil

[88] A.K. Marinoski, A.S. Silva, A.S. Vieira, and E. Ghisi, "Public acceptance of alternative water sources", In: *Water Conservation: Practices, Challenges and Future Implications.*. 1st Nova Science Publishers: Nova York, 2013, pp. 1-24.

[89] "Articulação Semiárido Brasileiro: Programa Um Milhão de Cisternas", *Available: http://www.asabrasil.org.br/acoes/p1mc#contexto-p1mc.*, 2016.Accessed: 8th Sep. 2016 Articulation Brazilian Semi-Arid: Programme One Million Cisterns.

[90] R.L.B. Nóbrega, C.O. Galvão, L.R. Palmier, and B.S.O. Cebalhos, "Aspectos político-institucionais do aproveitamento de água de chuva em áreas rurais do semiárido brasileiro", *Revista Brasileira de Recursos Hídricos,* vol. 17, no. 4, pp. 109-124, 2012. [Political and institutional aspects of rainwater harvesting in rural areas of the Brazilian semi-arid]. [Brazilian Journal of Water Resources]. [http://dx.doi.org/10.21168/rbrh.v17n4.p109-124]

[91] U.A.F. Gomes, L. Heller, and J.L. Pena, "A national programme for large scalerainwater harvesting: an individual or public responsibility?", *Water Resour. Manage.,* vol. 26, pp. 2703-2714, 2012. [http://dx.doi.org/10.1007/s11269-012-0041-1]

[92] M.I.C. Xavier, T.T. Saturno, M.A.F. Santos, and F.A.F. Souza, "Educação ambiental e meio rural: ações integradas em busca de uma melhoria na qualidade da água que abastece a comunidade de Lagoa de São João, município de Princesa Isabel/PB", *IX Simpósio Brasileiro de Captação e Manejo de Água de Chuva,* 2014 Feira de Santana: Brazil Environmental education and rural areas: integrated actions in search of a better water quality that supplies the community of Lagoa de São João, municipality of Princesa Isabel/PB. IX Brazilian Symposium on Capture and Rainwater Management [in Portuguese].

[93] M.J. Santos, D.C. Andrade, G.M. Santos, and P. Souza, "A educação ambiental crítica inserida no processo de execução do projeto P1+2 em escolas no semiárido de Sergipe", *IX Simpósio Brasileiro de Captação e Manejo de Água de Chuva,* 2014 Feira de Santana: Brazil The critical environmental education inserted into the project execution process P1+2 in schools in the semiarid of Sergipe. IX Brazilian Symposium on Capture and Rainwater Management [in Portuguese].

[94] J.L.A. Santos, and S.M.F. Dias, "Análise das estratégias educacionais presentes nos cursos de gerenciamento em recursos hídricos e sua relação com a melhoria da qualidade de vida", *IX Simpósio Brasileiro de Captação e Manejo de Água de Chuva,* 2014 Feira de Santana: Brazil Analysis of educational strategies presents in management courses in water resources and its relation to improving quality of life. IX Brazilian Symposium on Capture and Rainwater Management [in Portuguese].

[95] N.S. Oliveira, M.A.F. Santos, and J.L.A. Santos, "Ações de educação ambiental integradas entre o contexto de utilização de água de chuva para o consumo humano e o ensino profissionalizante na comunidade Caatinga do Vieira, Serrinha/BA", *IX Simpósio Brasileiro de Captação e Manejo de Água de Chuva,* 2014 Feira de Santana: Brazil Environmental education actions integrated between the rainwater harvesting context for human consumption and the professionalizing education in the Caatinga do Vieira community, Serrinha/BA. IX Brazilian Symposium on Capture and Rainwater Management [in Portuguese].

[96] A.P. Dias, T. Shubo, G.M.M. Gondim, and A.H.A.M. Neto, "Educação territorializada como estratégia para a melhoria da qualidade da água de cisterna e da saúde ambiental", *IX Simpósio Brasileiro de Captação e Manejo de Água de Chuva,* 2014 Feira de Santana: Brazil Territorialised education as a strategy for improving the cistern water quality and environmental health. IX Brazilian Symposium on Capture and Rainwater Management [in Portuguese].

[97] P.S.A. Junior, M.P. Lessa, J.O. Pessoa, and E. Cohim, "Percepção e utilização de duas fontes de abastecimento de água na zona rural do município de Várzea da Roça – Bahia", *IX Simpósio Brasileiro de Captação e Manejo de Água de Chuva,* 2014 Feira de Santana: Brazil Perception and use of two water supply sources in rural zone of municipality of Roça - Bahia. IX Brazilian Symposium on Capture and Rainwater Management [in Portuguese].

[98] G. Almeida, Metodologia para caracterização de efluentes domésticos para fins de reuso: estudo em Feira de Santana, Federal University of Bahia: Brazil, 2007. Methodology for characterizing domestic

wastewater for reuse purposes: a study in Feira de Santana [in Portuguese].

[99] K.L.G. Peixoto, H.S. Noguchi, A.L. Souza, and A.A. Santos, "Sensilibização a cerca do aproveitamento da água da chuva em escola", *IX Simpósio Brasileiro de Captação e Manejo de Água de Chuva,* 2014 Feira de Santana: Brazil Raising awareness about the rainwater harvesting in school. IX Brazilian Symposium on Capture and Rainwater Management [in Portuguese].

[100] "Brazil. Prefeitura Municipal de Curitiba", *Decreto nº 212, de 29 de março de 2007: Aprova o Regulamento de Edificações do Município de Curitiba e dá outras providências,* Prefeitura Municipal de Curitiba: Curitiba, . Municipal Government of Curitiba. Decree no. 212 of 29 March 2007: Approves the Regulations of Buildings of the Curitiba City and other measures, Municipal Government of Curitiba [in Portuguese].

[101] "Brazil. Prefeitura Municipal de Curitiba [City Hall of Curitiba]", *Decreto no 293, de 28 de março de 2006: Dispõe sobre os critérios do uso e conservação racional da água nas edificações e dá outras providências,* Curitiba, . Decree No. 293 of March 28, 2006: Regulates the criteria of rational use and conservation of water in buildings and other measures [in Portuguese].

CHAPTER 2

Rainwater Tanks to Save Water in Buildings: An Australian Perspective

M. Ashiqur Rahman, Md Mahmudul Haque, Amir Ahmed and **Ataur Rahman**[*]

School of Computing, Engineering and Mathematics, Western Sydney University, Sydney, NSW 2751, Australia

Abstract: Water supply is an essential service to buildings along with other services such as electricity supply and telephone. Many of the new cities in developing countries do not have 24-hour safe drinking water in their buildings, and water supply is often intermittent and is of poor quality. In contrast, developed countries use a much higher volume of potable water in their buildings, which can be reduced significantly. In urban areas, sustainability and water efficiency of buildings are of great significance as many water authorities have been struggling to meet the increasing water demand due to higher populations. The climate change is bringing more extremity in the climate systems such as long and frequent droughts and higher temperature episodes, which increase water demand significantly when water availability is the lowest. This chapter presents water recycling and reuse in buildings with a special focus on rainwater harvesting systems in Australia to reduce potable water demand within buildings. It has been found that rainwater harvesting is one of the most popular means of alternative water supplies in buildings; however, there are further scopes of development in the rainwater harvesting systems by bringing innovations into these systems to make future buildings more water-efficient.

Keywords: Green building, Greywater, Mains water, Rainwater tank, Rain water, Water supply, Water recycling, Water efficiency, Waste water, Storm water, Water conservation.

INTRODUCTION

A building needs a variety of services such as electricity, water, telephone, sewerage system and air-conditioning. Among these services, water supply is of special significance as its presence is a prerequisite to few other essential services such as sewage disposal system. In many developing countries, buildings do not

[*] **Corresponding author Ataur Rahman:** School of Computing, Engineering and Mathematics, Western Sydney University, Building XB2.48, Locked Bag 1797, Penrith, NSW 2751, Australia; Tel: 614-47360145; E-mails: a.rahman@westernsydney.edu.au; a.rahman@uws.edu.au

Enedir Ghisi (Ed.)

have continuous water supply *e.g.* water is supplied in morning in a day or once or twice per week. For example, in a megacity like Dhaka (the capital of Bangladesh with population over 16 million), although drinking water supply coverage was 87% in 2015, only 32% of the urban population lived in households that had piped water supply [1]. In 2011, rainwater harvesting for new houses was made mandatory towards solving water scarcity and flooding problems in Dhaka city [2]; however, it has not solved the water supply problem of Dhaka. The water supply in Dhaka is regarded as unsafe, and hardly anyone drinks this water without boiling. This is a typical problem to many large cities in the developing countries.

Presence of safe drinking water can be taken as the sign of 'balanced development' *i.e.* its absence in buildings can be regarded as lack of development. In fact, one can make a flushing building with sophisticated communication systems, but to provide clean drinking water and sewage disposal system is far more challenging as it needs much bigger investment and constant maintenance and monitoring. In a large city like Sydney, the approval of any new residential area is highly dependent on adequate water supply, which is not guaranteed. Even in peri-urban Sydney, many people depend on roof harvested rainwater to meet all their water demands as they are not connected to city's water mains.

Reduction of potable water demand in a building is highly sought in both developed and developing countries as cities are ever expanding with time, which increases water demand to a challenging level. Also, during drought conditions, water supply is stressed and authorities seek measures to reduce water demands *e.g.* application of mandatory water use restrictions. For the last twenty years or so, provision of alternative water supplies is well discussed and debated covering wastewater recycling, rainwater harvesting and greywater reuse among other sources. This chapter focuses how water demand in buildings can be supplemented by rainwater harvesting with a special focus to Australian condition.

WATER USE AND RECYCLING IN BUILDINGS

The use of potable water in buildings can be minimized significantly by pursuing two different water reuse techniques, namely rainwater harvesting and greywater reuse systems. These techniques can be adopted either separately or in a combination, especially for purposes that do not require potable water such as flushing toilets and gardening. The rainwater harvesting system mainly requires two elements - a collecting surface (usually a roof) and a water storage tank (above or below ground). In a rainwater harvesting system, possible water contamination may occur during the precipitation if the air itself is polluted (*e.g.*

acid rain) or if the collection surface contains accumulated pollutants during dry weather period. A first flush device is generally fitted with the rainwater harvesting system, which diverts first part of the runoff containing high level of pollution. In the greywater system, water is collected from less contaminated sources within a building such as bathing and washing. This water usually does not come into contact with high levels of contamination, such as sewage, and thus this water has the potential for reuse or recycling in a building. Hence, due to the low level of contamination, the harvested rainwater and greywater may be treated on-site with a simple treatment system, and has the potential to be used as potable water, which can ultimately reduce the water demand.

Australia has the second largest domestic water use rate with about 500 l/p/d (liter/person/day), followed by Italy (400 l/p/d) and Japan (380 l/p/d), while the USA has the highest water consumption rate (above 500 l/p/d) [3]. A large part of this water consumption is attributed to water use in buildings that consists of direct consumption by human beings, and water usage for various household purposes such as cooking, bathing, toilet flushing, cloth washing, general cleaning inside the households, car and hard surface washing, plant watering, filling swimming pools and recreational fountains and feeding pets/livestock.

Household sector is the second highest water user in Australia after agriculture sector [4]. Interestingly, only a small portion of residential water use belongs to potable needs. For example, potable use comprises only 9% of indoor water usage as reported in a Western Australian study [5]. It has been reported that the remaining 91% non-potable domestic water demand can be met by alternative water sources [6]. An important demographic study by Hurlimann [7] on domestic water use and satisfaction level with alternative water sources in Victoria, Australia revealed that garden watering had the highest usage among all the building water uses. Water from laundry has the potential to meet about 30% of garden water demand.

Table **1** represents water use for 2001 in twenty two largest cities in Australia. This table shows that 59% of total water use belongs to domestic water use - both inside and outside of a household. The end use of recycled water was found to be limited to landscape irrigation, car washing and toilet flushing. In Hong Kong, sea water is used for flushing about 70% of its toilets [8]. Saudi Arabia is an arid country where temperature is quite high, and there is no large river system and lake; in order to cope with the increasing water demand the Saudi Government has developed 33 desalination plants, and has emerged as the largest producer of desalinated water in the world [9]. Studies on application of sustainable architectural principles for developing water efficient residential buildings in Saudi Arabia have resulted in few modifications to traditional building design to achieve increased water savings in buildings.

Numerous researches have been conducted on rainwater harvesting and reuse of treated greywater to promote potable water savings within a building. These studies have suggested different water saving measures such as use of rainwater and reuse of greywater separately or in a combination of these two. There have been debates on the cost effectiveness of rainwater harvesting and greywater reuse. According to Rahman *et al.* [10], rainwater harvesting has received greater attention because collection of fresh rainwater from nature is relatively easy and it can be used readily for non-potable purposes without formal treatment. A study conducted in Queensland, Australia mentioned that roof harvested rainwater can be used safely for toilet flushing and hot water purpose [11]. Rainwater can also be used safely for cloth washing purpose [12].

Table 1. Water use in 22 largest cities in Australia [13] [#].

Water Use Type	Volume (ML)	Percentage of Total Water Use	Percentage of Residential Water Use
Residential garden watering	414,000	20.1%	34%
Toilet flushing	244,000	11.8%	20%
Laundry	183,000	8.9%	15%
Bathroom	317,000	15.3%	26%
Kitchen	61,000	3.0%	5%
Total residential use	**1,219,000**	**59.0%**	**100%**
Industrial & commercial	437,000	21.2%	
Local government, parks and fire fighting	139,000	6.7%	
System losses	221,000	10.7%	
Customer meter errors	49,000	2.4%	
Total	**2,065,000**	**100%**	

[#] (Note: combined population of these cities are 13.4 million. The primary source of these data is Water Services Association of Australia [14]).

A research by Ghisi and Ferreira [15] involving a multi-storied residential building in Brazil found that potable water saving from the rainwater harvesting system was 14.7%-17.7%, whereas it was 28.7%-34.8% for the reclaimed greywater. A combined use of rainwater and greywater was reported a water saving of about 42.7%. Similarly, the study by Zhang *et al.* [16] in rural areas of Western Australia showed that a water saving of 32.5% could be achieved from greywater reuse, while rainwater harvesting could achieve about 25.1% water saving. In contrast, Muthukumaran *et al.* [17] reported that use of rain water in a purpose-built household in regional Victoria, Australia could achieve up to 40% water saving.

Since 1990's, Australia has embraced Water Sensitive Urban Design (WSUD) in the urban planning and development that aims to save potable water [18]. The Building Sustainability Index (referred to as BASIX) has also been introduced by New South Wales Department of Planning in Australia [19]. BASIX is a web-based tool to measure the potential performance of new residential dwellings against a range of sustainability indices. BASIX requires a saving of at least 40% potable water in all the new houses in New South Wales by adopting a range of water saving measures including installation of rainwater tanks for toilet flushing and garden watering.

Notably, a number of researchers have focused on efficient toilet flushing in order to minimize potable water use within buildings. In contrast, many studies have evaluated water saving potential by using recycled water for toilet flushing. Fig. **1** shows how modern toilet in Japan has been designed with sanitary fittings that reutilize water from washing basin for toilet flushing. This type of toilet flushing needs only 4.8 liters water per flush [20]; moreover, TOTO's latest model, Neorest hybrid toilet series is fitted with Hybrid Ecology System that offers additional environmental benefits.

Proença and Ghisi [21] examined water end-uses in Brazilian office buildings, and found that toilet flushing alone comprised of 52-84% of total water consumption in ten buildings. It was suggested that installation of more efficient toilet flushing system is important to reduce the potable water demand along with use of rainwater and greywater to achieve a significant water saving target in new office buildings in Brazil. Taleb and Sharples [22] noted that some suggested modifications to their studied building such as low-flow tap aerators in kitchen, bathroom and shower along with dual-flush cisterns, efficient washing machine and a greywater system were capable of collecting 90% bath and shower wastewater to supply water to toilet cisterns. A combination can achieve approximately 222 l/p/d water saving. The potential daily water saving rate was estimated for low-flow tap at 21.2%, for a grey water system at 7.7% and for efficient washing machine at 7%.

Water saving potentials for treated and untreated greywater were addressed with sufficient details by many studies. The advantages of greywater are that it contains fewer pathogens and 90% less nitrogen in comparison with black water. This makes greywater treatment less complex and comparatively cheaper than wastewater [16]. Untreated greywater (storage time less than 24 hours to avoid anaerobic state) can readily be used for subsurface irrigation and acceptable by water authorities and residents in Australia [16]. On site grey water reuse for reducing urban water demand was also examined from socioeconomic perspective. A promising research by Friedler [23] estimated a penetration ratio at 19-31% by 2015 in Israel, which is equivalent to water saving of 30–51 $\times10^6$ m^3/y. It was shown in the study that on-site greywater reuse could be economically feasible while water price has been continually increasing. Public support was 85% positive towards greywater reuse in the Israeli case study.

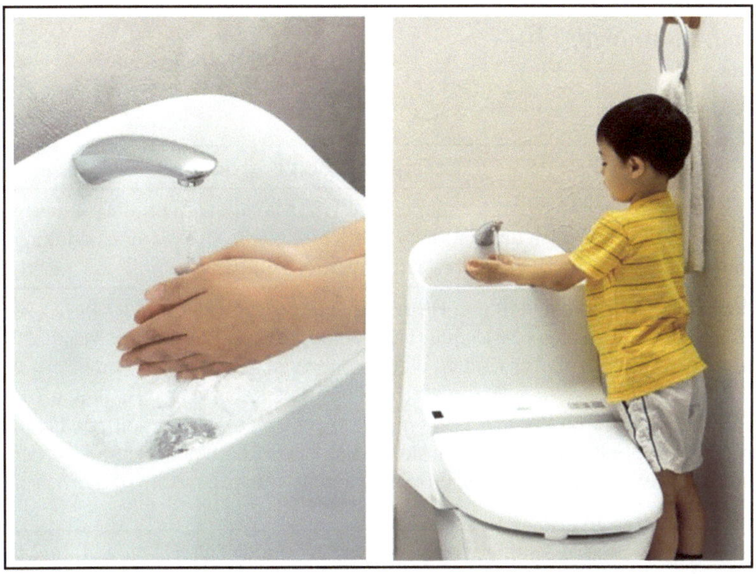

Fig. (1). A new water-saving toilet in Japan to reuse hand-washed water [20].

RAINWATER HARVESTING FOR REDUCING POTABLE WATER CONSUMPTION IN BUILDINGS

International Perspective

Rainwater, if harvested adequately, can be a major source of water in households. Recorded history suggests that sophisticated rainwater collection systems have been in use as early as 3,000 BC in Jordan [24]. In recent time, a number of scientific studies around the world have affirmed the fact that Rainwater Harvesting Systems (RWHS) can increase potable water savings in buildings through supplementing mains water with harvested rainwater [14, 25 - 28]. These studies have shown that water for toilet flushing, cloth washing, building air-conditioning, cleaning and gardening does not need to be potable, instead roof harvested rainwater could be a potential source to meet such non-potable water demands in a building.

The benefits gained from the RWHS have encouraged many countries to implement such systems in both residential and commercial buildings (Table **2**). Examples of the best practices from around the world reveal how RWHS in buildings can provide safe and reliable water supply, whilst reducing over exploitation of limited surface and groundwater resources [29]. Besides, some countries have started embracing strict policies to adopt rainwater use especially in their urban localities where acute water shortage persists. For example, India has made it mandatory in some of its megacities like New Delhi and Chennai to include a RWHS in the new building plan for getting necessary approval from the local government authority [30].

Table 2. Examples of RWHS in buildings.

Countries	Examples of Best Practice
China	Seventeen provinces have so far adopted RWHS. A total of 5.6 million rainwater tanks have been installed having a total capacity of 1.8 billion m^3. This provides drinking water for approximately 15 million people and irrigation facility to 1.2 million ha land area [31].
Germany	Bad Hersfeld Clinic building in Hessen is conserving 6,180 m^3 of water per year through RWHS resulting in saving of €13,500 (US$15,000) per year. During the years 2002 to 2006, tap water need has dropped around 17 percent due to increased use of rainwater [29]. In 1998, RWHS was introduced in Berlin where rooftops of 19 buildings (32,000 m^2) were used to collect and store rainwater in a 3,500 m^3 basement tank to use for non-potable purposes [32]. During 1990 to 1999, Germany has installed more than 100,000 decentralized rainwater storage tanks for service water purposes, providing a total storage capacity of more than 600,000 m^3 [12].
Japan	The 8,400 m^2 rooftop of a sumo-wrestling arena in Sumida City is being utilized for RWH since 1985. The collected rainwater is stored for non-potable water use in a 1,000 m^3 underground tank [33]. RWHS with a total roof top catchment area of 43,400 m^2 and a total tank volume capacity of 4,300 m^3 have been installed in several multipurpose stadiums located in Tokyo, Nagoya and Fukoka. These have resulted in significant economic savings in the stadiums [34]. To date, around 750 private and public buildings in Tokyo have installed RWHS [31].
Korea	Star City (4 apartment towers with 1310 units) in Seoul has installed a capacity to store 3,000 m^3 of water with a total floor area of 1,500 m^2 in one of its tower buildings. This facilitates the conservation of 40,000 m^3 of water per year. Realizing the benefit of this project, Seoul City government has passed an ordinance with incentive program to promote RWHS in new building projects [29].
Singapore	Singapore collects roof water from buildings for non-potable uses resulting in 4% water saving. Cost of roof water use has been calculated as S$0.96 per m^3 while the cost for main water is S$1.17 per m^3. Changi Airport uses rainwater for non-potable purposes which accounts for 28-30% of the total water usage, resulting in savings of approximately S$ 390,000 (US$ 290,000) per year [31].
UK	In the city of London, 2,200 m^2 roof area is connected to RWHS having a capacity of 14.56 m^3, which has resulted in 51 percent efficiency improvement of the system [35]. A water saving of 57% on average has been obtained from a RWHS fitted to a house in Nottingham [25].

An empirical assessment conducted by Ward *et al.* [28] on a UK office building revealed that a significant volume of potable water and cost savings (including energy costs for pumping) were attained through the use of RWHS. In a similar study, Villarreal and Dixon [36] postulated that buildings with large roof areas in Ringdansen, Sweden having the potential to install RWHS, which would contribute 30% of the mains water savings with a 40 m^3 tank when rainwater is utilized for toilet flushing and running washing machines.

A study performed by Marinoski [37] observed a potable water savings of about 40% in a multi-storey building in Florianópolis, Brazil. In a follow-up economic analysis on the use of either greywater or rainwater separately or in combination for the same multi-storey building, the three systems appeared to be cost effective as the payback periods were smaller than 8 years, while the greywater system became the most cost effective one (less than 5 years), followed closely by the rainwater [25]. Further, Ghisi *et al.* [38] estimated the average potential for potable water savings as 69%, with a range of 34% to 92% depending on the demand for potable water in Santa Catarina, southern Brazil. An empirical study conducted in five different regions of Brazil revealed a potable water savings between 48 to 100% from RWHS [39].

Logistical problems such as spacing for rainwater tanks or changes in plumbing system may, at times, hinder the promotion of RWHS in a particular location [40, 41]. However, these could be minimized by constructing underground reservoirs or storing water in the foundation of the building [42]. Also government rebate schemes or other monetary incentive-based mechanisms can address the relevant financial issues. Despite some limitations, evidence of the benefits from RWHS in buildings has been realized in many countries including both developed and developing nations.

Examples of best practices from around the world affirm that harvested rainwater can be effectively collected and utilized in green buildings for non-potable uses such as flushing toilets, watering gardens, hot water systems and/or cooling buildings which eventually make a significant economic saving by minimizing the consumption of potable water for these purposes. At the international level, acceptability and installation rate of RWHS are gradually increasing not only due to these economic benefits but also due to the potential environmental benefits to be gained from the RWHS. Chilton *et al.* [35] postulate that there are excellent reasons for installing a RWHS when the environmental benefits are considered, such as delay of building new surface water treatment system (*e.g.* building a large reservoir costs billions of dollars), reduced risk of flooding due to detention effects and storm water treatment due to first flush effects of a RWHS.

Australian Perspective

Australia has a long tradition of utilizing harvested rainwater [29, 43]. Most state governments in Australia give monetary incentives in the form of a rebate to the home proprietors to encourage them to install RWHS [44]. As an example, Sydney Water Corporation used to offer a rainwater tank rebate of up to $ 1,400 depending on the nature of water use and the tank size [10]. In March 2013, 34% or around 1.7 million Australian households (living in a dwelling suitable for a rainwater tank) reported to have a rainwater tank as a source of water (Fig. **2**). This statistics indicate an increase of RWHS from 32% in 2010 and 24% in 2007. This increase in RWHS usage could be attributed to water use restrictions imposed during the millennium drought, government rebates, favorable water policies and regulations and water pricing factors [45].

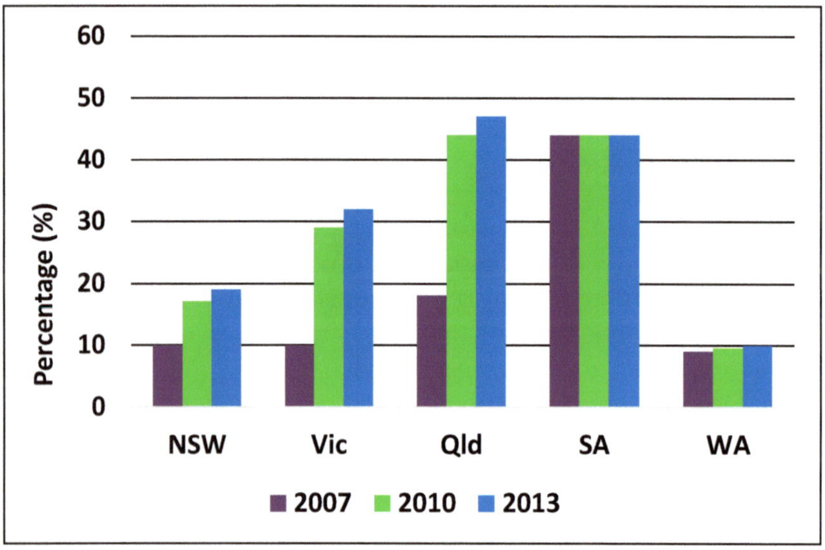

Fig. (2). Percentage of Australian households by States with rainwater tanks [45].

Numerous researches on RWHS in Australia have shown that they can provide significant mains water savings [12, 46 - 51]. In 2013, it was estimated that the domestic RWHS across NSW can provide water savings of up to 38 GL (Gigaliter) [52]. In a study conducted in Newcastle, Coombes *et al.* [46] reported that rainwater usage could promote potable water savings up to 60%. Besides, Muthukumaran *et al.* [17] showed that the use of rainwater inside a purpose-built home in regional Victoria can save up to 40 per cent of potable water use. Further, an empirical study on performance of rainwater tanks in South-East Queensland found that it is possible to achieve average water savings between 40 and 50 Kiloliter (KL)/household/year from internally plumbed rainwater tanks [53].

Several studies have been conducted to evaluate the cost-effectiveness of RWHS in residential buildings in Australia [16, 51, 54, 55]. In a study on four major cities of Australia, for RWHS in large buildings, it was found that Sydney had the shortest payback period (only about 10 years) followed by Perth, Darwin and Melbourne [16].

It has been a customary practice to use a tank size of 3 KL for collecting rainwater in detached residential houses in Greater Sydney locations; however, it is recommended to design the appropriate tank size based on local rainfall conditions, water demand and other relevant parameters such as climate change impacts in order to increase the reliability and optimal water savings of the RWHS [56, 57]. Rahman *et al.* [10] estimated that a 5 KL rainwater tank could save between 54 to 69.5 KL of water per year at a reliability range from 38-69% when the rainwater is used for toilet flushing, laundry and gardening.

Moreover, a number of studies conducted in Melbourne suggest that rather than having a fixed size of the tank, variable tank size should be adopted based on the local rainfall. Studies have found a considerable difference in required tank sizes ranging from 1.1 to 2.5 KL for a roof catchment area of 200 m^2 in detached houses to achieve 90% reliability for RWHS [50, 58]. There have been efforts in developing design curves for optimal rainwater tank sizing at any given location in greater Sydney region based on local rainfall, roof catchment size, desired water savings and reliability [56].

Nonetheless, in order to enhance financial viability of RWHS, it is imperative to increase the demand for rainwater rather than simply increasing the size of the rainwater tank [59]. This necessitates the importance of further enhancing public perception on the utilization of rainwater harvesting in Australia. Moreover, public education and awareness are also important, so that households understand not to use rainwater for drinking and cooking, especially where disinfection of rainwater is absent [60]. Otherwise, any epidemiological disease incidence caused by ingestion of rainwater may jeopardize the campaign of rainwater use in Australia.

AUSTRALIAN CASE STUDIES ON RAINWATER HARVESTING

Case Study in Brisbane

Cook *et al.* [61] reported a monitoring study of a communal RWHS in South East Queensland, Australia. They monitored the water demand and water supply data from the communal RWHS for the period of January 2010 to November 2011. The communal RWHS was placed in Capo di Monte (CDM), a 46 home development, located at Mount Tambourine in Greater Brisbane, Australia. CDM has its own decentralized water and wastewater systems as it lies outside of the service boundary of municipal water and wastewater systems. The communal rainwater system was used to meet household potable water demand (*i.e.* kitchen, laundry and bathroom). Non-potable water demands (toilet flushing and garden watering) were satisfied by a wastewater recycling scheme.

The CDM has a population of 75 people with average age of the population being 60 years. The nearby weather station indicated relatively wet years around the CDM with an average annual rainfall of about 1,318 mm/year during the period of 1982 to 2005. The communal RWHS storages rain water in two 200 KL tanks by collecting roof runoff through a pipe network which is dominated by gravitational flow. Total catchment area consists of about 10,700 m^2 to collect rainwater. The communal RWHS has a water treatment plant that is a combination of sand filtration, UV radiation and chlorination. The water treatment plant provides water to a 40KL balance tank to distribute potable water to each house and a small community center. The two 200 KL storage tanks are always kept at 50% capacity in order to allow for emergency fire-fighting capability. The water demand and rainwater supply monitoring data showed that the communal RWHS provided an average of 5 KL water per day and satisfied around 90% of water demand.

In an another case study, Muthukumaran *et al.* [17] demonstrated water savings of a residential house in Brisbane, Victoria that was built according to an ecologically sustainable design including rainwater harvesting and greywater reuse system. The residential house "Sharland Oasis" had water efficient taps, showerheads; and dual flush toilets were connected with RWHS. It had a total of 4 rainwater tanks including two above ground rainwater tanks with a capacity of 2.1 and 2.7 KL respectively, and two underground concrete tanks with a capacity of 4.5 KL each. Above ground tanks were used to supply garden watering and other outdoor water activities (*e.g.* car washing) and underground tanks were used to supply rainwater to the internal appliances (*e.g.* washing machine, dishwasher and toilets). There was a connection in pump supplying water from the underground rainwater tank to the internal appliances with mains water supply, which automatically switches when required. Sharland Oasis had 6 numbers of people including two adults and four children. Muthukumaran *et al.* [17] monitored water consumption and rainwater supply for the period of December 2006 to April 2008 using a data logging system. They found that total average daily water consumption for Sharland Oasis was 578 L/day with a combination of 95% indoor and 5% outdoor usage. They found that RWHS supplied 40% of total water usage during the monitoring period. They concluded that a significant water saving could be achieved through the use of RWHS in residential houses.

Case Study in South-Eastern Australia

Burns *et al.* [62] reported the performance of 12 domestic rainwater tanks installed in some houses in South-eastern Australia by monitoring the water demand and rainwater supply over a two year period. The rainwater tanks were connected to provide water for both indoor water demand (*i.e.* toilet flushing, clothes washing and hot water usage) and outdoor water demand (garden watering, car washing and swimming pool top-up). The 12 rainwater tank sizes were ranged from 3 KL to 28 KL, and each house had an average occupancy of 2.5 people. During the period April 2010 to March 2012, Burns *et al.* [62] collected water level data in the rainwater tanks through the installed depth logger for every 6 minutes. They also recorded water demands connected to rainwater tanks and garden. They compared mean daily rainwater usage with mean daily potable mains water usage for each household, and found that rainwater usage varied notably between the monitored houses. The reasons for variability were found to be variation in household occupancy, water usage appliances and behavioral factors. They found that more rainwater was used by the households that had more internal demand connected to the rainwater tanks. Some households considerably reduced their potable mains water use by utilizing water entirely from rainwater tanks for toilet flushing, clothe washing and hot water usage. Around 23 to 69% of water reductions in potable mains water usage were found for few houses, which used rainwater tanks for toilet flushing and clothe washing but not for hot water. Few other households reported only 10% water reduction in mains supply due to lesser demand (*e.g.* only one toilet flushing demand was connected to rainwater). Burns *et al.* [62] concluded that rainwater tanks would be an effective mechanism to save water by reducing pressure on the mains supply.

Case Study in Sydney

Water savings potential of residential RWHS in detached houses at 10 different locations in Greater Sydney, Australia was investigated by Rahman *et al.* [10]. They considered water demand of toilet, laundry and garden watering for 3 different tank sizes (*e.g.* 2 KL, 3KL and 5 KL) in their studies. Moreover, a number of assumptions were made to investigate the water savings potentials from the RWHS, such as (i) 4 persons in a household; (ii) each household had a site area of 450m^2 with roof, lawn and impervious areas of 200 m^2, 150 m^2 and 100 m^2, respectively; (iii) location of the tanks was on the ground; (iv) each tank would be supplemented by mains water when the water level in the tanks dropped below 20% of the capacity; (v) a pump for indoor water demand supply and gravity system for garden watering supply and (vi) installation of leaf eater devices and consideration of first flush of 150L to remove contaminants from the roof runoff.

They reported that on an average a 2 KL rainwater tank could meet the toilet and laundry demand for 86% of the days in a year and a 5 KL tank could meet 97.5% of the days in a year for the same purpose. For the combined use of toilet, laundry and garden watering use, a 5 KL tank could meet about 57% of the days in a year. They also reported that a 5 KL tank could save about 34 KL, 49 KL and 58 KL of water in a year for toilet and laundry, garden watering and combined water use, respectively. They concluded that the average water savings were strongly related with average annual rainfall, and a higher rainwater tank size was preferable instead of 2 KL and 3 KL tanks to achieve a greater water savings.

Case Study in Sydney, Newcastle and Wollongong

Eroksuz and Rahman [42] investigated the rainwater harvesting potential in multi-unit buildings in three Australian cities: Sydney, Newcastle and Wollongong. They considered four types of water demand in their investigation including toilet flushing (36 l/p/d), laundry (36 l/p/d), hot water (62 l/p/d) and outdoor garden (28 l/p/d). A number of tank sizes (10 to 100 KL) and roof sizes (*e.g.* 500 m^2 to 2500 m^2) were adopted in the analysis. They reported that a 10 KL tank offered around 21% water savings, whereas a 100 KL tank offered around 55% water savings. They also examined the water saving potentials from RWHS under drought condition. They considered two drought years (2002 and 2004) as the study period in Sydney when annual rainfall was around 71% and 83% of the long term annual average rainfall. They found that even in drought years, a good water savings (37 to 42%) could be achieved with a 70 KL tank in a multi-unit residential development.

Case Study in Melbourne

Performance of two rainwater tanks situated in the campus of Swinburne University of Technology, Melbourne, Australia were evaluated by Imteaz *et al.* [63] in terms of rainwater harvesting effectiveness and payback period. These tanks were placed underground and had a capacity of 185 KL (named as North tank) and 110 KL (named as South tank). Catchments areas for North and South tanks were 907 m^2 and 2,491 m^2, respectively, and the tanks were not connected. Water from both the tanks was used for irrigation purpose, and the water demand was considered to be 2.25 KL/day from North tank and 0.94 KL/day from South tank. The authors investigated the tanks under three different climatic conditions: dry year, average year and wet year. The dry, average and wet years were selected based on the 1st decile, 5th decile and 9th decile of annual total rainfall amount from historical rainfall data. They found that a significant water savings was achieved from RWHS. Cumulative annual water savings from the North tank was found to be 739 KL, 564 KL and 401 KL in wet, average and dry years, respectively. South tank saved around 440 KL, 427 KL and 437 KL of water in wet, average and dry years, respectively. They concluded that both the rainwater tanks were quite effective in saving water, and the tanks could have saved more water if an optimized design would have been adopted.

WATER SAVINGS IN GREEN BUILDINGS

As a part of sustainable development, water conservation is regarded as one of the most critical factors as it is becoming more important with continuing population growth and increased extremity in climate systems due to global warming. Many government authorities have been promoting the concept of water conservation. In this regard, the concept of green building has become popular [64]. Green potable water system, a part of green buildings, is a new concept, which is defined as a system that achieves at least 20% overall decrease in potable water use compared with a traditional building having no water conservation measure. In green water systems, water efficiency is an important element, which is achieved by a variety of means *e.g.* use of water efficient devices, greywater recycling, wastewater recycling, rainwater harvesting and reduction of water losses.

Installation of green water systems in residential homes and commercial buildings usually incur higher initial capital investment than typical buildings. However, Chang *et al.* [65] claims that the extra benefits of such systems in combination with energy saving strategies in buildings should offset the increase of initial capital cost by the end of the building's life. They have developed a grey stochastic programming model and have applied such a cost-benefit-risk trade-off model to a typical Florida residential building. This method is useful in identifying optimal design strategies for water conservation and energy savings for a green building based on various sources of uncertainties.

Further, Cheng *et al.* [66] presented a calculation procedure to estimate the quantitative volume of water saved by practical green buildings in Taiwan. They invested water savings in 1,320 green buildings from 2000 to 2013, and reported that these green buildings achieved an average of 37.6% water saving compared to the baseline water usage rate for these buildings. The savings came from adoption of water-saving appliances, building design promoting water savings and rainwater utilization.

Kontokosta and Jain [67] applied weighted robust regression and geographically weighted regression (GWR) models to identify the determinants and spatial patterns of water consumption for 2,300 multi-unit buildings in New York City, USA. They found that occupancy, building size, building age, ownership structure, neighborhood demographic and socioeconomic characteristics, and the energy use intensity of a building have statistically significant effects on the water use intensity of multi-unit buildings. Many large cities like Sydney in Australia has an increasing focus of constructing multi-unit buildings in satellite cities, which will need less gardening thus resulting in less water consumption as compared with detached houses. However, these buildings should be equipped with large-scale innovative RWHS and greywater reuse to achieve greater water savings as compared with traditional buildings.

RECENT DEVELOPMENTS IN BUILDING WATER EFFICIENCY

More recently, a holistic view of water use in urban environment is becoming popular such as multiple criteria analysis in water use studies and accounting for the water-energy-food nexus [68]. Non-traditional alternative water source has been proposed to reduce demand of mains water in buildings; for example, Antunes *et al.* [69] presented a study on storm water harvesting from porous pavement for use in buildings. They found that for the city of Florianópolis in Brazil (having paved road surface area of about 11,044,216 m^2), a potable water savings of up to 19%, 76% and 70% can be achieved for residential, public and commercial sectors, respectively. More recently, in Australia, the Water Efficiency Labelling and Standards Scheme (WELS) 2016-2019 strategic plan has been accepted by the federal and state governments to develop necessary strategies and steps to achieve higher level of water savings in Australian buildings [70]. In 2016, Ren *et al.* [71] developed an integrated methodology based on Monte Carlo simulation approach to estimate total energy and water consumption in households of Australia by applying modules of water use and their related energy consumptions, which would help achieving higher water efficiency in buildings. An *et al.* [72] examined use of a RWHS in reducing potable water use and temperature reduction in buildings in Hong Kong. A temperature drop of 1.3 °C was noted inside the building due to a rainwater layer in the rain garden of the building in addition to reduction of potable water demand by 40 to 50%.

In future, buildings can be made more sustainable in terms of water use. For example, non-potable water needs can be met fully from well-designed greywater recycling and rainwater harvesting systems. In this regard, irrigation systems can be underground using greywater, toilets can be flushed with water coming out of hand basins and/or rainwater tanks. Rainwater tanks should be larger, preferably located in the building's basements or along the property fence with innovative design to match the land scape. In this regard, further research is warranted for developing design guidelines that can be applied in designing sustainable buildings having over 70% of potable water savings. Running water tap should be minimally used, shower heads should have higher water use efficiency and bathing system should be timed for a maximum of 4 to 6 minutes with a digital display of water used during a shower. Public education is important in this respect as positive awareness can help in achieving greater water savings. Favorable regulations that encourage water savings in buildings are desirable. Community involvement is essential to achieve full green buildings requiring much less potable water in future.

CONCLUSION

World is getting urbanized at an accelerating rate demanding a variety of services including provision of water to buildings. Many of the buildings in developing countries do not have presence of continuous water that is safe to drink; furthermore, water supply is intermittent. In contrast, developed countries use a higher volume of potable water in urban areas, which can be reduced significantly through positive public awareness and innovations in water recycling and conservation. In urban areas, sustainability and water efficiency of buildings are of great significance as many water authorities have been struggling to meet the increasing water demand due to increasing population. The climate change is bringing more extremity in the climate systems such as long and frequent droughts and more frequent high temperature episodes, which increase water demand significantly when water availability is the lowest. In Australia, favorable regulations, a greater public awareness and active research have made rainwater harvesting as a significant alternative source of water for use in buildings and urban areas; however, there are further scopes of development in the rain water harvesting systems by bringing innovations into these systems such as designing rainwater harvesting systems to achieve the highest water savings at a given building based on local rainfall and projected water demand instead of a blanket design. The concept of green buildings and Water Efficiency Labelling and Standards Scheme (WELS) in Australia are expected to increase water use efficiency in Australian buildings.

CONFLICT OF INTEREST

The authors confirm that they have no conflict of interest to declare for this publication.

ACKNOWLEDGEMENTS

Authors acknowledge Mr Erhan Eroksuz, Mr Caleb Amos, Dr Marlene van der Sterren, Mr Joseph Dbais, Mr Joseph Keane and Mr Cameron Snook for their inputs to rain water harvesting research studies conducted in Western Sydney University under the supervision of Professor Ataur Rahman.

REFERENCES

[1] UNICEF (United Nations Children's Fund) and WHO (World Health Organization), *Progress on Sanitation and Drinking Water: 2015 update and MDG assessment.* New York, 2015.

[2] *Bangladesh National Building Code (BNBC),* Ministry of Housing and Public Works: Dhaka, Chapter 7, pp. 291-303, 2011.

[3] UNDP (United Nations Development Programme), *Human Development Report 2006-Beyond scarcity: power, poverty and the global water crisis,* 2006.Accessed: 16th August 2016 [Online]. Available: http://www.undp.org/content/dam/undp/library/ corporate/HDR/2006%20Global%20HDR/HDR-2006-Beyond%20scarcity-Power-poverty-and-the - global-water-crisis.pdf

[4] ABS (Australian Bureau of Statistics), *Year Book Australia 2012,* May 2012.[Accessed: 16th August 2016]cat.no. 1301.0 [Online]. Available: http://www.abs.gov.au/ausstats/abs@.nsf/mf/1301.0.

[5] M. Loh, and P. Coghlan, *Domestic Water Use Study in Perth, Western Australia 1998–2001.* Water Corporation: Perth, 2003.

[6] Y. Zhang, A. Grant, A. Sharma, D. Chen, and L. Chen, "Alternative water resources for rural residential development in Western Australia", *Water Resour. Manage.,* vol. 24, pp. 25-36, 2010. [http://dx.doi.org/10.1007/s11269-009-9435-0]

[7] A. Hurlimann, "Household use of and satisfaction with alternative water sources in Victoria Australia", *J. Environ. Manage.,* vol. 92, no. 10, pp. 2691-2697, 2011. [http://dx.doi.org/10.1016/j.jenvman.2011.06.007] [PMID: 21715083]

[8] L.T. Wong, and K.W. Mui, "Determination of domestic flushing water consumption in Hong Kong", *Facilities,* vol. 23, pp. 82-92, 2005. [http://dx.doi.org/10.1108/02632770510575929]

[9] P. Vincent, *Saudi Arabia: An environmental overview.* Taylor & Francis: London, 2008. [http://dx.doi.org/10.1201/9780203030882]

[10] A. Rahman, J. Keane, and M.A. Imteaz, "Rainwater harvesting in greater Sydney: water savings reliability and economic benefits", *Resour. Conserv. Recycling,* vol. 61, pp. 16-21, 2012. [http://dx.doi.org/10.1016/j.resconrec.2011.12.002]

[11] P.J. Coombes, Rainwater tanks revisited: new opportunities for urban water cycle management, PhD thesis, University of Newcastle: Australia, 2002.

[12] T. Herrmann, and U. Schmida, "Rainwater utilisation in Germany: efficiency, dimensioning, hydraulic and environmental aspects", *Urban Water,* vol. 1, pp. 307-316, 1999. [http://dx.doi.org/10.1016/S1462-0758(00)00024-8]

[13] D. Rathjen, P. Cullen, N. Ashbolt, D. Cunliffe, J. Langford, A. Listowski, J. McKay, A. Priestley, and J. Radcliffe, "Recycling water for our cities", In: *Report to Prime Minister's Science.* Engineering and Innovation Council, Federal Government of Australia: Canberra, 2003.

[14] WSAA, *Water Services Association of Australia facts,* Melbourne, 2001.[Accessed: 20th August 2016] [Online]. Available: https://www.wsaa.asn.au/

[15] E. Ghisi, and D.F. Ferreira, "Potential for potable water savings by using rainwater and greywater in a multi-storey residential building in southern Brazil", *Build. Environ.,* vol. 42, pp. 2512-2522, 2007. [http://dx.doi.org/10.1016/j.buildenv.2006.07.019]

[16] Y. Zhang, D. Chen, L. Chen, and S. Ashbolt, "Potential for rainwater use in high-rise buildings in Australian cities", *J. Environ. Manage.,* vol. 91, no. 1, pp. 222-226, 2009. [http://dx.doi.org/10.1016/j.jenvman.2009.08.008] [PMID: 19744767]

[17] S. Muthukumaran, K. Baskaran, and N. Sexton, "Quantification of potable water savings by residential water conservation and reuse: a case study", *Resour. Conserv. Recycling,* vol. 55, pp. 945-952, 2011. [http://dx.doi.org/10.1016/j.resconrec.2011.04.013]

[18] *WSUD, "Water Sensitive Urban Design in the Sydney Region", CD-ROM, Upper Parramatta River Catchment Trust, Sydney Coastal Councils Group.* Western Sydney Regional Organization of Councils: Sydney, 2003.

[19] New South Wales Department of Planning (NSWDP), *The Building Sustainability Index (BASIX),* 2005.[Accessed: 20th August 2016] [Online]. Available: http://www.basix.nsw.gov.au/information/index.jsp

[20] Nost Tokyo News, *Waterbesparende technologieën in Japan,* March 20, 2014.[Accessed: 20th August 2016] [Online]. Available: http://news.nost.jp/2014/03/waterbesparende-technologieen-in-japan/

[21] L.C. Proença, and E. Ghisi, "Water end-uses in Brazilian office buildings", *Resour. Conserv. Recycling,* vol. 54, pp. 489-500, 2010. [http://dx.doi.org/10.1016/j.resconrec.2009.10.005]

[22] H.M. Taleb, and S. Sharples, "Developing sustainable residential buildings in Saudi Arabia: A case study", *Appl. Energy,* vol. 88, pp. 383-391, 2011. [http://dx.doi.org/10.1016/j.apenergy.2010.07.029]

[23] E. Friedler, "The water saving potential and the socio-economic feasibility of greywater reuse within the urban sector – Israel as a case study", *Int. J. Environ. Stud.,* vol. 65, pp. 57-69, 2008. [http://dx.doi.org/10.1080/00207230701846697]

[24] R.A. AbdelKhaleq, and I.A. Ahmed, "Rainwater harvesting in ancient civilizations in Jordan", *Water Science and Technology: Water Supply,* vol. 7, no. 1, pp. 85-93, 2007.

[25] A. Fewkes, "The use of rainwater for WC flushing: the field testing of a collection system", *Build. Environ.,* vol. 34, no. 6, pp. 765-772, 1999. [http://dx.doi.org/10.1016/S0360-1323(98)00063-8]

[26] A. Appan, "A dual-mode system for harnessing roof water for non-potable uses", *Urban Water,* vol. 1, no. 4, pp. 317-321, 2000. [http://dx.doi.org/10.1016/S1462-0758(00)00025-X]

[27] L. Handia, J.M. Tembo, and C. Mwiindwa, "Potential of rainwater harvesting in urban Zambia", In: *Physics and Chemistry of the Earth,* vol. 28. 2003, no. 20-27, pp. 893-896.

[28] S. Ward, F.A. Memon, and D. Butler, "Performance of a large building rainwater harvesting system", *Water Res.,* vol. 46, no. 16, pp. 5127-5134, 2012. [http://dx.doi.org/10.1016/j.watres.2012.06.043] [PMID: 22840659]

[29] UNEP, *Rainwater Harvesting: A lifeline for human well-bein..* Stockholm Environment Institute, 2009.

[30] UN-HABITAT, *Meeting Development Goals in Small Urban Centres: water and sanitation in the world's cities.* Blue drop series on rainwater harvesting and utilization: Nairobi, Kenya, 2005.

[31] UNEP, *Examples of Rainwater Harvesting and Utilization around the World,* 2016.[Accessed: 28th July 2016] [Online]. Available: http://www.unep.or.jp/ietc/publications/urban/urbanenv-2/9.asp

[32] UNEP (United Nations Environment Program), *Rainwater Harvesting and utilization: An environmentally sound approach for sustainable urban water management- an introductory guide for decision-makers.* IETC Urban Environment Series 2, 2002.

[33] J. Gould, and E. Nissen-Petersen, *Rainwater catchment systems for domestic supply: Design, construction and implementation.* Intermediate Technology Publication: London, 1999. [http://dx.doi.org/10.3362/9781780445694]

[34] M. Zaizen, T. Urakawa, Y. Matsumoto, and H. Takai, "The collection of rainwater from dome stadiums in Japan", *Urban Water,* vol. 1, no. 4, pp. 335-359, 1999.

[35] J.C. Chilton, G.G. Maidment, D. Marriott, A. Francis, and G. Tobias, "Case study of a rainwater recovery system in a commercial building with a large roof", *Urban Water,* vol. 1, pp. 345-354, 1999.

[http://dx.doi.org/10.1016/S1462-0758(00)00032-7]

[36] E.L. Villarreal, and A. Dixon, "Analysis of a rainwater collection system for domestic water supply in Ringdansen, Norrköping, Sweden", *Build. Environ.,* vol. 40, pp. 1174-1184, 2005. [http://dx.doi.org/10.1016/j.buildenv.2004.10.018]

[37] D.L. Marinoski, E. Ghisi, and L.A. Gomez, "Aproveitamento de agua pluvial e dimensionamento de reservatorio para fins nao potaveis: estdo de caso em um conjunto residencial localizado em Florianopolis-SC [Rainwater harvesting and rainwater tank sizing: a case study in a multi-storey residential building located in Florianopolis, Southern Brazil]", *CLACS04-1 Conferencia Latino-Americana de Contrucao Sustentavel e ENTAC04-10^{0} Encontro Nacional do Ambienete Contuido,* 2004 São Paulo-SP CD Rom, (in Portuguese)

[38] E. Ghisi, A. Montibeller, and R.W. Schmidt, "Potential for potable water savings by using rainwater: an analysis over 62 cities in southern Brazil", *Build. Environ.,* vol. 41, pp. 204-210, 2006. [http://dx.doi.org/10.1016/j.buildenv.2005.01.014]

[39] E. Ghisi, "Potential for potable water savings by using rainwater in the residential sector of Brazil", *Build. Environ.,* vol. 41, pp. 1544-1550, 2006. [http://dx.doi.org/10.1016/j.buildenv.2005.03.018]

[40] O.O. Aladenola, and O.B. Adeboye, "Assessing the potential for rainwater harvesting", *Water Resour. Manage.,* vol. 24, pp. 2129-2137, 2010. [http://dx.doi.org/10.1007/s11269-009-9542-y]

[41] M.P. Jones, and W.F. Hunt, "Performance of rainwater harvesting systems in the southeastern United States", *Resour. Conserv. Recycling,* vol. 54, pp. 623-629, 2010. [http://dx.doi.org/10.1016/j.resconrec.2009.11.002]

[42] E. Eroksuz, and A. Rahman, "Rainwater tanks in multi-unit buildings: a case study for three Australian cities", *Resour. Conserv. Recycling,* vol. 54, pp. 1449-1452, 2010. [http://dx.doi.org/10.1016/j.resconrec.2010.06.010]

[43] EHAA (Environmental Heritage Aboriginal Affairs), *Rainwater tanks: their selection, use and maintenance.* Government of South Australia, 1999.

[44] CRCWQT (Cooperative Research Centre for Water Quality and Treatment), *Public Health Aspects of Rainwater Tanks in Urban Australia.* Occasional Paper 10, Salisbury, South Australia, 2005.

[45] ABS (Australian Bureau of Statistics), *Environmental issues: water use and conservation,* March 2013.[Accessed: 29th July 2016] [Online]. Available: http://www.abs.gov.au

[46] P.J. Coombes, J.R. Argue, and G. Kuczera, "Figtree place: a case study in water sensitive urban development (WSUD)", *Urban Water,* vol. 1, no. 4, pp. 335-343, 1999. [http://dx.doi.org/10.1016/S1462-0758(00)00027-3]

[47] P.J. Coombes, and G. Kuczera, "Analysis of the performance of rainwater tanks in Australian capital cities", *28th International Hydrology and Water Resources Symposium,* 2003 Wollongong

[48] A. Chanan, and P. Woods, "Introducing total water cycle management in Sydney: a Kogarah Council initiative", *Desalination,* vol. 187, pp. 11-16, 2006. [http://dx.doi.org/10.1016/j.desal.2005.04.063]

[49] R. Marks, R. Clark, E. Rooke, and A. Berzins, "Development through integration of local water resources", *Desalination,* vol. 188, pp. 149-161, 2006. [http://dx.doi.org/10.1016/j.desal.2005.04.112]

[50] A. Khastagir, and N. Jayasuriya, "Optimal sizing of rainwater tanks for domestic water conservation", *J. Hydrol. (Amst.),* vol. 381, no. 3-4, pp. 181-188, 2010. [http://dx.doi.org/10.1016/j.jhydrol.2009.11.040]

[51] V.W. Tam, L. Tam, and S.X. Zeng, "Cost effectiveness and trade off on the use of rainwater tank: an empirical study in Australian residential decision-making", *Resour. Conserv. Recycling,* vol. 54, pp.

178-186, 2010.
[http://dx.doi.org/10.1016/j.resconrec.2009.07.014]

[52] ABS (Australian Bureau of Statistics), *Water account, Australia, 2012-13,* November 2014.[Accessed: 9th March 2015]cat. no. 4610.0 [Online]. Available: http://www.abs.gov.au/AUSSTATS/ abs@.nsf/mf/4610.0

[53] C.D. Beal, A. Sharma, T. Gardner, and M. Chong, "A desktop analysis of potable water savings from internally plumbed rainwater tanks in South-East Queensland, Australia", *Water Resour. Manage.,* vol. 26, pp. 1577-1590, 2012.
[http://dx.doi.org/10.1007/s11269-011-9973-0]

[54] M.A. Imteaz, A. Shanableh, A. Rahman, and A. Ahsan, "Optimisation of rainwater tank design from large roofs: a case study in Melbourne, Australia", *Resour. Conserv. Recycling,* vol. 55, pp. 1022-1029, 2011.
[http://dx.doi.org/10.1016/j.resconrec.2011.05.013]

[55] A. Khastagir, and N. Jayasuriya, "Investment evaluation of rainwater tanks", *Water Resour. Manage.,* vol. 25, pp. 3769-3784, 2011.
[http://dx.doi.org/10.1007/s11269-011-9883-1]

[56] C. Snook, A. Rahman, M. van der Sterren, M.M. Haque, and E. Hajani, "Developing design curves for rainwater harvesting in greater Sydney", In: *9th International Water Sensitive Urban Design.* Sydney, 2015.

[57] M.M. Haque, A. Rahman, and B. Samali, "Evaluation of climate change impacts on rainwater harvesting", *J. Clean. Prod.,* vol. 137, pp. 60-69, 2016.
[http://dx.doi.org/10.1016/j.jclepro.2016.07.038]

[58] M.A. Imteaz, A. Rahman, and A. Ahsan, "Reliability analysis of rainwater tanks: a comparison between south-east and central Melbourne", *Resour. Conserv. Recycling,* vol. 66, pp. 1-7, 2012.
[http://dx.doi.org/10.1016/j.resconrec.2012.05.009]

[59] P.J. Coombes, and P. Kozarovski, "Development of a regional model to understand the hydrological and economic benefits of rainwater tanks across New South Wales", *The 29th Hydrology and Water Resources Symposium,* 2005 Canberra

[60] T. Gardner, and A. Vieritz, "The role of rainwater tanks in Australia in the twenty first century", *Archit. Sci. Rev.,* vol. 53, no. 1, pp. 107-125, 2010.
[http://dx.doi.org/10.3763/asre.2009.0074]

[61] S. Cook, A. Sharma, and M. Chong, "Performance Analysis of a Communal Residential Rainwater System for Potable Supply: A Case Study in Brisbane, Australia", *Water Resour. Manage.,* vol. 57, pp. 4865-4876, 2013.
[http://dx.doi.org/10.1007/s11269-013-0443-8]

[62] M.J. Burns, T.D. Fletcher, H.P. Duncan, B.E. Hatt, A.R. Ladson, and C.J. Walsh, "The performance of rainwater tanks for stormwater retention and water supply at the household scale: an empirical study", *Hydrol. Processes,* vol. 29, pp. 152-160, 2015.
[http://dx.doi.org/10.1002/hyp.10142]

[63] M.A. Imteaz, J. Taylor, M. Pateras, and A. Shanableh, "Effectiveness and payback period analysis of rainwater tanks constructed within Swinburne University of Technology", In: *Solutions for a Sustainable Planet.* Engineers Australia, 2009, pp. 330-339.

[64] C.J. Kibert, "Green buildings: an overview of progress", *Journal of Land Use,* vol. 19, pp. 491-502, 2004.

[65] N.B. Chang, B.J. Rivera, and M.P. Wanielista, "Optimal design for water conservation and energy savings using green roofs in a green building under mixed uncertainties", *J. Clean. Prod.,* vol. 19, no. 11, pp. 1180-1188, 2011.
[http://dx.doi.org/10.1016/j.jclepro.2011.02.008]

[66] C. L. Cheng, J. J. Peng, M. C. Ho, W. J. Liao, and S. J. Chern, "Evaluation of water efficiency in green building in Taiwan", *Water,* vol. 8, no. 6, no. 236, 2016. [http://dx.doi.org/10.3390/w8060236]

[67] C.E. Kontokosta, and R.K. Jain, "Modeling the determinants of large-scale building water use: implications for data-driven urban sustainability policy", *Sustainable Cities and Society,* vol. 18, pp. 44-55, 2015. [http://dx.doi.org/10.1016/j.scs.2015.05.007]

[68] C. Amos, A. Rahman, and J. Gathenya, "Economic analysis and feasibility of rainwater harvesting systems in urban and peri-urban environments: A review of the global situation with a special focus on Australia and Kenya", *Water,* vol. 8, no. 4, pp. 149-169, 2016. [http://dx.doi.org/10.3390/w8040149]

[69] L.N. Antunes, L.P. Thives, and E. Ghisi, "Potential for potable water savings in buildings by using stormwater harvested from porous pavements", *Water,* vol. 8, no. 4, p. 110, 2016. [http://dx.doi.org/10.3390/w8040110]

[70] Government of Australia, *Water Efficiency Labelling and Standards Scheme,* 2016.Accessed on 28/09/2016 http://www.waterrating.gov.au/consultation/2015-wels-scheme-review

[71] Z. Ren, W.Y. Chan, X. Wang, J. Anticev, S. Cook, and D. Chen, "An integrated approach to modelling end-use energy and water consumption of Australian households", *Sustainable Cities and Society,* vol. 26, pp. 344-353, 2016. [http://dx.doi.org/10.1016/j.scs.2016.07.010]

[72] K.J. An, Y.F. Lam, S. Hao, T.E. Morakinyo, and H. Furumai, "Multi-purpose rainwater harvesting for water resource recovery and the cooling effect", *Water Res.,* vol. 86, no. 1, pp. 116-121, 2015. [http://dx.doi.org/10.1016/j.watres.2015.07.040] [PMID: 26253864]

CHAPTER 3

Criteria for the Optimal Sizing of Rainwater Harvesting Systems in Europe

Ilaria Gnecco[1,*], Anna Palla[1], Alberto Campisano[2] and **Carlo Modica[2]**

[1] *Department of Civil, Chemical and Environmental Engineering, University of Genova, Genova, Italy*

[2] *Department of Civil Engineering and Architecture, University of Catania, Catania, Italy*

Abstract: Rainwater Harvesting (RWH) systems are recognized as a widely accepted solution to save potable water in buildings. In addition, RWH systems may play an important role in mitigating the impact of increasing imperviousness in urban areas by contributing to increase both at-source retention and detention of storm water runoff. The chapter provides an overview of methodologies for designing RWH systems, with specific focus on numerical models based on the long-term water balance simulation of the tank. An overview of metrics to evaluate RWH system performance is presented, with regard to the estimation of both water saving potential and storm water control benefits. The accuracy of the modelling results with reference to the length of the available rainfall series and to the selected resolution time step of the used model is discussed. Results of an application to six cities in southern Europe are also discussed in order to highlight the impact of different precipitation regimes as well as the influence of rainwater demands on the system design and performance. Finally, the German, British and Italian standards on RWH are analysed and compared to identify differences and common design approaches.

Keywords: Behavioural model, Detention time, European standards, Long-term simulation, Operational parameters, Precipitation, Rainwater harvesting, Rainwater demand, Runoff control, System performance, Tank sizing, Water saving, yield.

INTRODUCTION

Rainwater Harvesting (RWH) systems are recognized as a widely accepted solution to save water in buildings. RWH has gained increasing popularity and confidence in Europe due to the improved reliability of system design and implementation issues. The installation of rainwater harvesting systems, as well as

* **Corresponding author Ilaria Gnecco:** Department of Civil, Chemical and Environmental Engineering, University of Genova, Genova, Italy; Tel: +39-010-353 2485; Fax: +39-010-353 2481; E-mail: ilaria.gnecco@unige.it

Enedir Ghisi (Ed.)

the use of water saving devices, are supported by the European legislation [1]. A significant number of studies from the literature (*e.g.* [2 - 5]) confirms the potential of RWH as an alternative source of fresh water in the house.

Domestic RWH consists of the small-scale concentration, collection, storage and use of rainwater coming from rooftops and other impervious surfaces. Rainwater can be used to replace drinking water for many uses in the house such as the flushing of toilets and the garden watering [6]. Several studies carried out across the world over the years have focused on RWH for toilet flushing. In many cases, this use represents one of the largest components of the household water consumption without requiring necessarily high water quality requisites [7, 8].

RWH systems may also play an important role in mitigating the impact of urban storm water due to the increase of imperviousness in urban areas [9]: according to the Low Impact Development (LID) principles and applications aimed at promoting storage, infiltration and evapotranspiration [10], RWH may contribute to increase at-source retention and detention of storm water. Then, catchment scale implementation of rainwater tanks may help to reduce frequency, volume and peaks of storm water runoff in urban drainage systems [11 - 14]. Moreover, the synergy between different LID systems can contribute to maximize the benefits and provide a range of urban ecosystem services (including mitigation of floods, droughts, noise, air pollution and urban heat island effects, improvement of biodiversity, *etc.*) [15].

Within this framework, the main objective of this chapter is to provide an overview of common methodologies for designing RWH systems and assessing their performance. A specific section is devoted to the analysis of the impact of the European precipitation regimes on the management of rainwater tanks as well as on the influence of water demands on the system performance. Further, design criteria are compared with standards provided by the main guidelines available in Europe, such as the British, the German and the Italian technical recommendations.

OPERATIONAL AND DESIGN PARAMETERS

RWH systems are usually made of three main components namely, the catchment area, the storage system, and the conveyance system. Devices used in the various system components are filters, one or more storage tanks, a supply facility, a separate pipe network and an overflow unit. Each component can be described in a simple way in terms of operational parameters. In the following sub-sections, the design parameters of the RWH systems are illustrated in detail.

Operational Parameters

Catchment Area

Any RWH system is designed in order to collect rainfall precipitated over a specific catchment area. The amount and quality of collected rainwater depend on the area and type of the catchment. The main operational parameters describing the catchment consist of its surface area and typology. The catchment area is generally impervious in order to maximize the runoff volume. Collection areas are usually rooftop surfaces or other paved areas (*e.g.* parking lots, small road networks). In the latter case, the impact of the associated pollutant load may be significant [16] thus requiring at least to divert the first flush volume from the total collected rainwater. The diverted rainwater volume can be evaluated based on results provided by the literature. As examples, Khastagir and Jayasuriya [3] subtract the first 0.33 mm of daily rainfall while Basinger *et al.* [17] assume 0.4 mm to be the first flush occurring after 3 dry days. Other authors recommend first flush amounts up to 200 litres per 100 m^2 of rooftop (2 mm) [18].

In the most basic and common solutions, rooftop rainwater is conveyed to the storage unit through the system of gutters of the building. Conversely, the conveyance of storm water runoff from other paved areas requires a dedicated drainage network. As for rooftops, the amount of collected rainwater is usually evaluated by assuming a constant runoff coefficient as a measure of losses in the surface depression storage. Based on this simplification, the volume of rainwater is calculated by multiplying the surface area by the runoff coefficient and the rainfall depth. The runoff coefficient is generally assumed in the range 0.8-0.9, which are typical values for rooftops of residential districts.

Storage System

Tanks for harvesting rainwater may be either above or below ground reservoirs. Various types of rainwater storage facilities can be found in the practice: cylindrical tanks and battery tanks (*i.e.* interconnected tanks) made of pottery, cement or polyethylene materials. Precautions required to use storage tanks include provision of adequate devices to minimise contamination from human, animal or other environmental pollutants including a tight cover to prevent algal growth and the breeding of mosquitos. Open containers are not generally recommended to store rainwater for domestic use. Since the tanks are generally covered, evaporation losses from the RWH system as well as the incident precipitation over the tank can be neglected in the water budget estimation.

The main operational parameter of the storage device is the storage capacity of the system, *i.e.* the maximum tank water volume that can be supplied to the final use.

The minimum technical volume (that is required to pump from the tank) is not generally considered in the sizing criteria; indeed, such volume of water does not concur to the supplied volume. The minimum technical volume is generally not defined (neglected) for the storage device or it is set irrespective of the storage capacity.

Conveyance System

Conveyance systems are required *i*) to convey the collected rainwater to the storage tanks and *ii*) to allow the transfer of the stored water to the final use.

The transfer of rainwater to the tank is usually accomplished by creating appropriate connection of the tank to one or more downpipes/gutters. Specific devices have to be installed in the case of first flush diversion in order to convey the more polluted portion of collected rainwater to the sewer system. Safe management of rainwater tanks requires that excess water can overflow; the overflow is usually re-directed to the downstream drainage network or ex-filtrated into the soil. Furthermore, the storage tank is generally equipped with an underdrain (low level outlet) in order to allow maintenance.

A pump is usually installed to transfer stored water to the various uses in the house. Rainwater may be pumped directly to toilet cisterns, washing machines and outdoor taps for irrigation purposes. Alternatively it can be pumped to a header tank (where it is gravity fed to the appliances).

Operational parameters of the conveyance system are the hydraulic efficiency, the typology of the pump and the management rules of the overflow and low level outlet. The selection of the pump has implications in the energy saving/consumption, while management rules of the overflow (downstream drainage network or ex-filtrated into the soil) and low level outlet can partially impact the benefits in reducing storm water runoff.

Design Parameters

The main design parameters of the domestic RWH systems are the rainfall pattern, the catchment area, the tank storage capacity, and the water demand.

Rainfall Pattern

The rainfall pattern is described by the characteristics of the time series of local climate records. The mean annual depth is normally used as the main design parameter; indeed, it allows to preliminary size the RWH system. The mean values of the monthly rainfall depth and the corresponding standard deviation are generally considered in order to assess seasonal patterns that can affect the system performance. A limited monthly variability allows a simpler optimization of the RWH system design [19].

Daily rainfall data are commonly used as input data for mass balance equation-based models in order to estimate the rainwater volume supplied to the house. Furthermore, using daily rainfall data, the analysis of the rainfall conditions is generally carried out at the site of concern in order to evaluate the number of rainy days and the hydrologic characteristics at the event scale: depth, duration, average intensity and Antecedent Dry Weather Period (ADWP). The number of rainy days is a simple way to characterize the daily rainfall data series when compared with the event characteristics that require more complex hydrologic analysis. Regression analysis could be performed to identify relationships between rainfall event characteristics and RWH system performance. Palla *et al.* [20] demonstrated that the Antecedent Dry Weather Period (ADWP) is the main hydrologic parameter affecting the system behaviour, while other rainfall event characteristics (including event rainfall depth, intensity and duration) reveal minor correlation.

Sub-daily resolution timescales of rainfall data are generally used to assess the hydrologic benefit of RWH systems, including storm water runoff control benefits (*e.g.* [13]). In particular, Campisano and Modica [21] have shown that at least hourly time step resolutions are required for a reliable evaluation of the tank volumetric retention efficiency, while high time resolutions (sub-hourly time steps) become mandatory if the analysis of the benefits of the RWH system to reduce storm water peaks is carried out.

Rainwater Demand

Rainwater can replace potable water for several domestic uses such as toilet flushing, cloth washing and garden watering [22]. Like precipitation, rainwater demands may be described on annual, monthly or daily basis.

Literature studies show that the demand for toilet flushing occurs at an almost constant daily rate during the year; indeed, the demand time series generated by WC usage do not exhibit high daily variances [23, 24]. Therefore, the rainwater demand for toilet flushing can be properly described as the total volume per year or per day. The annual water demand for toilet flushing is usually evaluated by assuming a constant daily demand per person (*i.e.* the daily water demand for undrinkable use can be estimated between 40 to 50 l/d per person) as a function of the total number of inhabitants served with RWH systems (see *e.g.* [1, 6]). At a sub-daily resolution timescale, the rainwater-demand diagram is generally described with two or three different supply periods in order to reproduce the typical water consumption with well-defined peaks [21, 25].

The rainwater demand for irrigation purposes is dependent on the surface area to be irrigated, the local weather conditions and irrigation methods and it is not constant during the year. Therefore, the rainwater demand for irrigation can be properly described as monthly or daily volumes, while the demand on annual basis can be used solely for a general mass balance-based evaluations.

Non-Dimensional Approach

As already discussed, the performance of RWH systems is influenced by several site specific variables, *i.e.*, the local rainfall pattern, the roof type and surface area, the tank size, the demand for rainwater, the number of people belonging to the household, *etc.* Understanding the sensitivity of each variable may require high computational efforts. Therefore, non-dimensional approaches have been used in the scientific literature in order to consider different combinations of design parameters in the performance assessment and optimum sizing of RWH systems (*e.g.* [23]).

Fewkes [23] firstly defined two dimensionless parameters in order to describe different combinations of roof area, storage capacity and demand, namely the demand fraction (*D/Q*) and the storage fraction *(S/Q)*:

$$ {}^{D}\!/_{Q} = \frac{D}{A \cdot R} \tag{1} $$

$$ {}^{S}\!/_{Q} = \frac{S}{A \cdot R} \tag{2} $$

where D [L^3] is the annual rainwater demand, A [L^2] is the catchment surface area (roof), R [L] is the annual rainfall depth and S [L^3] is the storage capacity. Palla *et al.* [19] assumed to replace the product $A \cdot R$ with the annual inflow to the tank Q evaluated as:

$$ Q = \varphi \cdot A \cdot R \tag{3} $$

where φ [-] is the catchment runoff coefficient. Simulation results reported by Palla *et al.* [19] indicate that the demand fraction considerably affects the system performance. Indeed, only when this parameter is close to unity, the performance of the rainwater harvesting system can be improved by increasing the storage capacity of the tank. However, the storage fraction parameter affects the detention time of the stored rainwater, thus having an important role on the water quality within the system. Using another approach to generalize results, Mun and Han

[25] referred to specific daily rainwater demand and tank volume (per square meter of the catchment area). However, the use of this kind of approach may be useful to analyse the performance of RWH systems in locations where the rainfall is an input data and not a design parameter.

Campisano and Modica [26] developed a dimensionless approach in order to compare various scenarios of operation of domestic RWH. The authors proposed the following dimensionless ratios, namely demand fraction (d) and modified storage fraction (s_m):

$$d = \frac{D_d}{A \cdot R_d} \quad \textbf{(4)}$$

$$s_m = \frac{S}{\frac{D}{365} \cdot \frac{n_D}{n_R}} \quad \textbf{(5)}$$

where D_d [L^3] and R_d [L] are the average daily values of toilet water demand and rainfall; S is the tank capacity; n_D and n_R are the average numbers of dry and rainy days in a year, respectively. Parameter s_m was used instead of the ratio S/Q proposed by Fewkes [23], as it allows the intra-annual characteristics of the rainfall pattern to be taken into account. For such a prerogative, parameter s_m enables an improved analysis of the system performance clearly pointing out the role of the antecedent dry weather period in affecting the RWH system performance [20]. In this context, the ratio n_D/n_R can be considered as the average dry period (days) in the year per each rainy day. Therefore, the parameter s_m allows to relate the available storage capacity to the required water demand during the defined time period. It should be emphasized that s_m is easy to calculate being values of n_D and n_R normally provided as basic data for rainfall stations.

RWH SYSTEM MODELLING

Criteria for the design of RWH systems are commonly based on the evaluation of the appropriate tank capacity to achieve the desired level of performance. Methodologies for the tank design are documented in the literature [17, 27] and generally include simplified approaches based on user-defined relationships (*e.g.* [28]), continuous mass balance simulations (*e.g.* [23]), non-parametric approaches (*e.g.* [17]) and statistical methods (*e.g.* [29]). The most common methodology makes use of behavioural analysis. Such a methodology is based on the results of long-term simulations of the tank water balance to assess inflow, outflow and change in storage volume of the tank (*e.g.* [27, 30 - 32]). Used simulation schemes normally consider rainfall and household water demand as tank inflow and outflow, respectively.

Behavioural Models

Typically, mass balance models are made of a set of interrelated modules including a behavioural model to represent rainwater demand (D), an inflow (rainfall R) module to represent the available water volumes, and an evaluation module. The evaluation module solves the mass balance equation of the tank:

$$V_t = Q_t + V_{t-1} - Y_t - O_t \tag{6}$$

where Q_t [L^3] is the inflow volume associated to the rainfall, Y_t [L^3] is the rainwater volume supplied to the house, V_t [L^3] is the volume stored in the tank, O_t [L^3] is the overflow volume from the tank. Subscript t indicates the current time step, with the assumption that all variables are always calculated at the end of each time step. Since rainwater tanks are usually set up with closed cisterns, evaporation losses from the system can be neglected in the mass balance equation. The inflow volume Q_t is evaluated as follows:

$$Q_t = \varphi \cdot A \cdot R_t \tag{7}$$

where φ [-] is the runoff coefficient; A [L^2] is the catchment surface area for rainwater collection and R_t [m] is the rainfall depth. The application of the mass balance equation for each time step can be carried out using different operating rules.

The "yield after spillage" (YAS) and "yield before spillage" (YBS) [30] rules have been used in many RWH evaluator tools. The YAS rule is generally adopted since it is more conservative than the YBS, the yield being extracted after that the overflow volume is subtracted. This algorithm is simple to implement and it can be used to evaluate RWH system performance under a large spectrum of conditions [19, 23, 27]:

$$Y_t = min\begin{cases} D_t \\ V_{t-1} \end{cases} \tag{8}$$

$$V_t = min\begin{cases} V_{t-1} + Q_t - Y_t \\ S - Y_t \end{cases} \quad \textbf{(9)}$$

where D_t [L^3] is the water demand and S [L^3] is the tank capacity.

Conversely, the YBS rule assumes that the yield is extracted by the tank before the overflow volume:

$$Y_t = min\begin{cases} D_t \\ V_{t-1} + Q_t \end{cases} \quad \textbf{(10)}$$

$$V_t = min\begin{cases} V_{t-1} + Q_t - Y_t \\ S \end{cases} \quad \textbf{(11)}$$

Selection of the Resolution Time Scale

The long-term water balance simulation of the tank is also a tool to explore multiple benefits of RWH tanks [2, 27, 33 - 37]. Studies reveal that simulation results may be affected by the time step of the simulation. In particular, the resolution time-scale that is chosen for the analysis (thus the time step selected for computations) may influence the estimation of the tank design/reliability in a significant way [11].

In this context, results by Fewkes and Butler [30] indicate that simulations with monthly time steps may provide inaccurate evaluation of the water saving performance. Such results suggest to use the daily time step resolution for water balance calculations. However, the daily resolution may be still insufficient if the aim of the analysis is to evaluate the tank storm water retention potential. For this purpose, accurate estimations require increased time resolutions. High time resolutions become mandatory if the tank effect on storm water peak reduction is analysed [38].

The choice of the time resolution for modelling the tank operation strictly depends on the availability of rainfall and water demand data and has obvious repercussions on both the accuracy of the results and on the required computational efforts to treat extended data sets. Campisano and Modica [21, 38] defined the appropriate modelling time resolution, based on the purpose that drives the implementation of the RWH system and on the expected accuracy of the results. A model framework to evaluate inflow to the tank (rainfall), outflow from the tank (toilet water demand), and to solve the water balance was developed for simulations at different timescales of resolution. Four temporal scales (daily, 1 hour, 15 min and 5 min time steps) were chosen to assess potential benefits of RWH tanks in terms of both water saving and volumetric retention efficiency. Also, the versatility of the tank to reduce the rainfall intensity peak was explored. Simulations have shown that the daily scale may be reliably chosen to evaluate the tank water saving efficiency, as the differences in results among the daily and sub-daily scales are relatively modest (less than 5% for a large range of conditions). Differently, (at least) the hourly time step resolution is required for a reliable evaluation of tank volumetric retention efficiency. Otherwise, the use of the daily scale may affect result accuracy in a significant way, especially for small tanks and for high values of water demand (differences in efficiency up to about 17%). Moreover, results at the 5 min time step showed that the rainwater tank may help reduce the rainfall intensity peak for a relatively high number of rainfall events (50% peak reduction for at least 50% of events), depending on the tank storage and on the rainfall event characteristics.

Appropriate Length of the Rainfall Data Series

The length of the rainfall data series plays a crucial role in the assessment of the system performance. Due to the limited availability of high-resolution rainfall records in several countries, it clearly emerges the importance to evaluate how the length of the time series affects the reliability of the results of the water balance simulations.

Several studies (*e.g.* [19, 27, 32]) have addressed this issue. Results seem to show that 30-year rainfall data records are sufficient for accurate long-term simulations, although the estimation on the corresponding accuracy level seems to be case-sensitive (*i.e.* affected by the specific precipitation regime and/or rainwater demand). Ghisi *et al.* [39] concluded that it is possible to use short-term simulations for the analysis of RWH systems in buildings where the rainwater demand is lower than 30% of the potable water demand.

Liaw and Tsai [32] investigated the impact of the data series length; in particular continuous simulations were performed under different input rainfall data in Taiwan by varying the series length from 1 to more than 80 years. Results show that simulation periods of 50 years or longer reduces the variation of performance estimation within 3% with a 95% confidence level. Similarly, Mitchell [27] in Australia compared results of the simulations performed using 50, 10 and 1-year rainfall records according to the availability of rainfall data series for the region. The results of the study revealed that the initial condition affects the evaluations as the simulation period decreases. In addition, the authors found that a 10-year simulation period is representative of the long-term precipitation pattern for the sites of concern.

Palla *et al.* [19] examined the reliability of RWH system performance by using the centenarian time series (daily resolution) recorded at Genoa and Florence (respectively 148-years and 167-years climate records). A moving time window was used to extract realizations of successive 30 and 50 years of daily rainfall records from the original time series of Genoa and Florence in order to analyse the impact on long-term simulation results. Results indicate that a 30-year simulation period allowed assessing the RWH system performance with sufficient accuracy (within 5%) when compared with simulation results based on centenarian time series.

RWH SYSTEM PERFORMANCE

The suitability of RWH systems to reduce the demand for drinking water as well as to control runoff (thus limiting the hydraulic failure of the drainage system) is usually measured by means of dimensionless performance indexes. With reference to storage-yield techniques, different performance metrics are documented in the literature. This section presents an overview of the main used performance indexes under different climatic and operational conditions.

The performance of RWH tanks installed in different sites located in Southern Europe is also examined, as an example.

Performance Indexes

The performance of RWH systems is generally described in terms of *reliability, i.e.* the probability that the tank is able to meet the target demand at any interval of time of the simulation period [40]. The measure of reliability includes the time-based reliability and the volumetric reliability.

Time-based reliability considers the proportion of intervals (during the simulation period) during which the storage system can meet the target demand. According to McMahon *et al.* [40] a general expression for its estimation is:

$$R_t = \frac{N_s}{N} \quad 0 \leq R_t \leq 1 \qquad \textbf{(12)}$$

where R_t is the time-based reliability, N_s is the number of intervals that the target demand was fully met and N is the total number of intervals of the simulation period.

Differently, Khastagir and Jayasuriya [3] defined the reliability as the probability of the system being not empty expressed in percentage:

$$R_e = \frac{P}{N} \cdot 100 \quad 0 \leq R_e \leq 100 \tag{13}$$

where R_e is the time-based reliability, P is the number of time intervals (generally days) that the tank is not empty and N is the total number of intervals.

Liaw and Tsai [32] estimated the time-based reliability, as the fraction of time that the demand is fully met:

$$R_e = 1 - \frac{n}{N} \quad 0 \leq R_e \leq 1 \tag{14}$$

where n is the number of time units where demand exceeds the storage volume and N is the total number units in the historical records. Likewise, in order to compute the reliability, Basinger *et al.* [17] have introduced the failure events corresponding to the number of days during which the rainwater stored volume is insufficient to meet the water demand. Thus the reliability is calculated as the complement to unity of the failure occurrence probability.

Volumetric reliability R_v is defined as the volume of supplied water divided by the total target demand during the entire simulation period [40]. With reference to RWH systems, the volumetric reliability is generally known as water-saving efficiency E_T [41]:

$$E_T = \frac{\sum_{t=1}^{T} Y_t}{\sum_{t=1}^{T} D_t} \quad 0 \leq E_T \leq 1 \tag{15}$$

where Y_t represents the rainwater supply (yield) volume at each time step t, D_t is the rainwater demand volume at each time step, and T is the total number of time steps. Note that T may refer to monthly, annual, or entire simulation periods. The water saving efficiency assumes value 0 if there is no yield from the rainwater tank (for example, if only water supplied from the mains is used) and it assumes value 1 if only the stored rainwater is used by the devices connected to the tank. Note that E_T can be expressed also in percentage, thus ranging between 0 and 100%.

The volumetric reliability is less restrictive than the time-based reliability, since it takes into account also the time steps during which the water demand is only partially met. On the contrary, whenever the water demand is not fully satisfied, the failure event occurs according to a time-based reliability approach. Further, the estimate of the time reliability is more sensitive to the computational time-step, the length of data record and the system characteristics. Therefore, the use of volumetric reliability is usually preferred for the evaluation of RWH systems' performance (*e.g.* [4, 19, 23, 27, 38]).

A complementary index to measure the performance of RWH systems is the rainwater overflow ratio, O_T, *i.e.* the volume of rainwater exceeding the system capacity divided by the inflow during the selected time interval:

$$O_T = \frac{\sum_{t=1}^{T} O_t}{\sum_{t=1}^{T} Q_t} \quad 0 \leq O_T \leq 1 \tag{16}$$

where O_T is the overflow ratio, O_t represents the rainwater volume exceeding the system capacity at each time step t, Q_t is the tank inflow volume at each time step, and T is the total number of time steps.

When RWH systems are used as storm water control measures two dimensionless indexes can be defined to measure the volume and peak reduction operated by RWH systems at the household scale [21]: the volumetric retention efficiency and the peak retention efficiency. The volumetric retention efficiency, E_R, is a measure of the capability of the tank to retain rainwater volumes, thus reducing downstream runoff. The following equation is used to determine the tank volumetric retention efficiency on annual basis:

$$E_R = \left(1 - \frac{\sum_{t=1}^{T} O_t}{\sum_{t=1}^{T} Q_t}\right) \cdot 100 \tag{17}$$

where the second of the two terms in brackets provides the overflow ratio evaluated on annual basis. Equation 17 clearly shows that E_R tends to 0% when the sum of discharged volumes O_t tends to the sum of Q_t that may occur in those cases characterized by small tanks and/or reduced rainwater demands. Oppositely, if the tank is large and/or the household demand for rainwater is high, the tank may provide significant retention performances with E_R approaching value 100%.

The peak retention efficiency, E_{PR}, can be defined as:

$$E_{PR} = \left(1 - \frac{O_{peak}}{Q_{peak}}\right) \cdot 100 \quad \textbf{(18)}$$

where O_{peak} [L^3] and Q_{peak} [L^3] are the volumes discharged at peak time intervals as overflow and inflow of the storage tank, respectively. E_{PR} is a measure of how much the storage tank is able to reduce/retain the runoff peak at the rainfall event scale. Different from the previous indicators, E_{PR} is evaluated for each event separately and its evaluation may provide significant information only if appropriate (high) temporal resolutions are considered for the tank water balance.

Performance Analysis in Europe

The precipitation regime significantly affects the performance of RWH systems and consequently their optimal design. This section aims at illustrating the RWH performance across Europe, moving from the continental to the Mediterranean areas.

Palla *et al.* [20] analysed the performance of domestic RWH systems under various European climate zones using the Köppen-Geiger classification. This climate classification is the first and still the most common quantitative system to classify world climates, recently updated by using large global data sets of precipitation and temperature time series (*e.g.* [42]). As reported by Peel *et al.* [43], Europe is characterized by four over five main climate types. In particular, the continental climate, namely D, covers 44.4% of the land area, followed by the dry climate, namely B, the temperate climate, namely C, and the polar climate, namely E, that cover respectively 36.3, 17.0 and 2.3%. No tropical climate regions, namely A, are present in Europe.

Table **1** provides an example of the main rainfall features with respect to the different European climate zones. Mean and standard deviation of the annual depth, event depth and antecedent dry weather period are reported, thus providing rainfall characteristics at both the annual and event scales. Rainfall data refer to 46 sites in Europe, equally distributed among 5 main climate zones. Detailed description of the site selection criterion is provided by Palla *et al.* [20].

From rainfall data reported in Table **1**, it can be observed that the warm continental and warm temperate climates are characterized, on average, by similar precipitation regime at both annual and event scale. Further, these climate zones reveal frequent events, quite evenly distributed across the whole year. The hot temperate climate is characterized by the highest annual rainfall depth and by less frequent events. In addition, events show larger depths when compared to the events observed for the warm continental and temperate climates. In the Mediterranean climate, although the annual rainfall depth is comparable to the warm climates, on average, the precipitation characteristics are less consistent within this climate zone as confirmed by the standard deviation values. At the event scale, this precipitation regime shows a more remarkable inter-annual variability if compared with the previous climates. The semi-arid climate reveals analogous range of variation for the rainfall event characteristics, while the lowest annual rainfall depth is recorded.

Table 1. Main rainfall characteristics (annual depth, event depth and antecedent dry weather period – ADWP) of the 5 European climates according to the Köppen classification. The values of the mean and the corresponding standard deviation are reported. Within brackets, the corresponding Köppen-Geiger climate class is indicated.

Köppen-Geiger Climate Type	**Annual Rainfall Depth [mm]**		**Event Rainfall Depth [mm]**		**ADWP [d]**	
	μ	σ	μ	σ	μ	σ
Warm Continental (Dfb)	707	147	11.7	15.8	3.1	3.2
Warm Temperate (Cfb)	734	101	12.7	17.2	3.3	3.4
Hot Temperate (Cfa)	1066	440	21.3	29.5	4.8	5.2
Mediterranean (Csa)	727	274	20.3	33.0	7.5	11.5
Semi-arid (Bsk)	431	132	11.6	19.1	7.9	10.8

Focusing on the water saving efficiency, the results of the performance analysis reported in Palla *et al.* [20] are synthetized in Table **2**, where the water saving efficiency (mean and standard deviation values) are reported for the 5 European climates with respect to the corresponding storage fraction ranging between 0.01 and 1. Findings of continuous simulations emphasise three main typical behaviours that can be associated to the following climatic groups:

- the warm continental and temperate zones;
- the hot temperate zone;
- the Mediterranean and semi-arid temperate zones.

Table **2** shows that the warm continental and temperate zones reveal the highest quantitative performance across the whole range of storage fractions, mainly in the range 0.01-0.1. Such behaviour is affected by the precipitation regime indeed, northern Europe is characterized by frequent rainfall events and a moderate inter-annual variability (see also Table **1**). Again, the Mediterranean and semi-arid temperate zones that show comparable values in terms of ADWP and in particular less frequent event with a significant range of variation reveal similar performance: the smallest frequency of rainfall events, the lowest water saving efficiency, thus confirming the influence of inter-annual variability in affecting the system performance. Further it has to be noticed that since the performance analysis was carried out by means of the storage fraction (dimensionless approach), the difference in term of annual rainfall depth are accounted within the S/Q values.

Table 2. Water saving efficiency with respect to storage fraction for the 5 European climates according to the Köppen classification. The values of the mean and the corresponding standard deviation are reported for each climate zone.

Storage Fraction [-]	Water Saving Efficiency [-]									
	Warm Continental		Warm Temperate		Hot Temperate		Mediterranean		Semi-arid	
	μ	σ	μ	σ	μ	σ	μ	σ	μ	σ
0.01	0.54	0.03	0.54	0.04	0.43	0.05	0.33	0.06	0.34	0.05
0.02	0.67	0.03	0.66	0.03	0.53	0.04	0.43	0.06	0.42	0.06
0.05	0.79	0.01	0.80	0.03	0.72	0.04	0.57	0.07	0.56	0.07
0.08	0.83	0.02	0.85	0.02	0.78	0.03	0.63	0.07	0.65	0.06
0.1	0.87	0.02	0.88	0.02	0.82	0.04	0.67	0.08	0.70	0.06
0.3	0.95	0.01	0.95	0.01	0.91	0.02	0.83	0.06	0.84	0.05
0.5	0.97	0.01	0.97	0.01	0.94	0.01	0.90	0.03	0.91	0.04
1	0.98	0.01	0.98	0.01	0.96	0.02	0.95	0.02	0.95	0.03

The Mediterranean climate is characterized by a wide range of rainfall characteristics both at the annual and at the event scale that is translated into a corresponding wide range of variation in the water saving efficiency, if compared to northern Europe. Based on such considerations, the following section provides a detailed description of the RWH system performance focusing on southern Europe.

Test Sites in Southern Europe

Six sites in southern Europe have been selected in order to account for different precipitation regimes within the Mediterranean climate.

In Table **3**, the location and the main characteristics of the six Mediterranean sites are summarized. In particular, at each site, rainfall data are available with daily resolution. The length of the precipitation record is greater than 30 years to suitably assess the system performance [19]. The six sites differ in terms of total annual precipitation, with an average value ranging from 540 mm recorded in

Malaga to about 1280 mm in Genoa (Table **3**). For the same cities, the characteristics of rainfall events in terms of depth, intensity, duration and ADWP are reported in Table **4**. Fig. **1** reports the cumulative frequency distribution of the rainfall event characteristics for the six precipitation series to briefly illustrate the main climate features of each site.

Table 3. Location of and main characteristics of the Barcelona, Malaga, Marseille, Corfu, Genoa and Lisbon weather stations. As for the annual precipitation, the mean and standard deviation values are reported.

Weather Station	Latitude (° 'N/S)	Longitude (° 'E/O)	Elevation (m a.s.l.)	Observation Period	Record Length (years)	Annual Precipitation (mm)
Barcelona	41°25' N	02°07' E	175	1950-1997	47	639±214
Malaga	36°40' N	04°29' O	7	1943-2009	66	540±211
Marseille	43°26' N	05°13' E	5	1947-2010	63	536±149
Corfù	39°37' N	19°55' E	11	1955-2001	46	1090±232
Genoa	44°24' N	08°55' E	55	1833-2008	175	1276±310
Lisbon	38°43' N	09°09' O	77	1941-1999	58	738±195

Table 4. Rainfall event characteristics of the precipitation time series of Barcelona, Malaga, Marseille, Corfu, Genoa and Lisbon. The mean and standard deviation values are reported for each rainfall event characteristics.

Station Name	Rainfall Event Characteristics		
	Rainfall Fepth (mm)	**Rainfall Duration (d)**	**ADWP (d)**
Barcelona	16.90 ± 28.77	1.64 ± 1.02	7.72 ± 8.45
Malaga	20.69 ± 34.93	1.86 ± 1.39	11.42 ± 19.04
Marseille	12.34 ± 17.89	1.79 ± 1.22	6.65 ± 7.26
Corfù	27.10 ± 39.43	2.53 ± 2.24	5.57 ± 7.71
Genoa	24.16 ± 41.47	2.43 ± 1.96	4.33 ± 4.43
Lisbon	20.15 ± 33.24	3.06 ± 2.99	6.80 ± 9.96

Results show that the rainfall event characteristics generally differ for the six sites. Indeed, only the behaviour of Genoa and Corfu is similar, while the other sites are generally comparable for one specific characteristic only, *e.g.* Malaga and Marseille for the rainfall event duration or Malaga and Lisbon for the rainfall event depth. In particular, Malaga and Genoa reveal the longest and the shortest ADWP, respectively, consistently with their total annual precipitation.

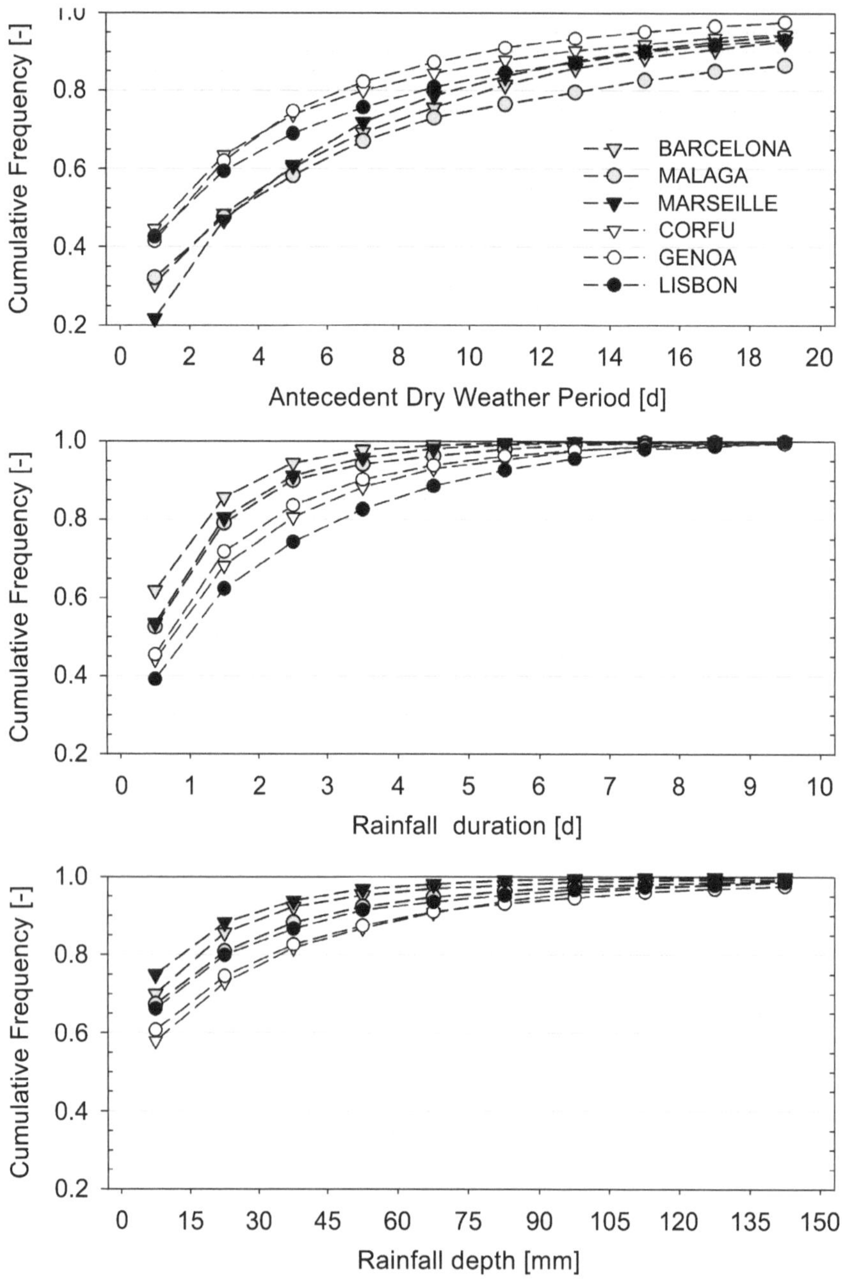

Fig. (1). Cumulative Frequency Distribution of the Antecedent Dry Weather Period (ADWP), the rainfall event duration and depth for the precipitation time series of Barcelona, Malaga, Marseille, Corfu, Genoa and Lisbon.

Simulation Scenarios

Water Demand Fraction

The performance of the rainwater harvesting system was analysed under varying conditions and system characteristics. The water demand for toilet flushing was calculated for the six sites in order to obtain the demand fraction equal to 1. Note that the optimum sizing of the tank is crucial only when the demand fraction is close to unity indeed at high/low demand fraction; the total water-saving efficiency keeps fairly constant almost, irrespective of the storage volume [19].

Storage Fraction

The storage capacity was assumed to range between 0.5 and 110 m^3, thus resulting in storage fractions, S/Q, varying from 0.01 to 1 at each site. Such an assumption complies with [30] for which storage fractions greater than 0.01 are required to enable accurate implementation of the behavioural model at a daily temporal resolution. In Table **5**, the storage fraction and the corresponding storage capacity (S) are listed for Barcelona, Malaga, Marseille, Corfu, Genoa and Lisbon.

Table 5. Storage fraction (S/Q) and the corresponding storage capacity (S) for the simulation scenarios of Barcelona, Malaga, Marseille, Corfu, Genoa and Lisbon.

Simulation Scenarios	Storage Fraction; *Storage Capacity* (-); (m^3)					
	Barcelona	**Malaga**	**Marseille**	**Corfù**	**Genoa**	**Lisbon**
1	0.01; *0.5*	0.01; *0.5*	0.01; 0.5	0.01; *1.0*	0.01; *1.0*	0.02; *1.0*
2	0.02; *1.0*	0.02; *1.0*	0.02; *1.0*	0.02; *2.0*	0.02; *2.0*	0.03; *2.0*
3	0.06; *3.0*	0.05; *2.0*	0.05; *2.0*	0.06; *5.0*	0.05; *5.0*	0.05; *3.0*
4	0.10; *5.0*	0.07; *3.0*	0.07; *3.0*	0.08; *7.0*	0.08; *8.0*	0.09; *5.0*
5	0.11; *5.5*	0.12; *5.0*	0.12; *5.0*	0.12; *10*	0.*10*;*10*	0.*10*; *6.0*
6	0.29; *15*	0.35; *15*	0.35; *15*	0.29; *25*	0.29; *30*	0.34; *20*
7	0.49; *25*	0.46; *20*	0.58; *25*	0.46; *40*	0.49; *50*	0.51; *30*
8	0.98; *50*	1.16; *50*	1.17; *50*	0.92; *80*	*1.08*; *110*	*1.02*; *60*

Performance Assessment of the RWH System

In Fig. **2**, the total water-saving efficiency and the total overflow ratio are plotted *versus* the storage fraction for the precipitation time series of the six selected cities. At demand fractions close to the unity, both the water-saving efficiency and overflow ratio indices are significantly affected by the storage fraction: storage fractions in the range 0.01-1 result in the increase in total efficiency from an average 0.3 to 0.95 and a reduction of the total overflow ratio from 0.7 to 0.05. By analysing only the quantitative performance, it is concluded that at high storage fractions the water-saving efficiency is maximized and the outflow is minimized.

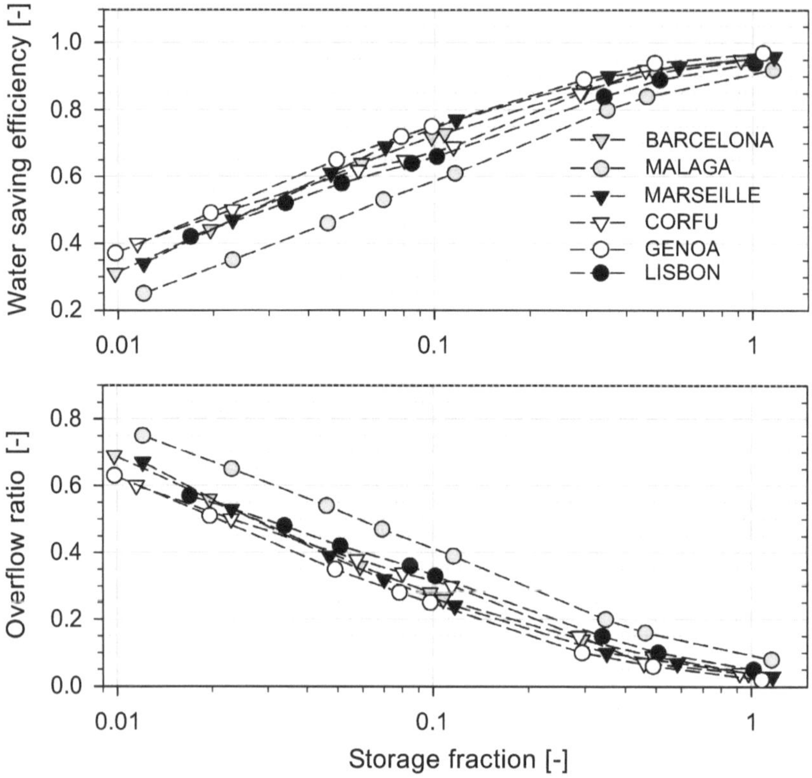

Fig. (2). Efficiency and Overflow *vs.* the storage fraction for the precipitation time series of Barcelona, Malaga, Marseille, Corfu, Genoa and Lisbon.

However, the suitable design of rainwater harvesting systems requires controlling the water quality degradation in the tank and, consequently, the water detention time. Based on results of a previous research dealing with grey water reuse [44], detention time below 48 hours is recommended to avoid significant water quality degradation. No similar research was carried out in the case of rainwater use although some water quality degradation during storage is expected. As to the six selected cities, the detention time was evaluated by assuming complete mixing of the inflow and stored water volume thus the detention time at each time step is equal to the age of the water stored in the tank after spillage [19]. Results indicate that low storage fractions (S/Q <0.1) should be adopted to minimize water quality degradation or rather to provide a detention time generally lower than 10 days thus limiting the required water treatment processes.

Although the estimation of the detention time has to be considered as a rough measure of the water quality degradation, these results point out that rainwater harvesting systems designed for supplying the optimum water demand ($D/Q \approx 1$) should be sized in the vicinity of storage fractions equal to 0.1 in order to provide optimal performance both in terms of water-saving efficiency and water quality.

By comparing the system quantitative performance corresponding to the different precipitation scenarios, it clearly emerges that at high storage fractions ($S/Q>0.3$), the total water-saving efficiency and the total overflow ratio are respectively limited in the 0.8-1.0 and 0.0-0.2 ranges almost irrespective of the precipitation regime (Fig. **2**). In the opposite case, at very low storage fractions ($S/Q\leq 0.02$), the total water-saving efficiency gradually increases from 0.35 for Malaga to 0.5 for Genoa (for $S/Q= 0.02$) or the overflow gradually increases from 0.65 for Malaga to 0.5 for Genoa (Fig. **2**). This different behaviour between Malaga and Genoa cannot be related to the difference in the annual precipitation, indeed the system performance is examined as a function of two non-dimensional parameters that are normalized with respect to the annual inflow [19]. Note that a similar discrepancy is observed also for Marseille and Malaga that show almost the same annual precipitation (see Table **3**). With reference to this latter case, the lower performance of Malaga (with respect to Marseille) is related to the specific precipitation regimes. In particular, the precipitation events in Malaga are less frequent and more intense than in Marseille, as confirmed by the average value of ADWP equal to 11.5 and 6.7 days, respectively (see Table **4**) thus producing a reduction of the compensation effect provided by the tank.

By comparing the system qualitative performance associated to different precipitation scenarios in the suggested range of storage fractions ($S/Q < 0.1$) it can be observed that the detention time is minimized for Malaga. Results reveals that a median detention time lower than 10 days is obtained for a storage fraction equal to 0.12, while 15 days are required for Marseille, Corfu and Lisbon. In the framework of the quality water degradation, the precipitation event characteristics (ADWP) positively impact on the system performance. Note that the contrasting goals of maximizing the water-saving efficiency and minimizing the detention time could be partially solved by selecting relatively higher storage capacity for precipitation conditions characterized by long ADWP.

However results are generally consistent across the six examined precipitation scenarios, therefore the design criterion $S/Q \approx 0.1$ seems to be valid for a generic installation in the Mediterranean climate.

GUIDELINES AND STANDARDS FOR THE DESIGN OF RWH SYSTEMS

Much of the available guidelines and standards for the design of RWH systems apply to households and commercial or industrial companies as well as public organizations, in which rainwater is used for non-potable uses. The guidelines and standards illustrate specifications for the design, installation and maintenance of the rainwater harvesting systems. The main guidelines available in Europe are briefly described and discussed in the following sub-sections.

German Standard (DIN 1989-1: 2002-04)

The German standard [45] is the first published standard in Europe and was prepared in 2002 with the general goal of achieving ecological and sustainable water management. The standard provides specific qualitative and quantitative aspects that have to be taken into account for the design for RWH systems with regard to: the catchment areas (see *collection areas* - Section 5 of the standard); the treatment operation (Section 6); the storage system (*rainwater reservoir* - Section 7); the conveyance system (pump - Section 8; supplemental feed system - Section 9; controls and pipe systems - Sections 10 to 12; overflow system - Sections 13, 14).

A specific section is dedicated to the design of the storage capacity (see section 16) where three sizing procedures are illustrated to meet different practical requirements. In particular the *abridged procedure* is reported for the sizing of small systems (*e.g.* one and two-family houses); the *simplified procedure* for all systems; and the *differentiated procedure* for large systems.

As for the *abridged procedure,* the storage capacity is calculated as a specific volume for connected catchment area in the range between 25 l/m^2 and 50 l/m^2 to be compared with a specific volume for user in the range between 800 and 1000 l per person. Note that in this case the collecting areas should be rooftops. The domestic use is considered constant over the entire year and the annual precipitation depth should range between 500 and 800 mm.

As for the *simplified procedure,* a dedicated calculation form is reported in the Appendix A of the standard. The input data are clearly indicated in the procedure as well as the relative evaluation methods: the precipitation has to be defined on annual basis; the size of the catchment is calculated as base area irrespective of the roof shape and slope; the values of the runoff coefficient (namely *yield coefficient* in the standard) are listed in table as a function of the area typology and vary in the range 0.3-0.8; a specific hydraulic efficiency of the treatment systems has to be considered (generally assumed equal to 0.9); and finally the water requirements are evaluated on annual basis as a function of use typology. Therefore, the *annual rainwater yield* (E_R), expressed in liters per year, is calculated as follows:

$$E_R = A_A \cdot e \cdot h_N \cdot \eta \tag{19}$$

where A_A is the collecting area expressed in square meters; e is the yield coefficient; h_N is the annual rainfall depth expressed in mm, and η is the hydraulic filter efficiency. Subsequently, the annual rainwater requirement (namely *process water requirement* in the standard) is calculated as follows:

$$BW_a = P_d \cdot n \cdot 365 \tag{20}$$

$$BW_a = A_{Bew} \cdot BS_a \tag{21}$$

where BW_a expressed in litres per year is alternately calculated based on the person-related or area-related data. P_d is the daily per person requirement; n is the number of users; A_{Bew} is the watering areas and BS_a is the specific annual requirement expressed as litres per square meters. Finally, the storage capacity is evaluated as the 6% of the lesser volume between E_R and BW_a.

As for the *differentiated procedure,* the optimization of the storages system capacity is proposed for large systems with individual consumption structures. The optimization strategy requires the simulation of the system behaviour at the daily temporal scale. The simulation period should cover at least 5 to 10-year interval of time; the longer the simulation period the more precise reliability of the system performance and thereby the optimum storage system size.

British Standard (BS 8515:2009)

The British standard [46] is defined as a code of practice that sets standards for the design, installation, quality of water, maintenance, and risk management of rainwater harvesting systems for both new-build projects and retro-fitted systems. The standard covers the storage tank, whether located above or below ground.

Focusing on the design section, the calculation methods to sizing the system are based on the procedures given in the German standard DIN 1989-1. The three methods are similarly named: *simplified approach*, *intermediate approach* and *detailed approach*. Furthermore, the British standard is the solely normative that indicates that the RWH systems can provide storm water control and that states specific methods to estimate the required additional storage volume when storm water control is to be integrated into the RWH systems (provided in the Annex A of the standard).

The *simplified approach* allows estimating tank capacity by using storage sizing graphs determined for specific numbers of household. The capacity is reported in ordinate as a function of roof plane area (in abscissa) and annual depth (regression lines). Similarly to the abridged procedure of the German standard the *simplified approach* is not suitable for commercial installations since the assumption relating to the demand (*i.e.* constant daily domestic use) is not applicable. Furthermore the additional storage capacity can be calculated by using a graph (provided in the Annex A.1) in order to integrate the sizing approach for storm water runoff purpose.

The *intermediate approach* provides a couple of equations in order to estimate the tank capacity based on a more flexible and accurate method. The equations are similar to the ones provided in the German simplified procedure and allows to calculate the annual rainwater yield (Y_R) and the annual non-potable water demand (D_N). Similarly to the German method, the storage capacity should be calculated as the lesser of 5% of the annual rainwater yield or 5% of the annual non-potable water demand. The selection of the 5% (that equates to 18 days in a year) instead of the 6% (that equates to 22 days in a year) of the German method is discussed and justified in order to take account of the daily rainfall variability in the hydrological regions of UK. Furthermore the additional storage capacity (*SC*) for storm water runoff purpose should be calculated, in the case of systems where supply is lower than demand, as follows:

$$SC = R_d \cdot A - (D_N - Y_R) \cdot 0.5 \qquad if\ D_N > Y_R \tag{22}$$

where R_d is the design storm event rainfall depth expressed in mm. As for systems where supply is greater that demand, the storm water control capacity is generally neglected, however in some cases it is possible to obtain some storm water control benefits, such as when the collected area is a green roof or a surface which results in lower runoff coefficient with respect to the hypothesized one (in the range between 0.7 and 0.8); when the tank is equipped with an outlet throttle to discharge excess flows; when the tank is automatically pumped out or drained after the event; when the tank is connect to an ex-filtration system for excess flows. In these latter cases, in order to size the storage capacity, it is recommended the use of the *detailed approach*.

The *detailed approach* should be used to calculate the storage size for the optimum design of large and complex RWH systems; in particular the optimum storage capacity is estimated by implementing a model of yield and demand at a daily temporal scale which is based on a continuous rainfall time series for a minimum of 3 years and preferably 5 years. The additional storage capacity needed for storm water control has to be evaluated according with three specific methods (illustrated in the Annex A.3): *analysis of 20+ extreme events; 100+ year extreme stochastic series* and *probability analysis with a 5-year time series*. The first two methods recommend of calculating the system and downstream network performance in an hourly temporal scale respectively for at least 20 observed extreme event or for all significant events of a more than 100 year series. In the third method a relevant 5-year time series has to be used in order to simulate the RWH behaviour and the maximum volume of water stored for 20% of the time should be assessed as measure of the storm water control capacity. This latter method may be applied for seasonal analysis if necessary.

Italian Guidelines (UNI/TS: 11445)

The Italian guidelines [47] illustrate the technical requirements for the design, installation (materials) and maintenance of RWH systems. The guidelines are organized into 17 sections differentiated for specific arguments (see *e.g.* Design – Section 5, Pump systems – Section 10; Controls and Measurement .- Section 15, *etc.*) and one Annex related to the technical elements for the RWH system design in the simplified method.

The Italian guidelines assumes for the system sizing two methods: namely the *simplified method* and the *analytical method*, similarly the other European standards.

The *simplified method* is based on the evaluation on two terms: the annual inflow, Q, and the annual water demand, R. In particular the annual inflow is evaluated by multiplying the collected area with the annual runoff depth and the latter is determined by multiplying the annual rainfall depth with the runoff coefficient of the corresponding collected area. The annual water demand for toilet flushing is evaluated by assuming a constant daily rate of 40 l/d/p. The annual water demand for irrigation of private green areas is evaluated by assuming a specific annual rate of 300 l per square meters (see the Annex A.2 of the Guidelines) The useful volume (V_u) of the tank is then assumed as the 6% of the lesser between the inflow and water demand volumes on annual basis. The Italian guidelines assume the same percentage of the German method but include the possibility of enlarge the useful volume by multiplying it for 1.5 in order to optimize the storage capacity in case of a significant variability of the precipitation and rain water demand.

Furthermore graphs for the dimensioning of the 4-person or 8-person household systems are also provided in the Annex A.3 in order to easily evaluate the storage capacity.

The *analytical method* can be adopted for the sizing of all RWH systems since it allows determining the storage capacity as a function of the desired system performance. The use of simulation model that implement a mass balance equation is required. As for the evaluation of the mass balance terms, the YAS operating rule is suggested since it allows a conservative estimate of the system performance irrespective of the model time scale. It is recommended to set up an initially empty storage and to use a daily temporal scale for storage fraction values greater than 0.01 otherwise hourly or sub-hourly temporal scales are required. The continuous simulation should be based on rainfall time series for a minimum of 30 years in order to obtain reliable performance.

The Simplified Approach and RWH System Reliability

By examining the European standard for the RWH design, it emerges that simplified approaches (namely *Simplified* in German and Italian standards and *Intermediate* in British ones) are available to easily design the tank capacity based on a limited number of parameters. Such approaches are strongly recommended in applications at the household level (small scale) and when the water demand is constant over the year. Conversely, if the tank capacity has to play a role in storm water runoff control (thus providing suitable retention/detention capacity), a detailed modelling approach is required to combine the RWH system performance with the expected hydrological benefits.

In order to evaluate the reliability of a RWH system designed with a simplified approach, the methodology is applied to a hypothetical household located in Genoa (Italy). Indeed, based on the long-term simulation results presented in the previous section, it is possible to assess the corresponding performance of the RWH system whose tank capacity is design according the *Simplified* method of the Italian standard.

As an example, it is assumed that the RWH system collects runoff from a pitched roof characterized by a surface area of 100 m^2; referring to the Italian standard, the runoff coefficient of 0.8 is considered. Then, the annual inflow Q can be easily computed by multiplying the annual rainfall depth of Genoa (see Table **3**) by the runoff coefficient and by the collected area, thus resulting in annual inflow volume Q of 102 m^3. The long-term simulation have been performed with a dimensionless approach by considering a demand fraction equal to 1 in term of water demand scenario, therefore the annual water demand, D, for the site of concern is assumed equal 102 m^3. Since the water demand to be supplied by rainwater is limited to toilet flushing (*i.e.* 40 l/d/p according to the Italian standard), the corresponding number of users is 7. Based on the simplified approach, the size of the tank is the 6% of the lesser between the annual inflow and water demand volumes, therefore the storage fraction, *S/Q* and the tank capacity, *S* result equal to 0.06 and 6 m^3, respectively.

Looking at RHW system performance illustrate in Fig. **2** with respect to the Genoa site (white dots), the water-saving efficiency and the overflow ratio corresponding to a storage fraction of 0.06 are respectively 0.69 and 0.31 based on long-term simulation scenario. Modelling results confirm that even designing the RWH system with a simplified approach, satisfactory performance in terms of rainwater demand are provided.

The simplified approach needing simply annual-basis parameters tends to oversize the storage capacity, on the other hand it allows sizing RWH systems characterised by high water saving efficiency performance. Taking into account the variability of the precipitation regime (mainly in terms of inter-event variability - ADWP) and how these rainfall characteristics affect the performance of RWH systems, it clearly emerges that only a detailed modelling approach is able to guarantee the optimal sizing of the RWH system. As discussed in the previous section, such behaviour is limited in northern Europe where the precipitation regime is characterized by frequent rainfall events distributed across the year.

CONCLUSION

While the practice of rain water harvesting (RWH) is quite old, technology development and increased water pressure has boosted its implementation in many countries [48].

The most common methodologies for designing rainwater harvesting systems are illustrated with a specific focus on models based on the tank long-term water balance simulation. An overview of metrics to evaluate RWH system performance is presented, with regard to the estimation of both water saving potential and runoff mitigation benefits. The accuracy of results of model simulations with reference to the length of the available rainfall series and to the selected resolution time step is discussed. Findings of research studies indicate that at least 30-year daily rainfall data records are required to reliable assess the RWH system performance while sub-daily time scale resolution is needed to investigate the hydrologic benefits.

Focusing on the impact of precipitation regime on the RWH system performance, long-term simulations are performed with respect to six cities in southern Europe characterized by a Mediterranean climate. With reference to a RWH system designed to meet constant water demand over the entire year (used for toilet flushing) and equivalent to a demand fraction equal to 1, modelling results indicate that storage tank should be sized in the vicinity of storage fractions equal to 0.1 in order to provide good performance both in terms of quantitative ($E_T > 0.6$ and $O_T < 0.4$) and qualitative performance (detention time < 10 days). Furthermore, results suggest a critical correlation between the rainfall event characteristic, ADWP, and the system performance: the water saving efficiency is in inverse proportion while the water quality degradation (expressed as detention time) is directly proportional to the ADWP. Aiming at the optimal sizing of the RWH system, such inverse behaviour could help the system designer in the contrasting goals of maximizing the water-saving efficiency and minimizing the detention time.

Finally, German, British and Italian standards have been compared to identify differences in the design approaches adopted by the three countries. Despite research studies as well as recommendations in force are organized in the perspective of water conservation objective, future common development could consider including benefits associated to the runoff retention of RWH systems [49]. Multi-criteria optimisation or other methods may be implemented in order to address and balance the contrasting targets dealing with water resource management in the urban environment.

The drafting of common guideline at the European level could effectively support the implementation of the RWH systems at the urban catchment scale. In the framework of sustainable urban water management strategies, it is argued that the widespread implementation of low impact development (LID) solutions (including rainwater harvesting) is needed to maximize the hydrological benefits.

CONFLICT OF INTEREST

The authors confirm that they have no conflict of interest to declare for this publication.

ACKNOWLEDGEMENT

Declared None.

REFERENCES

[1] C. E., "Study on water performance of buildings", *Reference report 070307/2008/520703/ETU/D2*, 2009.

[2] A. Campisano, I. Gnecco, C. Modica, and A. Palla, "Designing domestic rainwater harvesting systems under different climatic regimes in Italy", *Water Sci. Technol,* vol. 67, no. 11, pp. 2511-2518, 2013. [http://dx.doi.org/10.2166/wst.2013.143] [PMID: 23752383]

[3] A. Khastagir, and N. Jayasuriya, "Optimal sizing of rain water tanks for domestic water conservation", *J. Hydrol,* vol. 381, no. 3-4, pp. 181-188, 2010. [http://dx.doi.org/10.1016/j.jhydrol.2009.11.040]

[4] Y. Zhang, D. Chen, L. Chen, and S. Ashbolt, "Potential for rainwater use in high-rise buildings in Australian cities", *J. Environ. Manage,* vol. 91, no. 1, pp. 222-226, 2009. [http://dx.doi.org/10.1016/j.jenvman.2009.08.008] [PMID: 19744767]

[5] E. Ghisi, D. Lapolli Bressan, and M. Martini, "Rainwater tank capacity and potential for potable water savings by using rainwater in the residential sector of southeastern Brazil", *Build. Environ,* vol. 42, no. 4, pp. 1654-1666, 2007. [http://dx.doi.org/10.1016/j.buildenv.2006.02.007]

[6] US EPA, "Guidelines for Water Reuse", In: *EPA Report 625-R-04-108.* US Environmental Protection Agency: Washington, DC, USA, 2004.

[7] D. Glist, "Rainwater down the toilet", *Pollut. Eng,* vol. 37, no. 4, pp. 40-41, 2005.

[8] A. Mukhopadhyay, A. Akber, and E. Al-Awadi, "Analysis of freshwater consumption patterns in the private residences of Kuwait", *Urban Water,* vol. 3, pp. 53-62, 2001. [http://dx.doi.org/10.1016/S1462-0758(01)00016-4]

[9] M.J. Burns, T.D. Fletcher, H.P. Duncan, B.E. Hatt, A.R. Ladson, and C.J. Walsh, "The performance of rainwater tanks for stormwater retention and water supply at the household scale: an empirical study", *Hydrol. Processes,* vol. 29, pp. 152-160, 2015. [http://dx.doi.org/10.1002/hyp.10142]

[10] L.M. Ahiablame, B.A. Engel, and I. Chaubey, "Effectiveness of low impact development practices: Literature review and suggestions for future research", *Water Air Soil Pollut,* vol. 223, no. 7, pp. 4253-4273, 2012. [http://dx.doi.org/10.1007/s11270-012-1189-2]

[11] V.G. Mitchell, D.T. McCarthy, A. Deletic, and T.D. Fletcher, "Urban stormwater harvesting - sensitivity of a storage behaviour model", *Environ. Model. Softw,* vol. 2, no. 6, pp. 782-793, 2008.

[12] A. Gerolin, R.B. Kellagher, and M.G. Faram, "Rainwater harvesting systems for stormwater management: Feasibility and sizing considerations for the UK", *Proc. 7th International Conference on Sustainable Techniques and Strategies for Urban Water Management (Novatech 2010),* pp. 1-10, 2010.

[13] G. Petrucci, J.F. Deroubaix, B. De Gouvello, J.C. Deutsch, P. Bompard, and B. Tassin, "Rainwater harvesting to control stormwater runoff in suburban areas. An experimental case-study", *Urban Water J,* vol. 9, pp. 45-55, 2012. [http://dx.doi.org/10.1080/1573062X.2011.633610]

[14] M.J. Burns, T.D. Fletcher, C.J. Walsh, A.R. Ladson, and B.E. Hatt, "Flow-Regime Management at the Urban Land-Parcel Scale: Test of Feasibility", *J. Hydrol. Eng,* vol. 20, no. 12, p. 04015037, 2014. [http://dx.doi.org/10.1061/(ASCE)HE.1943-5584.0001002]

[15] E. Rozos, C. Makropoulos, and Č. Maksimovic, "Rethinking urban areas: an example of an integrated blue-green approach", *Water Sci. Technol. Water Supply,* vol. 13, no. 6, pp. 1534-1542, 2013. [http://dx.doi.org/10.2166/ws.2013.140]

[16] I. Gnecco, C. Berretta, L.G. Lanza, and P. La Barbera, "Quality of stormwater runoff from paved surfaces of two production sites", *Water Sci. Technol,* vol. 54, no. 6-7, pp. 177-184, 2006. [http://dx.doi.org/10.2166/wst.2006.610] [PMID: 17120648]

[17] M. Basinger, F. Montalto, and L. Upmanu, "A rainwater harvesting system reliability model based on nonparametric stochastic rainfall generator", *J. Hydrol,* vol. 392, no. 3-4, pp. 105-118, 2010. [http://dx.doi.org/10.1016/j.jhydrol.2010.07.039]

[18] T. Abbasi, and S.A. Abbasi, "Sources of pollution in rooftop rainwater harvesting systems and their control", *Crit. Rev. Environ. Sci. Technol,* vol. 41, no. 23, pp. 2097-2167, 2011. [http://dx.doi.org/10.1080/10643389.2010.497438]

[19] A. Palla, I. Gnecco, and L.G. Lanza, "Non-dimensional design parameters and performance assessment of rainwater harvesting systems", *J. Hydrol,* vol. 401, no. 1-2, pp. 65-76, 2011. [http://dx.doi.org/10.1016/j.jhydrol.2011.02.009]

[20] A. Palla, I. Gnecco, L.G. Lanza, and P. La Barbera, "Performance analysis of domestic rainwater harvesting systems under various European climate zones", *Resour. Conserv. Recycling,* vol. 62, pp. 71-80, 2012. [http://dx.doi.org/10.1016/j.resconrec.2012.02.006]

[21] A. Campisano, and C. Modica, "Appropriate resolution timescale to evaluate water saving and retention potential of rainwater harvesting for toilet flushing in single houses", *J. Hydroinfor,* vol. 17, no. 3, pp. 331-346, 2015. [http://dx.doi.org/10.2166/hydro.2015.022]

[22] V. Lazarova, S. Hills, and R. Birks, "Using recycled water for non-potable, urban uses: a review with particular reference to toilet flushing", *Water Sci. Technol. Water Supply,* vol. 3, no. 4, pp. 69-77, 2003.

[23] A. Fewkes, "Modelling the performance of rainwater collection systems: towards a generalised approach", *Urban Water,* vol. 1, pp. 323-333, 2000. [http://dx.doi.org/10.1016/S1462-0758(00)00026-1]

[24] A.S. Silva, and E. Ghisi, "Uncertainty analysis of daily potable water demand on the performance evaluation of rainwater harvesting systems in residential buildings", *J. Environ. Manage,* vol. 180, pp. 82-93, 2016. [http://dx.doi.org/10.1016/j.jenvman.2016.05.028] [PMID: 27208997]

[25] J.S. Mun, and M.Y. Han, "Design and operational parameters of a rooftop rainwater harvesting system: definition, sensitivity and verification", *J. Environ. Manage,* vol. 93, no. 1, pp. 147-153, 2012. [http://dx.doi.org/10.1016/j.jenvman.2011.08.024] [PMID: 22054581]

[26] A. Campisano, and C. Modica, "Optimal sizing of storage tanks for domestic rain water harvesting in Sicily", *Resour. Conserv. Recycling,* vol. 63, pp. 9-16, 2012. [http://dx.doi.org/10.1016/j.resconrec.2012.03.007]

[27] V.G. Mitchell, "How important is the selection of computational analysis method to the accuracy of rainwater tank behaviour modelling?", *Hydrol. Processes,* vol. 21, no. 21, pp. 2850-2861, 2007. [http://dx.doi.org/10.1002/hyp.6499]

[28] S. Ward, F.A. Memon, and D. Butler, "Rainwater harvesting: model-based design evaluation", *Water Sci. Technol,* vol. 61, no. 1, pp. 85-96, 2010. [http://dx.doi.org/10.2166/wst.2010.783] [PMID: 20057094]

[29] Y.P. Guo, and B.W. Baetz, "Sizing of rainwater storage units for green building applications", *J. Hydrol. Eng,* vol. 12, no. 2, pp. 197-205, 2007. [http://dx.doi.org/10.1061/(ASCE)1084-0699(2007)12:2(197)]

[30] A. Fewkes, and D. Butler, "Simulating the performance of rainwater collection systems using behavioural models", *Build. Serv. Eng. Res. Tech,* vol. 21, no. 2, pp. 99-106, 2000. [http://dx.doi.org/10.1177/014362440002100204]

[31] T. Herrmann, and U. Schmida, "Rainwater utilisation in Germany: efficiency, dimensioning, hydraulic and environmental aspects", *Urban Water j,* vol. 1, pp. 307-316, 1999. [http://dx.doi.org/10.1016/S1462-0758(00)00024-8]

[32] C.H. Liaw, and Y.L. Tsai, "Optimum storage volume of rooftop rain water harvesting systems for domestic use", *J. Am. Water Resour. Assoc,* vol. 40, no. 4, pp. 901-912, 2004. [http://dx.doi.org/10.1111/j.1752-1688.2004.tb01054.x]

[33] E. Ghisi, and D.F. Ferreira, "Potential for potable water savings by using rainwater and greywater in a multi-story residential building in southern Brazil", *Build. Environ,* vol. 42, no. 4, pp. 2512-2522, 2007.
[http://dx.doi.org/10.1016/j.buildenv.2006.07.019]

[34] I.M. Brodie, "Stormwater reuse treatment requirements and screening-level risk assessment at two urban spatial scales", *J. Water Reuse Desalin,* vol. 2, no. 1, pp. 13-21, 2012.
[http://dx.doi.org/10.2166/wrd.2012.072]

[35] A. Campisano, and C. Modica, "Regional scale analysis for the design of storage tanks for domestic rainwater harvesting systems", *Water Sci. Technol,* vol. 66, no. 1, pp. 1-8, 2012.
[http://dx.doi.org/10.2166/wst.2012.171] [PMID: 22678193]

[36] C.M. Agudelo-Vera, K.J. Keesman, A.R. Mels, and H.H. Rijnaarts, "Evaluating the potential of improving residential water balance at building scale", *Water Res,* vol. 47, no. 20, pp. 7287-7299, 2013.
[http://dx.doi.org/10.1016/j.watres.2013.10.040] [PMID: 24238880]

[37] I. Leusbrock, T.A. Nanninga, K. Lieberg, C.M. Agudelo-Vera, K.J. Keesman, G. Zeeman, and H.H. Rijnaarts, "The urban harvest approach as framework and planning tool for improved water and resource cycles", *Water Sci. Technol,* vol. 72, no. 6, pp. 998-1006, 2015.
[http://dx.doi.org/10.2166/wst.2015.299] [PMID: 26360761]

[38] A. Campisano, and C. Modica, "Selecting time scale resolution to evaluate water saving and retention potential of rainwater harvesting tanks", *Procedia Eng,* vol. 70, pp. 218-227, 2014.
[http://dx.doi.org/10.1016/j.proeng.2014.02.025]

[39] E. Ghisi, K.A. Cardoso, and R.F. Rupp, "Short-term versus long-term rainfall time series in the assessment of potable water savings by using rainwater in houses", *J. Environ. Manage,* vol. 100, pp. 109-119, 2012.
[http://dx.doi.org/10.1016/j.jenvman.2011.12.031] [PMID: 22377369]

[40] T.A. McMahon, A.J. Adeloye, and S.L. Zhou, "Understanding performance measures of reservoirs", *J. Hydrol,* vol. 324, no. 1-4, pp. 359-382, 2006.
[http://dx.doi.org/10.1016/j.jhydrol.2005.09.030]

[41] A. Dixon, D. Butler, and A. Fewkes, "Computer simulation of domestic water reuse system: investigating greywater and rainwater in combination", *Water Sci. Technol,* vol. 38, no. 4, pp. 25-32, 1999.
[http://dx.doi.org/10.1016/S0273-1223(99)00083-9]

[42] M. Kottek, J. Grieser, C. Beck, B. Rudolf, and F. Rubel, "World Map of the Köppen-Geiger climate classification updated", *Meteorol. Z,* vol. 15, no. 3, pp. 259-263, 2006.
[http://dx.doi.org/10.1127/0941-2948/2006/0130]

[43] M.C. Peel, B.L. Finlayson, and T.A. McMahon, "Updated world map of the Köppen -Geiger climate classification", *Hydrol. Earth Syst. Sci. Discuss,* vol. 4, pp. 439-473, 2007.
[http://dx.doi.org/10.5194/hessd-4-439-2007]

[44] A. Dixon, D. Butler, A. Fewkes, and M. Robinson, "Measurement and modelling of quality changes in stored untreated greywater", *Urban Water j,* vol. 1, no. 4, pp. 293-306, 2000.
[http://dx.doi.org/10.1016/S1462-0758(00)00031-5]

[45] "DIN 1989-1: 2002-04, Rainwater harvesting systems – Part I: planning, installation, operation and maintenance", *DIN Deutsches Institut für Normung e.V,* 2004.

[46] "BS 8515:2009, Rainwater harvesting system- code of practice", *British Standard Institution (BSI),* 2009.
[http://dx.doi.org/10.3403/30171876]

[47] UNI/TS, "Rainwater harvesting systems for non-potable uses. Design, installation and maintenance", In: *UNI Ente Italiano di Unificazione.* 2012. (In Italian)

[48] A. Campisano, D. Butler, S. Ward, M.J. Burns, E. Friedler, K. DeBusk, L.N. Fisher-Jeffes, E. Ghisi, A. Rahman, H. Furumai, and M. Han, "Urban rainwater harvesting systems: Research, implementation and future perspectives", *Water Res,* vol. 115, pp. 195-209, 2017.

[49] A. Palla, I. Gnecco, and P. La Barbera, "The impact of domestic rainwater harvesting systems in storm water runoff mitigation at the urban block scale", *J. Environ. Manage,* vol. 191, pp. 297-305, 2017. [http://dx.doi.org/10.1016/j.jenvman.2017.01.025]

CHAPTER 4

Economic, Environmental, and Social Criteria Evaluation of Rainwater Harvesting System Options for an Office and Lab Building on the University of Utah Campus

Ghazaleh Vaseghi[1,*], **Ilke Celik**[2], **Steven Burian**[3] and **Defne Apul**[2]

[1] *Chemical Engineering Department, University of Toledo, Ohio, USA*

[2] *Civil Engineering Department, University of Toledo, Ohio, USA*

[3] *Civil Engineering Department, University of Utah, Utah, USA*

Abstract: This chapter presents an approach for using multi-criteria decision analysis (MCDA) to study the tradeoffs of rainwater harvesting (RWH) system designs. Three scenarios have been designed and evaluated. Each one captures rainwater from the rooftops of the connected Civil and Materials Engineering (CME) and Hedco buildings on the University of Utah campus. Scenario 1 utilizes two separate storage cisterns; one large underground, and one smaller placed on the top of the CME building. Scenario 2 utilizes only one large underground cistern. It uses a pressure-regulated pump to supply a constant flow of water to each toilet in the building. Scenario 3 utilizes only one large cistern placed on top of the CME building. The design includes a pump to convey collected runoff from the Hedco building to the rooftop cistern. From there, toilets are flushed from the gravity-fed water. MCDA was used to integrate environmental impacts, cost, water use and social impacts of the proposed scenarios. GaBi software was used for environmental impact analysis; TRACI and ReCiPe were used as assessment tools. After assessment of the designs, it is suggested that Scenario 3 represents the most favorable option for any change to the toilet-flushing system in the CME.

Keywords: Concrete cistern, Environmental impacts, Economics, Fiberglass, GaBi, Global Warming Potential, Irrigation, Multi-criteria, Rainwater harvesting, Social impacts, Toilet flush.

INTRODUCTION

According to Utah foundation report in 2014, Utah's population has increased almost three times since 1970, and is estimated to be doubled by 2050 [1]. As

* **Corresponding author Ghazaleh Vaseghi:** Chemical Engineering Department, University of Toledo, Toledo, OH, USA; Tel: +1(575)405-6713; Email: ghazaleh.vaseghi@rockets.utoledo.edu

Enedir Ghisi (Ed.)

population grows, water demand also increases, and the cities with inadequate water supplies can experience the negative effects of water scarcity. Rainwater harvesting (RWH) offers a possible solution to this problem by capturing rainwater runoff from impervious surfaces (mainly rooftops) and storing it for later use. RWH has been practiced for centuries but was abandoned as cities developed extensive water distribution systems. Yet, as of the past decade, a renaissance of RWH has occurred with many studies from many parts of the world (*e.g.* Africa [2]; Asia [3 - 5], Australia [6], Europe [7, 8], North America [9 - 11]).

RWH systems have typically been assessed in the literature according to their energy consumption, water saving potential, and greenhouse gas emissions [12 - 16]. Some detailed life-cycle environmental impact assessment reports that discuss acidification, eutrophication, toxicity, ozone depletion, fossil depletion *etc.* impacts of the RWH systems, have also been published recently [17 - 20]. However, in addition to technical and environmental evaluations, economic and social indicators should be incorporated in evaluations of RWH systems as part of sustainable design.

Sustainable design of buildings to include RWH systems can be addressed *via* a multi-criteria decision analysis (MCDA). MCDA has been used to analyze a variety of problems; remediation [21], energy solutions [22], environmental policy [23] or integrated risk assessment [24]. This study presents an approach for using MCDA to integrate environmental impact, water use, life cycle cost and social impact from three proposed RWH scenarios that can be implemented in an educational building on the University of Utah campus (the Civil and Materials Engineering (CME) building and the Hedco lab building, which connected by a covered walkway). The background objective of this study was to establish analytical basis to evaluate high performance RWH systems that would enable designers to base choices on comprehensive evaluation.

METHODS

Site Description

The focus for this study consists of two connected buildings, the CME building and Hedco, on the University of Utah campus in Salt Lake City, Utah. The CME building is 4 stories tall with extra height for mechanical equipment. Hedco is 3 stories, but approximately 30 feet lower. The building footprints are slightly offset, as shown by the plan view drawing in Fig. (**1**). The CME building contains faculty and student offices, while Hedco contains an array of laboratories and shops for machining, materials processing, and chemical analysis. The total roof area of CME and Hedco is 28,824 ft^2. There is great interest to implement water conservation and storm water management features on the University of Utah campus and this building provides a potential location for RWH.

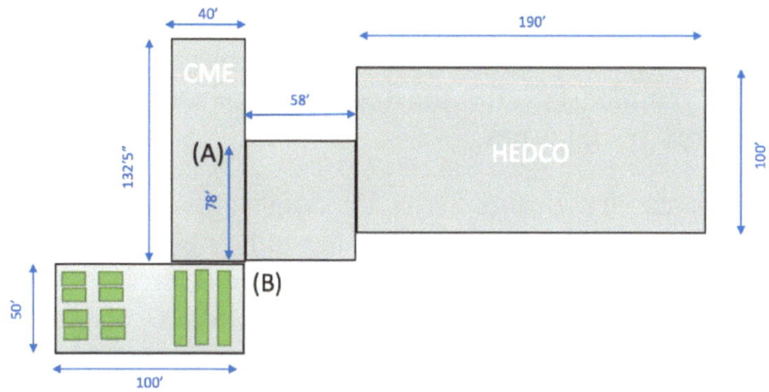

Scenario	1	2	3
Tank/Location	(A) (B)	(B)	(A)
Material	Fiberglass Tank A, Concrete Tank B	Concrete Tank B	Fiberglass Tank A
Capture area	CME for Tank A , CME and Hedco for Tank B	CME and Hedco	CME for Tank A
System aids	• Pump to feed A • Reinforcement • Water treatment • Educational Garden	• Pump to feed each toilet flushing	• Pump to redirect harvested water from Hedco to CME • Gravity fed • Reinforcement • Educational Garden

Fig. (1). System layout for three scenarios.

Scenarios

Overview of Scenarios

RWH systems are tanks/cisterns placed atop or on ground level of a site or building to collect and store rainfall or runoff. The location of the tank/cistern affects required equipment (*e.g.* pump, pipeline, *etc.*), and cost and energy usage. The work described in this paper studies three RWH configurations with different cistern storage volumes and locations (Fig. **1**). The first scenario has two storage cisterns; one large underground, and one smaller placed atop the CME building. The idea is to use both locations and see if maximizing the amount of collected rainwater can maximize the benefits of the RWH system. The roof tank is smaller as there is already one large cistern underground; thus, it works as a complementary combination. Scenario 2 uses one large underground cistern, and has a pressure-regulated pump to supply a constant flow of water to each toilet in the building. Scenario 3 uses one large cistern placed on top of the CME building. The volume of roof cistern is the maximum for a high reliability. Runoff from the Hedco building is pumped into the rooftop cistern. From there, toilets would be flushed from the gravity-fed water.

Underground cisterns are designed to be constructed of concrete. From an environmental perspective, concrete has the advantage of being a reusable material. Different applications and structures can afford the use of recycled concrete while maintaining the same standards. Since concrete is produced locally, the environmental and financial cost of transportation is relatively low. Economically, concrete may be one of the cheaper options to consider [25, 26]. The rooftop equipment has to be made of lighter materials to accommodate the weight constraints of the roof structure. Therefore, fiberglass is specified for the roof cisterns.

Scenario 1 – Two Cisterns

First system design uses two cisterns. In this scenario, a majority of the rainwater runoff from both the CME and Hedco building rooftops are collected and gravity fed into a large underground cistern. A pump raises a portion of the runoff into a smaller cistern located on the roof of the CME building with overflow being delivered into the municipal storm drain system. Water is, then, gravity fed through plumbing in the CME building to each of the toilets for flushing. The underground cistern is placed nearby the point of use, on the south side of the CME building. Municipal water supply is available to replenish the cistern if the storage passes a threshold, ensuring that the system does not run dry. There will be a mass flow controller installed to the system to provide drip irrigation for an educational garden located on the southeast side of the Hedco building.

The second smaller fiberglass cistern is located in a cavity near the roof of CME building, above the fourth floor, between the ductwork and aligned with the toilets in the building. It gravity feeds the harvested rainwater into the toilets through the building's plumbing. When the water level of roof cistern falls below 15%, water will be pumped from the underground cistern to the roof cistern. The required pump will only run to replenish the roof cistern, and could be programmed to run at night when power costs are lower which will cut back on operation cost. It is anticipated that when required, the pump will be capable of replenishing the rooftop cistern with 1,500 gallons of water in approximately 15 minutes which means the duration of pump runtime will be short. In total, Scenario 1 needs two pumps and two UV disinfection filters. A water treatment system is needed at the point of water delivery from the cistern to the toilets.

Some disadvantages of this system are that it requires additional structural support in the ductwork area near the roof in order to hold an additional 20,000 pounds, and will also require pump and water treatment. Its advantage is that with the capacity of the large underground cistern the need for municipal water can most likely be completely eliminated. It also requires smaller pump capacity that is not in frequent demand. An educational garden can be located to the southwest of CME building and be gravity-fed by the water collected on the roof to increase the positive impacts of the RWH system in this design.

Scenario 2 – One Cistern Underground

Scenario 2 places one large concrete cistern underground in the area south of the CME building. Runoff from both the CME and the Hedco buildings is diverted into the cistern and a pressure-regulated pump provides pressurized water to the plumbing system for flushing toilets.

The design of one underground large-capacity concrete cistern requires no additional structural support to CME. It will require a high pressure pump to pump the underground water up 50 feet to the toilets each time a toilet is flushed. The underground cistern will remain connected to the municipal system for when the cistern runs below 15% and needs to augment its water supply. Scenario 2 requires two pumps and one UV disinfection system. Construction of educational garden is not recommended in this scenario as the water cannot be gravity-fed and thus it will require an extra pump to irrigate the garden.

Scenario 3 – One Cistern Rooftop

Scenario 3 includes one large cistern on the roof of the CME building. As rainwater drains from the Hedco building it is immediately pumped into the rooftop cistern atop the CME. The water would then be gravity fed into the building's plumbing for flushing of toilets and into an educational garden adjacent to the CME building.

Although the single large cistern atop the roof of the CME is made of fiberglass, which is much lighter than concrete, it would approximately add up to 125,000 lb of weight on the structure (review System Design section). Thus, substantial structural reinforcement is required for the building. The roof cistern is connected to the municipal water system in order to augment its water supply when the cistern runs below 15%.

Despite the disadvantages of designing one cistern on the roof, which is the requirement of reinforcement, the advantages include gravity fed water to the sources—toilets and an educational garden on the southeast side of Hedco building. Moreover, the pump would only be required to run when it rains, which is infrequent during the summer months in Salt Lake City. Scenario 3 does not require any pumping to feed the toilets, but requires a pump to redirect Hedco harvested rainwater up to the higher elevation of the CME building rooftop for storage. Also, it needs one UV disinfection system.

System Design

Two main factors are considered for optimizing RWH systems: local precipitation patterns and site characteristics such as landscape, roof and underground area, *etc* [27]. The two parameters along with capture efficiency were used to calculate captured water volume ($V_{captured}$) in eq. (1).

$$V_{captured} = A \times i \times F \quad \textbf{(1)}$$

where A is the capture area (ft^2), i is the precipitation intensity (in), and F is capture efficiency (%). The capture area was the same for the three scenarios and equals the total surface area of Hedco and CME Building (28,824 ft^2). Local precipitation intensity data was extracted from Country Studies and National Climatic Data Center database (Table **1**) [28, 29]. Rainfall values for a 30-year period (1960-1990) were reviewed for dry, wet and average years in regard to rainfall amounts. With an ever-changing climate, the monthly precipitation volumes were analyzed according to dry weather patterns. Therefore, we used averages of the dry year rainfalls in the $V_{captured}$ equation to size for the extreme condition. Roof capture efficiency is assumed to be 0.75 (75% of the rainfall on the rooftop is captured in cistern).

Required water from the municipal water system is calculated with eq. (2) as follow:

$$V_{municipal} = V_{demand} - V_{captured} \quad \textbf{(2)}$$

where $V_{municipal}$ is the water supply from municipal water system (ft^3), and V_{demand} is the water demanded indoor and outdoor of the building (ft^3). To quantify water demand from municipal water system, the indoor and outdoor water demand of the CME and Hedco buildings are calculated. The indoor toilet demand is determined by counting the number of offices and factoring in the users that may come from the laboratories. This resulted in a building occupancy estimate of 80-100 people. To allow for future growth and increased users, it was determined that an overestimate would be better rather than underestimating the demand and relying on the municipal water system. Therefore, the water demand is based on 100 people in the building using the toilet every day. Using Economic and Environmental Analysis of Sanitation Technologies (EEAST) as a reference, the toilet flushing for educational buildings and office buildings for toilet flushing was considered as 500 gallons per day [19].

Irrigation is an additional benefit if there is a surplus of water. Outdoor water demand was considered as the water demand for a 1,000 square feet vegetable garden (1,000 square feet irrigation area, but about 5,000 square feet footprint), which is 2 inches per week per square feet. For more detailed explanation of the function and process of the spreadsheet, see Appendix D.

Table 1. Monthly rainfall depth and volume quantity [28, 29].

Month	Dry Year (in)	Wet Year (in)	Average Year (in)	Monthly Captured Volume (gal)
January	0.17	1.59	1.21	2,967
February	1.31	1.90	1.25	22,864
March	0.80	3.08	1.79	13,963
April	2.37	4.50	1.99	41,364
May	0.90	5.18	1.95	15,708
June	1.60	1.02	0.98	27,925
July	0.34	0.36	0.61	5,934
August	0.19	1.52	0.69	3,316
September	0.50	0.36	1.21	8,727
October	3.48	1.88	1.52	60,738
November	1.17	1.25	1.45	20,420
December	0.47	1.40	1.41	8,203

For the proposed Scenario 1 the size of fiberglass roof cistern and underground concrete cistern are determined to be 2,500 and 30,000 gallons, respectively. Similarly, Scenario 2 requires a single concrete cistern with capacity of 30,000 gallons. And, Scenario 3 uses a 30,000-gallon roof cistern (see Appendix A). However, this latest design is only a consideration if structural analysis shows it is possible to introduce this dead/live load on the roof without further reinforcement. A customized spreadsheet tool, Rainwater Harvester 2.0, and Storm Water Management Model (SWMM) each confirmed the size of the fiberglass roof cistern and concrete underground cistern for all scenarios (Appendix D).

Life Cycle Assessment

The goal of this study was to evaluate and compare the environmental, economic, energy, water usage and social impacts of three alternative RWH systems implemented for the CME and Hedco buildings in Salt Lake City, Utah. Environmental LCA system boundaries of the study was cradle to end-of-use, included raw material processing and construction, transportation and use phases.

For this LCA, the functional unit is the gallons of water provided for both indoor (toilet flush) and outdoor (irrigation) use for over 600,000 people using CME building during 30 years. The reference flow is determined to be 100,000 gallons of water per year. The lifetime of the three RWH systems were the same and assumed to be 30 years. The Life Cycle Impact Assessment (LCIA) is conducted using the GaBi 6.0 and EcoInvent 2.0 database.

The results of this project may be useful for any other commercial, industrial, residential, and/or educational buildings, which have numerous visitors daily. However, the results depend on demand and supply which may change from building to building. Therefore, it is necessary to scale them according to the specified number of visitors, toilets and/or sizes of the building before they can be used.

Life Cycle Inventory

Impacts resulting from each and every stage of the designs are summed up for each scenario separately. Total values for different impact categories, total energy consumption and total tap water usage, and benefits of social impacts are interpreted and compared among the three scenarios in order to identify the option with the lowest environmental impact.

Various datasets are needed for environmental impact analysis in GaBi, including the specification of raw material (applications, countries of production, *etc.*), total weights of raw materials for each design (pounds or kilograms), energy needed for molding raw materials (pounds or kilograms), and total use of tap water for each scenario during the complete lifetime of 30 years (pounds or kilograms).

Raw Material Processing and Construction Phase

The material inventory of this study consisted of concrete (large underground cistern), fiberglass (small/large roof cistern), polyvinyl chloride or PVC (pipeline), stainless steel (pumps and UV disinfection) and steel (reinforcement). Table **2** shows the type of raw materials selected for all scenarios, and the total required weights of materials in pounds. US-based data was not available for all materials. Therefore, data for Germany was selected for the analysis. In one case data for Germany was not available in the GaBi database (fiberglass), therefore US data was used. And, in two others neither Germany nor U.S. were available (stainless steel and tap water). Hence, Europe-based dataset, which is very close to Germany data, was selected for both stainless steel and tap water. Since a significant fraction of water resources in Salt Lake City, Utah is groundwater, and in order to reduce the variation among chosen materials and real scenarios, the tap water was assumed to be from groundwater resources.

Total weights of materials were calculated from the size and number of equipment, density of the material and number of replacements during a complete life cycle of the design, which is both operating and maintenance phases (Table **2**).

Table 2. Total weights and specifications of raw materials for each scenario in Gabi simulations (lb).

Material	Specifications on Gabi		Total Weights (lb)		
	Country	**Type**	**Scenario 1**	**Scenario 2**	**Scenario 3**
Concrete (Underground Cistern + Pad)	Germany	Ready-mix C30/37-agg*	3.37E+05	4.12E+05	0.00E+00
Fiberglass (Roof Cistern)	USA	Duct Wrap-agg	3.80E+02	0.00E+00	7.02E+03
Steel (Reinforcement)	Germany	Steel Sheet-HDG-agg	3.50E+03	0.00E+00	3.50E+03
PVC (Pipeline + Diverter)	Germany	PVC Granulate-Plastic Production-agg	5.15E+03	4.84E+03	4.27E+03
Stainless Steel (UV Filter + Pump)	Europe	White Hot Rolled-agg	4.02E+03	2.23E+03	1.79E+03
Tap Water	Europe	Water Purification-from Groundwater	NA	NA	NA

*agg stands for "aggregated", means that this option considers all preliminary steps necessary to provide the material

The number of new disinfection units added to the system through maintenance phase is also considered in the calculation of total molding energy requirements for stainless steel.

The following calculations show the procedure of finding energy requirements for molding up one pound of each of the materials:

Ready-mix concrete [30, 31]:

$$\text{Energy} = 0.963\ \frac{\text{MBTU}}{\text{yd}^3} \times \frac{1}{0.7646}\frac{\text{yd}^3}{\text{m}^3} \times \frac{1}{2300}\frac{\text{m}^3}{\text{kg}} \times \frac{1}{2.2}\frac{\text{kg}}{\text{lb}} \times 10^6\frac{\text{BTU}}{\text{MBTU}} \times 0.0002931\frac{\text{kWh}}{\text{BTU}} = 0.072955\frac{\text{kWh}}{\text{lb}}$$

Steel [32]:

$$\text{Energy} = \text{range of } (446.43 - 535.71)\frac{\text{kWh}}{\text{ton}}$$

$$\text{Average Energy} = 491.07\ \frac{\text{kWh}}{\text{ton}} \times \frac{1}{2000}\frac{\text{ton}}{\text{lb}} = 0.246\ \frac{\text{kWh}}{\text{lb}}$$

PVC [33]:

$$\text{Energy} = \frac{1027}{13974}\frac{\text{kWh}}{\text{lb}} = 0.073\ \frac{\text{kWh}}{\text{lb}}$$

Stainless Steel [34]:

$$\text{Energy} = 12\ \frac{\text{kJ}}{\text{bend}}$$

One bend is defined as a 20 × 20 mm stainless steel sheet with a thickness of 3 mm, therefore:

$$\text{Volume of one bend} = 20 \times 20 \times 3 = 1200\ \text{mm}^3 = 1.2\ \text{cm}^3$$

$$\text{Mass of one bend} = 1.2\ \text{cm}^3 \times 7.955\ \frac{\text{mg}}{\text{cm}^3} = 9.546\ \text{g} \times 0.0022\ \frac{\text{lb}}{\text{g}} = 0.021\ \text{lb}$$

Therefore,

$$\text{Energy} = 12\ \frac{\text{kJ}}{\text{bend}} \times \frac{1}{0.021}\frac{\text{bend}}{\text{lb}} = 570.207\ \frac{\text{kJ}}{\text{lb}} \times 0.00028\ \frac{\text{kWh}}{\text{kJ}} = 0.159658\ \frac{\text{kWh}}{\text{lb}} \sim 0.16\ \frac{\text{kWh}}{\text{lb}}$$

Fiberglass [32]:

$$\text{Energy} = \text{range of } (242.25 - 289.28)\ \frac{\text{kWh}}{\text{ton}}$$

$$\text{Average Energy} = 265.87\ \frac{\text{kWh}}{\text{ton}} \times \frac{1}{2000}\frac{\text{ton}}{\text{lb}} = 0.133\ \frac{\text{kWh}}{\text{ton}}$$

Transportation Phase

For the transportation step, the total weights of all materials were summed separately for each scenario and multiplied by the distance to provide the value of kilogram/kilometer of the diesel used. The average of 30 kilometers of distance was assumed for transportation of all materials. USA based diesel was selected for the fuel of the truck transport.

Operation & Maintenance Phase

For the life cycle inventory of the operation phase, electricity consumption of pump and disinfection systems were factored in. Scenario 1 needs two pumps and two UV disinfection units. The pump supplied 20 minutes of operation during a day. The UV disinfection unit is placed before the cistern; thus, the rainwater is disinfected before entering the tank. Therefore, UV systems work whenever it rains. Based on the precipitation data, in 30 years, there are 2,188 days of rain. Assuming average rainy day is 4 hours, there would be total of 8,752 hours per 30 years meaning average of 0.9 hours per day during which the pumps and filters are working. Scenario 2 requires a high pressure pump to pump the underground water up to 50 feet to the toilets each time a toilet is flushed. It also requires two pumps and one UV disinfection system. The pump has to be on almost all the day (24 hours) in order to provide the required water to the third floor. The filter works the same hours per day for the three scenarios; therefore, the total working hours in 30 years for Scenario 2 is half of that for Scenario 1. Scenario 3 does not require any pumping to feed the toilets, but requires a pump to re-direct Hedco harvested rainwater to CME. However, it needs one disinfection system working on the same trend with Scenarios one and two.

The maintenance assumptions of each scenario are different: routine maintenance for Scenarios 1 and 3 was assumed twice in a year while it was assumed annually for Scenario 2. This phase includes roof washing, cleaning inflow filters, tank inspections and disinfection, and replacement of filters. All steps use water except for disinfection in which UV disinfection system was used about 4 hours each time for each system. Therefore, total disinfection operation of 480 hours for Scenario 1 (2 system, 60 times, 4 hours), 120 hours for Scenario 2 (1 system, 30 times, 4 hours) and 240 hours for Scenario 3 (1 system, 60 times, 4 hours) were considered in the maintenance phase and added up to the total hours of UVs working in operating phase. In addition, total weight of stainless steel includes the mass used in UV disinfection systems for maintenance. The tap water usage was also calculated for each scenario. The calculation steps for municipal water usage are presented in Appendix C.

Life Cycle Impact Assessment

Impact categories were studied with GaBi using the two most common methods: TRACI and ReCiPe. Global warming potential (GWP, kg CO_2 eq.), acidification (AP, kg SO_2 eq.), eutrophication (EP, kg N eq.), ozone depletion potential (ODP, kg CFC -11 eq.), and human health cancer (HHC, CTUc) were modeled in TRACI because it is the most commonly used impact assessment method in the US. Additionally, fossil depletion (FD, kg oil eq.) and water depletion (WD, m^3) were modeled in ReCiPe. The seven mid-point impact categories are selected based on the most important concerns about human's health and life. As different impact categories have different units, normalization of data was necessary before summing for overall environmental impact comparison. It was done by dividing the values of each category of each system to the related ones from the system that provides the lowest value of that category. For instance, if GWP is 10, 20, and 30 kg CO_2 eq. for Scenarios 1, 2, and 3, respectively; then, all values are going to be divided by GWP of Scenario 1, which is the lowest. Thus, it will be unit one for the first system, and Scenario 2 and Scenario 3 can be compared to this system as they provide two and three times higher GWP than Scenario 1, respectively.

The environmental impacts from business as usual (BAU) was also studied and compared to those generated from each scenario. BAU represented a complete system providing water from municipal sources without rainwater harvesting.

Life Cycle Cost

Life cycle cost analysis was conducted by comparing total cost and net present value of the systems. These calculations included two separate cisterns, two pumps, one flush diverter, one ultraviolet light disinfection unit, monthly maintenance (high) and discount rates of 5.50% for 30 years.

During the lifetime of the project, it was assumed that there would be no increase in water and energy costs and no technological advancement. Values were estimated based on researched information and scaled to the design parameter to give a more realistic cost evaluation.

Each of the designs and construction parameters were entered into the spreadsheet with their respective unit cost. The unit costs were determined through average costs of the units in the web resources, and for parameters that have to be evaluated by a structural engineer, such as installation, the costs were assumed. Some parameters were omitted due to long-term insignificant cost or a lack of data. Example cost inputs for Scenario 1 and maintenance can be found in Appendix B.

To calculate the cost of the entire system for Scenario 1, it initially accounted only for the cost of the fiberglass required to construct, install and structurally reinforce the building for the 2,500-gallon roof cistern. The design was assumed to require monthly inspection because of its sophistication. The two cisterns will require disinfection, flush, and inspection. The single underground cistern, Scenario 2, was the simplest design alternative and only requires an additional cost of a pump for irrigation. The design was assumed to require only annual maintenance and inspection because of its simple design and single cistern with little structural hazard concerns. To introduce a 250,000 lb load on the roof in Scenario 3 without requiring reinforcement would be unlikely; hence, this life cycle cost analysis included an assumed value for structural reinforcement of $8,400 to $11,200 for that load.

Social Impact Assessment

Even though there exists no exact definition of social impact assessment in the literature, this type of assessment usually involves a systematic analysis of social effects of a project or policy on individuals, groups of people or society as a whole [35]. The social evaluation criteria for three RWH systems were defined as departmental income, number of jobs and visual impacts of the systems (Table **3**). These criteria were used to evaluate reduction in water cost, boost in local economy, to the ability to attract and keep people in the region. As a visual and educational social impact, it was decided that the excess water would create an opportunity for a community garden and educational awareness station. The vegetable garden is designed to replace 5,000 square feet of grass and provide local vegetables for students and those who live close. Within the garden would be information boards to explain the system and provide educational value.

Table 3. Evaluation criteria for social impact assessment.

Criteria	Effects and Expectations	Criterion Scores
Cost/water bill reduction	Reduction in water cost	Funding for educational/entertaining events for students
Increase in number of jobs	Boost the local economy	To offer temporary job
	To keep and attract people in the region	To offer permanent job
Visual impact	Encourage people to visit the department	Educational garden
Educational impact	Informing visitors about water shortage and introducing RWH systems as one of alternatives	Posting flyers and notes around the building and/or inside bathrooms

Criterion scores were evaluated by considering the expectations of RWH. The Department's income will increase based on the reduction in water bill *via* water saving. This saving can be offered as funding for students. Increases in temporary jobs were evaluated regarding the main material market. For example, Scenario 2 was assumed to provide social benefits of providing temporary jobs in 5 different local industries including concrete, fiberglass, steel, PVC and stainless steel. On the other hand, Scenario 3 has potential to affect three industries such as concrete, PVC and stainless steel (see Table **2** to see the details of material inventory). Effect on permanent jobs was evaluated by dependence of a gardener to maintain the educational gardens. Visual impacts were also considered for the existence of an educational garden.

RESULTS AND DISCUSSION

Life Cycle Impact Assessment

The comparison of total normalized impacts of three scenarios is shown in Fig. (**2**). Total environmental impacts come from the construction phase (production of elements and shipping and transportation to the land), operating phase (energy consumption), maintenance and additional tap water usages.

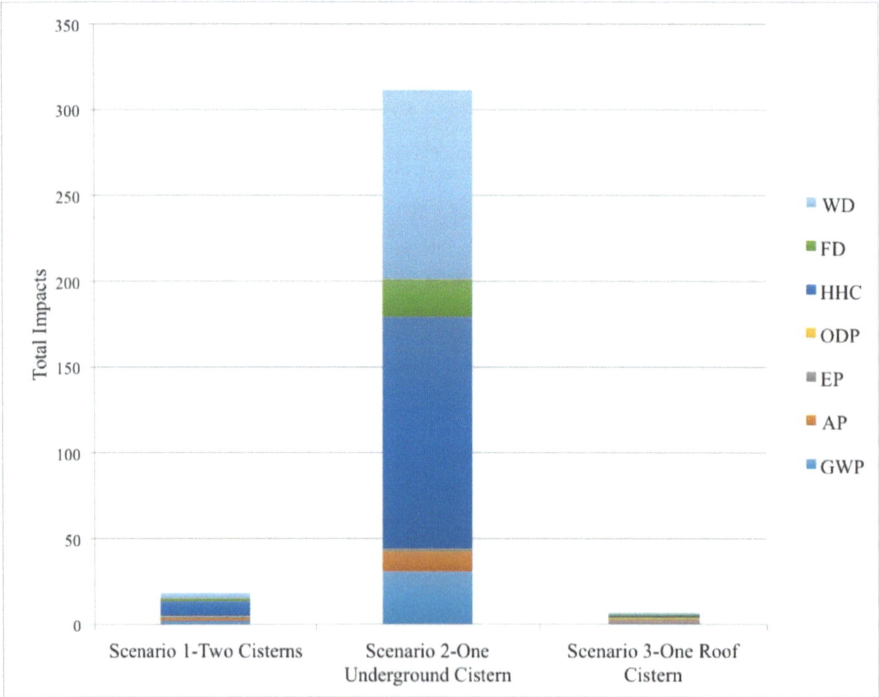

Fig. (2). Comparison of total normalized environmental impacts of the scenarios.

Although different impact categories have different units, each impact category was normalized to Scenario 3 (review Life Cycle Impact Assessment section), as the results show that Scenario 3 is the most environmentally friendly option in regard with all impact categories. Then, all impact categories are summed up and provide an overall environmental impact comparison. Scenario 3 has 2 to 30 times lower impacts than Scenario 1, and Scenario 2, respectively. Scenario 2 has largest impact for all the presented impact categorizes. 24-hour running of pump and consequently, electricity consumption is the primary reason for such high impacts of this scenario.

Since in earlier discussions Scenario 2 showed release of the highest level of impact categories, especially, GWP, HHC, and WD, it is important to determine the main element causing this impact. Hence, analysis of different elements for each scenario is provided in Fig. (**3**). As illustrated in this figure, 50% of GWP in the first scenario is caused by the storage part, which is reasonable because there are two tanks used in this alternative. Filtering material, which is stainless steel, contributes significantly in producing HHC (65%), ODP (62%), and AP (50%). In addition, PVC pipeline plays an important role in causing WD (50%) and FD (40%). Therefore, to reduce the impacts of interest, one can change the most contributed material in the related part of the design layout.

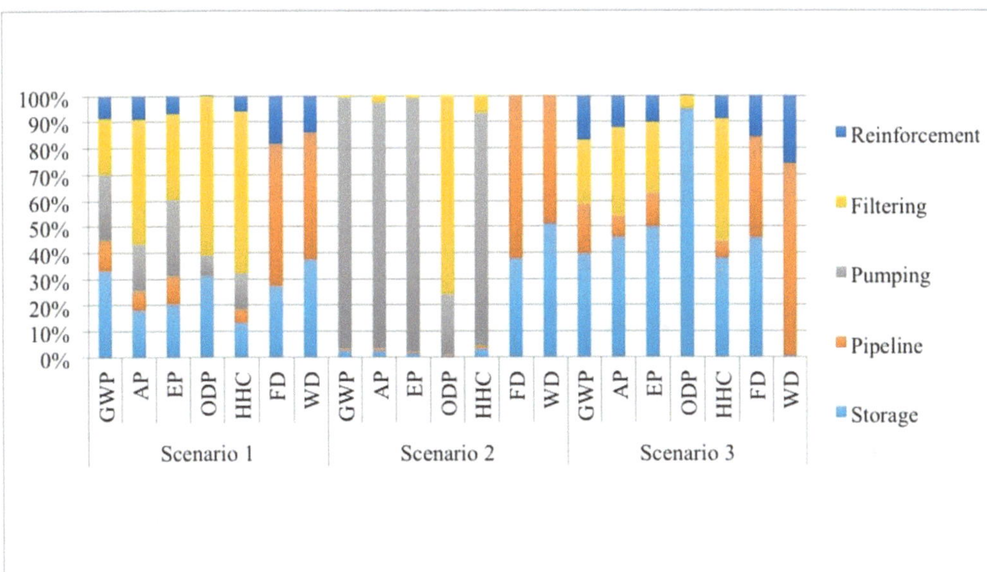

Fig. (3). Contribution of elements in environmental impacts.

High values of impacts from Scenario 2 are those from GWP, AP, HHC, FD, and WD (Fig. **2**). Fig. (**3**) demonstrates that for this design, storage covers the highest part of GWP; AP is, also, mainly caused by storage and filtering systems. Filtering covers the highest percentage of producing HHC. And, FD and WD are mostly caused by the pipeline part of this layout. Each scenario can be modified by using different pipeline pathway; therefore, the length and amount of PVC used in the system would change. Also, tanks and filters with different materials can be considered to reduce the amount of some impact categories.

Furthermore, Fig. (**3**) represents that more than 40% of the GWP for Scenario 3 as a result of the storage part, which is made of fiberglass. ODP, which is another impact category highly produced by Scenario 3, is greatly caused by the storage part (more than 90%). Additionally, pipeline construction of this scenario contributes in high values of Water Depletion. Therefore, in order to improve this scenario, one should mainly focus on improving the storage and pipeline parts.

Results from comparing three RWH alternatives with BAU outputs imply that, in regards with environmental impacts, BAU is always the best option. However, GaBi simulation for tap water (BAU) only includes the operational phase of potable water usage, which makes the comparison not very reasonable. Therefore, in a different approach, the sum of impact categories resulted from operation and maintenance phases of RWH systems were calculated and compared to each other and to the BAU system. Then, all data were normalized to Scenario 3, which has the minimum amounts, and the normalized data are plotted as Fig. (**4**).

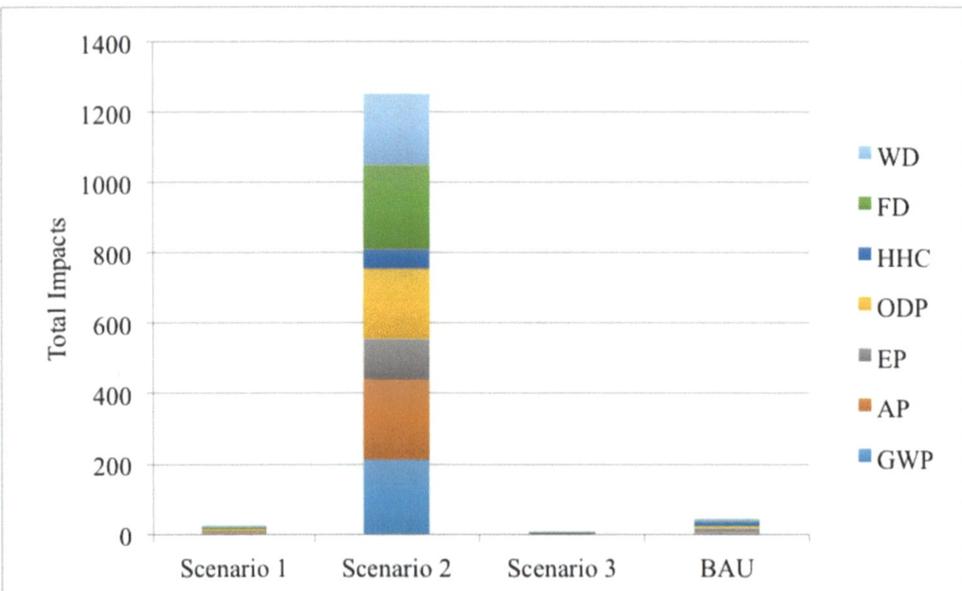

Fig. (4). RWH scenarios *vs.* BAU (data normalized to Scenario 3).

As visually shown in Fig. (**4**), there is a substantial difference between the impacts caused by RWH Scenario 2 and the others. Considering Scenario 2 data as outliers, BAU always show higher impact categories compared to RWH Scenario 1. Accordingly, Fig. (**4**) proves the advantage of using RWH system designs 1 or 3 instead of BAU.

Energy consumptions during the production and operation phase of the RWH systems are presented in Table **4**. The energy comparison of the three scenarios demonstrates that Scenario 3 has the best energy consumption performance. Energy consumption results show that both the production and operation phase of Scenario 2 are larger than those of Scenarios 1 and 3. Scenario 2 requires the largest amount energy, which is around 20 to 200 times higher than Scenario 1 and 3, respectively. Electricity consumption of the pump during the operation phase of Scenario 2 primarily accounts for such high energy consumption since it is assumed to be running 24-hours for toilet flushing. Energy consumption of Scenarios 1 and 3 in operation phase contribute 42% and 48% to overall, respectively.

Table 4. Total energy consumptions for each scenario.

Energy consumption (kWh)			
	Scenario 1-two Cisterns	**Scenario 2-one Underground Cistern**	**Scenario 3-one Roof Cistern**
Production Phase			
Concrete	2.46E+04	3.01E+04	0.00E+00
Fiberglass	5.54E00	0.00E+00	9.33E+02
Steel	8.61E+02	0.00E+00	8.61E+02
PVC	3.76E+02	3.54E+02	3.11E+02
Stainless Steel	6.43E+02	3.56E+02	2.86E+02
Operation Phase			
Pump	1.48E+04	1.07E+06	0.00E+00
Filter	4.56E+03	2.22E+03	2.22E+03
Total	4.58E+04	1.11E+06	4.61E+03

Life Cycle Cost Analysis

Table **5** presents the total cost and net present value of each scenario. The results show Scenario 1 is the most expensive design while Scenario 2 is the least costly one.

The cost of each design is over $100,000. Although the environmental and cost data do not match, there is a good explanation behind the results. The highest cost in Scenario 1, valued close to $300,000, is due to its high maintenance. If the system could be designed to require only semi-annual maintenance (medium), the cost would be significantly reduced to closely match the other two scenarios. Also, Scenario 1 includes one extra cistern compared with the other two scenarios. A smaller size cistern could be selected and utilize the low cost of municipal water, the cost of the design could decrease as well.

Table 5. Life cycle cost of scenarios.

Scenario	Total Cost ($)	Net Present Value ($)
1	503,680	290,664
2	117,765	101,901
3	189,602	130,261

The total monthly inspection of Scenario 1 has an annual cost of $13,558. Depending on the price of steel to reinforce the structure, the cost was estimated to be $8,443 to $8,843*. Assuming the higher price of steel used to reinforce the structure, the value is added to the capital cost of the underground cistern to get a total cost of the system to be $503,680 and the net present value is $290,664.

The total annual maintenance cost of the system for Scenario 2 is $345. Therefore, the total cost of the system is $117,765 and the net present value is $101,901. The values generated include tank installation and excavation costs into a generic installation cost. The net present value was based on one cistern, two pumps, one flush diverter, one ultraviolet light disinfection unit, annual maintenance (low) and discount rates of 5.50% for 30 years, the lifetime of the project.

The total cost of semi-annual maintenance for the third system is $980. Without the structural reinforcement, the total cost of the design with a discount rate of 5.50% for 30 years would be $110,103 and $89,834 for the total cost on the system and the net present value, respectively. Assuming a worst-case scenario, the addition of the structural reinforcement would be $189,602 for total cost on the system and $130,261* for net present value. Semi-annual maintenance in Scenario 3, which means twice inspection as Scenario 2, causes the 60% difference in total costs of the two systems. The net present value was based on one cistern, no pump, one flush diverter, one ultraviolet light disinfection unit, semi-annual maintenance (medium) and discount rates of 5.50% for 30 years – the life of the project.

*Based on using steel to reinforce the roof where it does not specify weight and uses only square footage—in this case, 7 pounds of steel per square foot and $3 to $4 per pound of steel. The dimension of the system is assumed to be 36 x 8 x 6.5 feet, using the largest square footage [36].

The value generated is questionable because retrofitting a building is typically more expensive due to relocation and redesign of equipment and loading to the building roof, values up to the $150,000 to $200,000 for a green roof [36].

Life Cycle Water Usage

The total municipal water needed and the total days of pumping for each scenario, affect the LCC and LCA results, which were discussed earlier. Table **6** summarizes the water performance of each scenario. The analysis determined that there are points where the cisterns are empty due to extreme long periods of drought, but over a 30-year time frame, rainwater harvesting has proven to be sufficient enough to meet the water demand of the CME building.

Table 6. Life cycle water usage data for each scenario.

	Scenario 1	Scenario 2	Scenario 3
Total days of pumping to CME tank on the roof	1608	-	-
Total days of municipal backup supply	212	264	419
Total overflow water out of system (gallons)	2,424,285	2,450,668	1,437,343
Total municipal water needed for dry periods (gallons)	424,000	434,000	531,000
Percentage of water saving (outdoor)	92.11%	91.93%	90.12%
Total water used from tanks (gallons)	4,138,000	4,138,000	4,138,000
Strictly rainwater	3,714,000	3,704,000	3,607,000
Percentage rainwater used	89.75%	89.51%	87.17%

Among the presented aspects, Scenario 1 shows the best environmental performance. For instance, Scenario 1 requires municipal backup water supply for 212 days (in period of 30 years), while Scenario 2 and 3 required 264 and 419 days, respectively. The slightly shorter supply time of Scenario 1 (20%) than Scenario 2 is an expected result since the only difference between these two scenarios is small cistern in the top of CME building (2,500 ft^3). However, the total backup time of Scenario 3 is almost twice as much as Scenario 1. Besides having an extra 2,500 ft^3 storage in Scenario 1 (resulting in 20% shorter supply time), in Scenario 3 the amount of water to enter storage is reduced due to evapo-transpiration from Hedco to CME. For the same reason, total overflow water out of system is close in Scenarios 1 and 2, and almost twice Scenario 3. Total municipal water requirements of Scenario 1 and 2 during dry season are also close to each other and around 25% lower than Scenario 3.

A significant result of this study is all these cisterns provided around 90% water saving in municipal water usage. The current amount of water that is used for indoor and outdoor uses over a 30-year time frame is estimated to be about 8,896,000 gallons. This is all water that comes from the municipal water system in BAU. Therefore, with the reduction of water usage, the energy consumption to move the water is expected to be dramatically decreased. These results show the importance of RWH systems in educational building, and are consistent with the literature [35 - 38].

Social Impacts

Social impacts of the three RWH system scenarios were presented earlier in this chapter (Table **3**). Table **7** represents this evaluation with scores to identify the scenarios with the most significant contribution to social benefits. Existence of each positive impact was scored as 1, while those lacking was set to 0.

Results of this type of evaluation show that the positive social impacts of Scenarios 1 and 3 nearly doubled that of Scenario 2. The lower social performance of Scenario 2 mainly comes from different number of offered job and visual impacts. Local fiberglass and steel markets are not boosted in Scenario 2. Therefore, Scenario 2 does not help with making permanent job opportunities. In addition, it was determined that the excess water would create opportunity for a community garden and educational awareness station. The vegetable garden is designed to replace 5,000 square feet of grass and provide local vegetables for students and those who live close by. Within the garden would be information boards to explain the system and provide the educational awareness.

Table 7. Evaluation of social impacts.

	Scenario 1	Scenario 2	Scenario 3
Department income	1	1	1
Number of temporary jobs	5	3	5
Number of permanent jobs	1	0	1
Visual impact	1	0	1
Educational impact	1	1	1
Total score evaluation	9	5	9

Unlike Scenarios 1 and 3, gravity-fed educational rain garden was not opportune for Scenario 2. Thus, this design does not include visual attractiveness (educational garden).

Final Design to Maximize Sustainability

In order to maximize the sustainability and potential benefits, it is decided to enhance the network and surrounding features according to four bottom lines. This means, improving the economic, environmental, and social aspects by reducing water demand for the indoor toilet usage and applying extra water to irrigation. Reducing municipal water demand has many environmental benefits and decreases the environmental stresses that come from delivering municipal water on a centralized system.

Fig. (**5**) represents ranking of Scenarios regarding each criterion. The lowest impacts in all the aspects (*i.e.* environment, cost, water usage, and social) translates that the scenario performs better. For this purpose, a similar approach to Lingov and Seager [24] was used. The worst scenario in each category was normalized to 1 by dividing the highest impact in each category (data was extracted from Fig. (**2**), Table **4** and Fig. (**6**) for environment, cost and water use categories, respectively). In contrast, the social impact assessment was normalized by dividing the lowest impact (is meant to be the worst) among three to the individual impact. For instance, the worst scenario was Scenario 2, and the corresponding value was calculated as 5 in Table **7**. The normalized score for Scenario 1 and 3 were calculated by dividing their scores to the Scenario 2's score. LCA analysis shows that Scenarios 1 and 3 are both very plausible designs for increasing the environmental stewardship. Scenario 2 is the most cost effective option. Analysis of environmental impact introduces Scenario 3 as the best alternative, while Scenario 1 is the best choice for the highest saving in potable water usage. Social impacts are equally improved by Scenarios 1 and 3, as the best options according to this criterion.

Fig. (5). Ranking of scenarios.

CONCLUSION

The aim of this study was to evaluate RWH system designs based on a quadruple bottom line (economic, environmental, social, and water usage) to replace the currently-used potable water to flush toilets in the CME building of University of Utah. Based on the normalized scoring used, Scenario 3 is the most viable RWH option. Despite the promise of low cost that Scenario 2 offers, its total impact outcompetes with the other two systems as a result of the difference in environmental impact criteria of the systems. Environmental impact of Scenario 2 was found to be around 15 to 30 times higher due to the corresponding impact of high energy intensity of the system. Scenario 1 appears to be the most costly both in terms of installation and lifetime operation and maintenance while Scenario 2 appears to have the lowest life cycle cost. However, it should be noted that sufficient appraisal has not been given to the Scenario 3 requirement to retrofit the CME with additional structural support. From social aspects points of view, both Scenarios 1 and 3 are ready to give the higher benefit by offering lower impacts. Water use criteria of the scenarios are relatively close to each other - the difference is only 20% maximum - and Scenario 1 performs the best.

CONFLICT OF INTEREST

The authors confirm that they have no conflict of interest to declare for this publication.

ACKNOWLEDGEMENTS

This research was funded by the National Science Foundation's Environmental Sustainability (Grant no. 1236660 and 1235855). We also thank Kathy Ho, Steven Arhart, and Daniel Herbert who laid the groundwork for this manuscript.

APPENDIX A – SPREADSHEET MODEL DESCRIPTION

The developed spreadsheet to size the system and analyze changes to the layout was based on a water balance. This appendix describes the flow of water through the system according to the assumptions and design layout for Scenario 1, as it is the most complex scenario with two cisterns.

Precipitation data from the Salt Lake International Airport NCDC rain gage was input into the spreadsheet. This data set the time frame for the analysis of the model, which was January 1st, 1960 to January 1st, 1990. Within the 30 years time frame of data, there were several outliers removed, and the data for 1964 was missing. System assumptions that control the model are the Hedco roof area (*i.e.* 23,524 square feet), the CME roof area (*i.e.* 5,300 square feet), indoor toilet demand of 500 gallons per day (50 gallons per day on weekends), 2 inches water per square feet per watering session, 1,000 square feet of irrigation area, and use of 640 gallons per 1,000 square feet by grass for each watering session for the purpose of irrigation comparison. Assumptions are presented in Table **8**.

Table 8. Assumptions and parameters for spreadsheet model tool.

Indoor usage (gallons/day)	500
Outdoor usage (gallons/day)	1,247
Irrigation area (sq. ft)	1,000
Plant irrigation demand (inch/week/sq. ft)	2
Grass irrigation demand (gallons/1000 sq. ft)	640
Minimum volume of CME tank (gallons)	500
CME tank* volume (gallons)	2,500
Minimum volume of Hedco tank (gallons)	2,000
Hedco tank* volume (gallons)	30,000
Pumping volume from Hedco to CME (gallons)	2,000

*Hedco Tank refers to the large underground tank and CME Tank refers to the smaller roof tank

Based on the assumptions and model parameters a mass balance was created as a flow chart:

When it rains the runoff is added to either the Hedco Tank or CME Tank dependent of the roof runoff. If the CME Tank overflows, the water goes down to the Hedco underground tank. If the Hedco tank overflows, the water is lost to the municipal storm drain. When the CME Tank runs low, water is pumped from the Hedco Tank to the CME Tank. If the system in its entirety runs dry, the municipal water will fill the Hedco tank with a certain volume. Hence, the water can be pumped to the CME Tank, and used for the indoor and outdoor demands. As mentioned in the assumptions, 500 and 50 gallons are withdrawn for the indoor demand for weekdays and weekends. Irrigation demand is an additional withdrawal depending on the irrigation schedule.

APPENDIX B – LIFE CYCLE COST EXAMPLE INPUTS FOR SCENARIO 1

Most of the inputs for Scenario 1 were repeated for Scenarios 2 and 3, which means the pump capacity were over-designed for some of the scenarios. Prior to adding values to the spreadsheet shown in Fig. (**8**), the user of the spreadsheet has to input storage requirement data, system characteristics for cistern materials and usage, maintenance level, and discount rate for the net present value calculations as seen in Fig. (**6**).

Once the parameters were determined in Fig. (**6**), another spreadsheet tab was selected (Fig. **7**). Scenario 1 inputs were added as shown in Fig. (**7**). Some of the values were combined and input into the spreadsheet. For example, the piping from roof to cistern and parallel plumbing for interior reuse distribution had a total square footage of 1,700 square feet and the cost of the item is $2.44 per foot based on FlexPVC company's website. The roof modification was mentioned in the LCC portion and added on the spreadsheet separately. A comprehensive disinfection package with the capacity of 15 gallons per minute and an inflow filter were selected to meet the building code for toilet flushing use. The cistern refill system from municipal source in drought was the water that will be used to augment the 179,500 gallons of annual toilet flushing water. The proposed pump for this system is the AnsiPro AP96-ANS, but no pricing was available online for the pump. Therefore, the default pump cost of $1,188 was selected. The tank installation cost was a default value with the two cisterns size and cost input into the spreadsheet.

Design & Maintenance Options

STORAGE REQUIREMENTS	Unit	Model Default	User	Chosen Option
Impervious Drainage Area, DA (often roof area)	sq ft	5,000	22000.00	22,000
Max Design Rainfall Event	in	2	2	2
Precipitation Volume Generated per Event	gallons	6,233		27,427
Total Storage Needed	gallons	6,300	30000	30,000

SYSTEM CHARACTERISTICS	Unit	Model Default	User	Chosen Option
Type of Tank Desired (*P=Plastic, M=Metal, F=Fiberglass, C=Concrete. See sizing cost chart below*)	-	C	C	C
Primary Use *(I = Indoor *, Non-potable; O = Outdoor Irrigation.)*	-	O	I	I
Height of Building (Used to calculate 'Indoor' Use costs)	story	3	3	3
Number of Fixtures per Floor (toilets, used to calculate 'Indoor' Use costs)	ea	10	4	4

* Local health codes strictly control indoor water use. Before in stalling any indoor rainwater use system, check with local regulations to insure it is permitted. Regulations may require rainwater be treated before indoor use; this model does not estimate costs associated with treatment as systems may vary significantly. If the cost of the treatment system is know, the user may enter this cost is the "other" row in either Capital Cost Method A or B

DESIGN & MAINTENANCE OPTIONS	Unit	Model Default	User	Chosen Option
Choose Level of Maintenance, Irrigation *("H"=high; "M"=medium; "L"=Low)*	-	M	L	L

WHOLE LIFE COST OPTIONS	Unit	Model Default	User	Chosen Option
Discount Rate	%	5.50	5.5	5.5

Fig. (6). Scenario 1 input for the design and maintenance options.

Total Facility Base Costs	Unit	Unit Cost	Quantity	Cost
Tank (Materials Only)	Concrete	$ 2	30000	$ 49,800
Roof Modifications	None	$ 8,843	1	$ 8,843
Inflow/Collection Structures, Channeling on Roof*				$ -
Inflow Filters		$ 80	1	$ 80
External Conveyance Systems/Gutters*				$ -
Piping from Roof to Cistern*		$ 2	1700	$ 4,148
Overflow Structure (off-drain to storm drain)				$ -
Pump System		$ 1,188	2	$ 2,376
Paraller Plumbing for Interior Reuse Distribution				$ -
Irrigation System*				$ -
Cistern Refill System from Municipal Source in Drought		$ 568	1	$ 568
Other	Fiberglass Cistern	$ 1	2500	$ 3,326
Other				$ -
Total Facility Base Costs				**$ 69,140**
Associated Capital Costs	**Unit**	**Unit Cost**	**Quantity**	**Cost**
Mobilization/Delivery				$ -
Excavation/Tank Site Preparation				$ -
Haul/Dispose of Excavated Material				$ -
Tank Installation		$ 29,880	1	$ 29,880
Disinfection System		$ 2,329	1	$ 2,329
Permitting				$ -
Construction Inspection				
Other				$ -
Other				$ -
Total Associated Capital Costs				**$ 32,209**
Total Facility Costs				**$ 101,349**

Fig. (7). Scenario 1 input for the capital cost.

Year	Discount Factor	Capital & Assoc. Costs	Regular Maint. Costs	Corrective Maint.	Total Costs	Present Value of Costs	Cumulative Costs		
							Cash	Present Value	Discounted Cost per Year
Cash Sum ($)					$ 210,868	$ 134,725			
0	1,000	$ 96,934			$ 96,934	$ 96,934	$ 96,934	$ 96,934	$ 134,725
1	0,948	$	$ 345	$ -	$ 345	$ 327	$ 97,279	$ 97,261	$ 37,791
2	0,898	$	$ 345	$ -	$ 345	$ 310	$ 97,624	$ 97,571	$ 37,464
3	0,852	$	$ 345	$ 270	$ 615	$ 524	$ 98,239	$ 98,095	$ 37,154
4	0,807	$	$ 345	$ -	$ 345	$ 278	$ 98,584	$ 98,374	$ 36,630
5	0,765	$	$ 345	$ 10,301	$ 10,646	$ 8,146	$ 109,230	$ 106,519	$ 36,351
6	0,725	$	$ 345	$ 270	$ 615	$ 446	$ 109,845	$ 106,965	$ 28,206
7	0,687	$	$ 345	$ -	$ 345	$ 237	$ 110,190	$ 107,202	$ 27,760
8	0,652	$	$ 345	$ -	$ 345	$ 225	$ 110,535	$ 107,427	$ 27,523
9	0,618	$	$ 345	$ 270	$ 615	$ 380	$ 111,150	$ 107,807	$ 27,298
10	0,585	$	$ 345	$ 10,301	$ 10,646	$ 6,232	$ 121,796	$ 114,039	$ 26,918
11	0,555	$	$ 345	$ -	$ 345	$ 191	$ 122,141	$ 114,231	$ 20,686
12	0,526	$	$ 345	$ 270	$ 615	$ 323	$ 122,756	$ 114,554	$ 20,494
13	0,499	$	$ 345	$ -	$ 345	$ 172	$ 123,101	$ 114,726	$ 20,171
14	0,473	$	$ 345	$ -	$ 345	$ 162	$ 123,446	$ 114,889	$ 19,999
15	0,448	$	$ 345	$ 10,571	$ 10,916	$ 4,890	$ 123,362	$ 119,779	$ 19,836
16	0,425	$	$ 345	$ -	$ 345	$ 146	$ 134,707	$ 119,926	$ 14,946
17	0,402	$	$ 345	$ -	$ 345	$ 139	$ 135,052	$ 120,064	$ 14,799
18	0,381	$	$ 345	$ 270	$ 615	$ 235	$ 135,667	$ 120,299	$ 14,661
19	0,362	$	$ 345	$ -	$ 345	$ 125	$ 136,012	$ 120,424	$ 14,426
20	0,343	$	$ 345	$ 10,301	$ 10,646	$ 3,649	$ 146,658	$ 124,072	$ 14,301
21	0,325	$	$ 345	$ 270	$ 615	$ 200	$ 147,273	$ 124,272	$ 10,653
22	0,308	$	$ 345	$ -	$ 345	$ 106	$ 147,618	$ 124,378	$ 10,453
23	0,292	$	$ 345	$ -	$ 345	$ 101	$ 147,963	$ 124,479	$ 10,347
24	0,277	$	$ 345	$ 270	$ 615	$ 170	$ 148,578	$ 124,649	$ 10,246
25	0,262	$	$ 345	$ 10,301	$ 10,646	$ 2,792	$ 159,224	$ 127,441	$ 10,076
26	0,249	$	$ 345	$ -	$ 345	$ 86	$ 159,569	$ 127,527	$ 7,284
27	0,236	$	$ 345	$ 270	$ 615	$ 145	$ 160,184	$ 127,672	$ 7,198
28	0,223	$	$ 345	$ -	$ 345	$ 77	$ 160,529	$ 127,749	$ 7,053
29	0,212	$	$ 345	$ -	$ 345	$ 73	$ 160,874	$ 127,822	$ 6,976
30	0,201	$	$ 345	$ 10,571	$ 10,916	$ 2,190	$ 171,790	$ 130,012	$ 6,903

Fig. (8). Scenario 1 output for the whole life costs.

There is a "maintenance and cost" tab for the spreadsheet tools, in which it goes into greater detail what low, medium and high maintenance entails. Since the level of maintenance was determined in the "design and maintenance option" of AnsiPro A96 series pump brochure [39], the "whole life cycle tab" should be selected to view the total cost and the net present value as shown in Fig. (**8**).

The spreadsheet tool is for a 50-year project life and changes to the spreadsheets are locked. Since the lifetime of this study was 30 years, the outputs for each year had to be summed up to get the total costs and net present costs.

APPENDIX C – PUMP DESIGN FOR SCENARIO 1

Given:
Water @ 50 ˚C:

gamma = 62.4 lb/ft^3
spec. gravity = 1.0

Q = ~100 gpm

Vs = ~3 ft/s
Vd = ~15 ft/s

PSIA (Salt Lake City) = ~12.5 PSI

Use Schedule 40 Steel Pipe

Suction Line

- Approximately 5 ft of Schedule 40 steel pipe.

- Include foot valve with strainer (K = 3.0) at the pipe inlet to prevent debris from entering the pipe and potentially damaging pump and to eliminate need to require a self-priming pump.

- Choose 3-inch nominal diameter Schedule 40 steel pipe to produce velocities between 3 ft/s and 7.5 ft/s for design flow of 100 gpm.

Discharge Line

- Approximately (10 ft + 55 ft + 75 ft) = 140 ft of Schedule 40 steel pipe.

- Include two 90° flanged bends.

- Include check valve (swinging type) a few feet downstream of pump to prevent backflow and elevated pressures from harming pump at shutoff and non-operation periods.

- Choose 2.5-inch nominal diameter Schedule 40 steel pipe to produce velocities between 7.5 ft/s and 15 ft/s for design flow of 100 gpm.

Based on the notion of a 2,500-gallon tank, once it reaches 15% (275 gallons), capacity will refill to 75% capacity (1,875 gallons). This means the pump has to fill 1,500 gallons. A good, even number for the flow rate to do this is 100 gpm, meaning the 1,500 gallons would take about 15 minutes to refill.

Det'm pressure head (hp) for design Q of 100 gpm.

$h_{l,total} = 2.02469 \times 10^{-3} (Q)^{1.852} + 2.1825 \times 10^{-5} (Q)^2$ (take Q = 100 gpm)

$h^p = (65 - 0) + h_{l,total}$

$h_p = 65 + 2.02469 \times 10\text{-}^{3} (100)^{1.852} + 2.1825 \times 10^{-5} (100)^2$ □(aka: Total Dynamic Head, TDH; h_a)

Total Head in the system is calculated to be 75.5 ft.

Using the pump curve sizing tool at Engineered Software Incorporation website, a great pump fit is the AnsiPro AP96-ANSI (Fig. **9**). The specifications for the pump are listed in Table **9**:

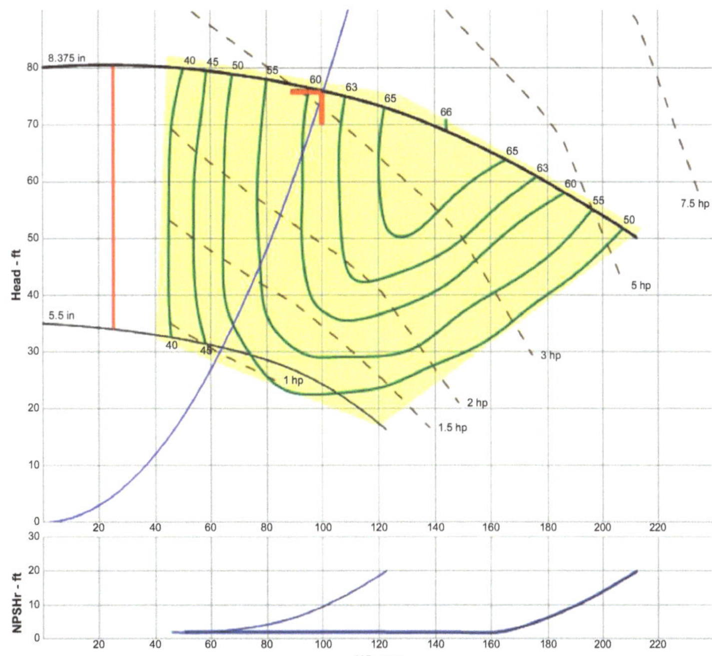

Fig. (9). AnsiPro AP96-ANSI pump curve [39].

Table 9. AnsiPro AP96-ANSI pump characteristics [39].

Size	3x2-8 A60
Curve	AP96-1807
Speed	1750 rpm
Diameter	8.375 in
Head	75.8 ft
Efficiency	61.2%
NPSHr	2
Power	3.13 hp
Min. Flow	25 gpm

Scenario 1

Based on precipitation data, each year, the tanks are getting below the specified volume for about 14 days. This results to 424 total times in 30 years. Each time, the municipal water fills the tank to almost 1000 gallons. Therefore:

$$\text{Municipal Water Usage} = 14\,\frac{\text{days}}{\text{year}} \times 30\ \text{years} \times 1000\,\frac{\text{gallons}}{\text{day}} = 424{,}000\ \text{gallons}$$

The same calculations were done for the other two scenarios.

Scenario 2

$$\text{Municipal Water Usage} = 14.5\,\frac{\text{days}}{\text{year}} \times 30\ \text{years} \times 1000\,\frac{\text{gallons}}{\text{day}} = 434{,}000\ \text{gallons}$$

Scenario 3

$$\text{Municipal Water Usage} = 17.7\,\frac{\text{days}}{\text{year}} \times 30\ \text{years} \times 1000\,\frac{\text{gallons}}{\text{day}} = 531{,}000\ \text{gallons}$$

APPENDIX D – SYSTEM SIZING

For the purpose of this study, a spreadsheet has been created to calculate the volume of water in a tank after a monthly indoor use. It was determined that the indoor use can have a large variance depending on the amount of people in the building; therefore, different quantities of users were analyzed in this spreadsheet. The final output is shown in Table **10**.

Table 10. Monthly storage tank volume for different building occupancy quantities.

	Tank Volume at the Beginning of the Month Depending on Occupancy (Number of People)					Inflow (Gallons)
	75	**100**	**125**	**150**	**175**	
Demand (gal/day)	**360**	**480**	**600**	**720**	**840**	
January	14,400	14,400	14,400	14,400	14,400	2,967
February	6,567	2,967	633	4,233	7,833	22,864
March	18,631	11,431	4,231	2,969	10,169	13,963
April	21,794	10,994	194	10,606	21,406	41,364
May	52,358	37,958	23,558	9,158	5,242	15,708
June	57,266	39,266	21,266	3,266	14,734	27,925
July	74,391	52,791	31,191	9,591	12,009	5,934
August	69,525	44,325	19,125	6,075	31,275	3,316
September	62,042	33,242	4,442	24,358	53,158	8,727
October	59,968	27,568	4,832	37,232	69,632	60,738
November	109,906	73,906	37,906	1,906	34,094	20,420
December	119,526	79,926	40,326	726	38,874	8,203
Average Tank Volume	55,531	35,731	15,931	3,869	23,669	

From the average tank volumes it was determined that for 100-person occupancy the tank should be sized around 35,000 gallons. Fig. (**10**) shows the monthly tank volumes depending on different building occupancies. The variance in tank demand over a year, and how some tanks go dry if the indoor demand is under estimated, can be noted in this figure.

Using another spreadsheet, a continuous model was created to simulate different scenarios for 30+ years of precipitation and water demand. For Scenario 1, it was determined to size the cisterns at 2,500 and 30,000 gallons for the roof and underground cistern, respectively. Similarly, Scenario 2 was determined to use a ground cistern sized at 30,000 gallons. And, a 30,000-gallon tank on the roof was considered for Scenario 3. However, Scenario 3 is only a consideration if structural analysis shows that this is possible to introduce this dead/live load on the roof. To further illustrate the feasibility of the system design, outdoor irrigation can be completely eliminated, and if the system boundary remained within the building, the municipal water supply that will have to augment the harvested rainwater for a 30-year period is 3,000 gallons.

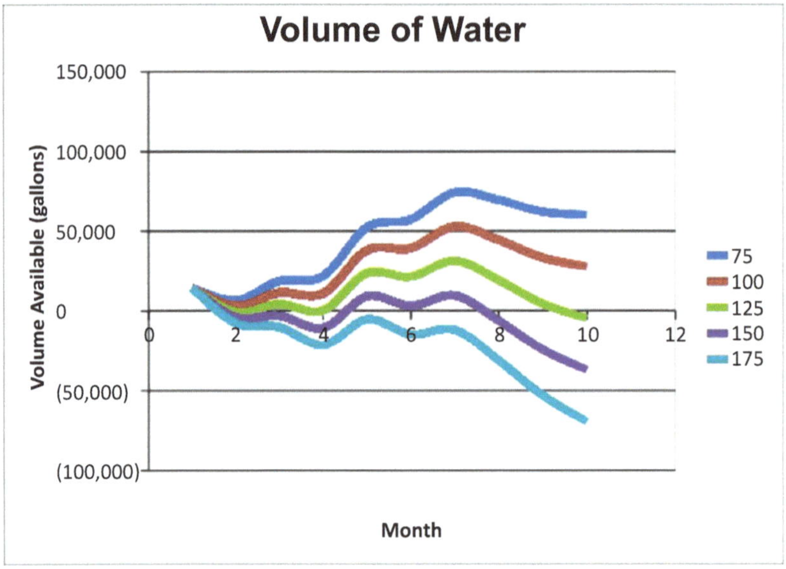

Fig. (10). Volume of water in a tank depending on building occupancy.

Scenario 1 was too complex to be modeled in SWMM, as there are two tanks at different elevations with conditional formats. When the tank on the roof (2,500 gallons) gets below 500 gallons, the pump turns on that night to add 2,000 gallons from the tank on the ground. Similarly, the two tanks have overflows not to flood surrounding areas.

Scenarios 2 and 3 were modeled in SWMM to show the amount of overflow from the tank, as well as the storage capacity. These results were compared to that in Table **6**. However, using SWMM some assumptions or design parameters could not be modeled. For instance, the fact that municipal water fills the 30,000-gallon tank when it reached 1,500 gallons or less was not accounted in the model. Therefore, the number of days below 1,500 gallons is much higher than the actual number of these days happening. However, the total gallons of municipal water used, and the precipitation data (in.) are within a percent of what is modeled in the spreadsheet.

REFERENCES

[1] D. Harbeke, B. Garbett, D. Matsumori, and S.J. Hershey Kroes, *A snapshot of 2050; an analysis of projected population change in Utah,* 2014.*Research Report, Rep. 720,* 2014.[Accessed Nov. 2014] [Online] Available: http://www.utahfoundation.orghttp://www.utahfoundation.org.

[2] J.M. Kahinda, A.E. Taigbenu, and R.J. Boroto, "Domestic rainwater harvesting as an adaptation measure to climate change in South Africa", *Phys. Chem. Earth,* vol. 35, no. 13-14, pp. 742-751, 2010. [http://dx.doi.org/10.1016/j.pce.2010.07.004]

[3] M.Y. Han, and J.S. Mun, "Particle behaviour consideration to maximize the settling capacity of rainwater storage tanks", *Water Sci. Technol.,* vol. 56, no. 11, pp. 73-79, 2007. [http://dx.doi.org/10.2166/wst.2007.778] [PMID: 18057644]

[4] C.J. Glendenning, and R.W. Vervoort, "Hydrological impacts of rainwater harvesting in a case study catchment: The Arvari River, Rajasthan, India. Part 1: Field-scale impacts", *Agric. Water Manage.,* vol. 98, no. 2, pp. 331-340, 2010. [http://dx.doi.org/10.1016/j.agwat.2010.09.003]

[5] R. Alam, G. Munna, M.A. Chowdhury, M.S. Sarkar, M. Ahmed, M.T. Rahman, F. Jesmin, and M.A. Toimoor, "Feasibility study of rainwater harvesting system in Sylhet City", *Environ. Monit. Assess.,* vol. 184, no. 1, pp. 573-580, 2012. [http://dx.doi.org/10.1007/s10661-011-1989-7] [PMID: 21416213]

[6] V.W. Tam, L. Tam, and S.X. Zeng, "Cost effectiveness and tradeoff on the use of rainwater tank: an empirical study in Australian residential decision-making", *Resour. Conserv. Recycling,* vol. 54, no. 3, pp. 178-186, 2010. [http://dx.doi.org/10.1016/j.resconrec.2009.07.014]

[7] S. Lassaux, R. Renzoni, and A. Germanin, "Life cycle assessment of water: From the pumping station to the wastewater treatment plant", *International Journal of Life Cycle Assessment,* vol. 12, no. 2, pp. 118-126, 2007. [http://dx.doi.org/10.1065/lca2005.12.243]

[8] R. Farreny, T. Morales-Pinzón, A. Guisasola, C. Tayà, J. Rieradevall, and X. Gabarrell, "Roof selection for rainwater harvesting: quantity and quality assessments in Spain", *Water Res.,* vol. 45, no. 10, pp. 3245-3254, 2011. [http://dx.doi.org/10.1016/j.watres.2011.03.036] [PMID: 21492898]

[9] E. Ghisi, D.D. Tavares, and V.L. Rocha, "Rainwater harvesting in petrol stations in Brasilia: Potential for potable water savings and investment feasibility analysis", *Resour. Conserv. Recycling,* vol. 54, no. 2, pp. 79-85, 2009. [http://dx.doi.org/10.1016/j.resconrec.2009.06.010]

[10] S.J. Burian, and D. Jones, "National assessment of rainwater harvesting as a stormwater best management practice: Challenges, needs, and recommendations", *International Low Impact Development Conference,* , 2010pp. 842-852 [http://dx.doi.org/10.1061/41099(367)74]

[11] R.B. Thomas, M.J. Kirisits, D.J. Lye, and K.A. Kinney, "Rainwater harvesting in the United States: a survey of common system practices", *J. Clean. Prod.,* vol. 75, pp. 166-173, 2014. [http://dx.doi.org/10.1016/j.jclepro.2014.03.073]

[12] C. Anand, and D.S. Apul, "Cost, energy, and CO2 emissions analysis of standard, high efficiency, rainwater Flushed, and composting toilets", *J. Environ. Manage.,* vol. 92, no. 3, pp. 419-428, 2011. [http://dx.doi.org/10.1016/j.jenvman.2010.08.005] [PMID: 21056531]

[13] W. Ahmed, A. Vieritz, A. Goonetilleke, and T. Gardner, "Health risk from the use of roof-harvested rainwater in Southeast Queensland, Australia, as potable or nonpotable water, determined using quantitative microbial risk assessment", *Appl. Environ. Microbiol.,* vol. 76, no. 22, pp. 7382-7391, 2010. [http://dx.doi.org/10.1128/AEM.00944-10] [PMID: 20851954]

[14] L.C. Proenca, E. Ghisi, D.F. Tavares, and G.M. Coelho, "Potential for electricity savings by reducing potable water consumption in a city scale", *Resour. Conserv. Recycling,* vol. 55, no. 11, pp. 960-965, 2011. [http://dx.doi.org/10.1016/j.resconrec.2011.05.003]

[15] T. Morales-Pinzón, R. Lurueña, and J. Rieradevall, "C. M., Gasol, and X. Gabarrell, “Financial feasibility and environmental analysis of potential rainwater harvesting systems: a case study in Spain", *Resour. Conserv. Recycling,* vol. 69, pp. 130-140, 2012.

[http://dx.doi.org/10.1016/j.resconrec.2012.09.014]

[16] M.V. Vargas-Parraa, G. Villalbaa, and X. Gabarrell, "Applying exergy analysis to rainwater harvesting systems to assess resource efficiency", *Resour. Conserv. Recycling,* vol. 72, pp. 50-59, 2013. [http://dx.doi.org/10.1016/j.resconrec.2012.12.008]

[17] S. Angrill, R. Farreny, C.M. Gasol, X. Gabarrell, B. Viñolas, A. Josa, and J. Rieradevall, "Environmental analysis of rainwater harvesting infrastructures in diffuse and compact urban models of Mediterranean climate", *International Journal of Life Cycle Assessment,* vol. 17, no. 1, pp. 25-42, 2012. [http://dx.doi.org/10.1007/s11367-011-0330-6]

[18] J. Devkota, H. Schlachter, and D. Apul, "Life cycle based evaluation of harvested rainwater use in toilets and for irrigation", *J. Clean. Prod.,* vol. 95, pp. 311-321, 2015. [http://dx.doi.org/10.1016/j.jclepro.2015.02.021]

[19] J. Devkota, H. Schlachter, C. Anand, R. Phillips, and D. Apul, "Development and application of EEAST: a life cycle based model for use of harvested rainwater and composting toilets in buildings", *J. Environ. Manage.,* vol. 130, pp. 397-404, 2013. [http://dx.doi.org/10.1016/j.jenvman.2013.09.015] [PMID: 24141064]

[20] S.R. Ghimire, J.M. Johnston, W.W. Ingwersen, and T.R. Hawkins, "Life cycle assessment of domestic and agricultural rainwater harvesting systems", *Environ. Sci. Technol.,* vol. 48, no. 7, pp. 4069-4077, 2014. [http://dx.doi.org/10.1021/es500189f] [PMID: 24605844]

[21] I. Linkov, F.K. Satterstrom, G. Kiker, T.P. Seager, T. Bridges, K.H. Gardner, S.H. Rogers, D.A. Belluck, and A. Meyer, "Multicriteria decision analysis: a comprehensive decision approach for management of contaminated sediments", *Risk Anal.,* vol. 26, no. 1, pp. 61-78, 2006. [http://dx.doi.org/10.1111/j.1539-6924.2006.00713.x] [PMID: 16492181]

[22] W.F. Wu, Y. Hsu, Y.J. Chang, and Y.C. Chen, "Selection of renewable energy based on life cycle assessment and multi-criteria decision making", *Open J. Energy Effic.,* vol. 5, no. 5, pp. 1-11, 2016. [http://dx.doi.org/10.4236/ojee.2016.51001]

[23] A. Rabl, and M. Holland, "Environmental assessment framework for policy applications: Life cycle assessment, external costs and multi-criteria analysis", *J. Environ. Plann. Manage.,* vol. 51, no. 1, pp. 81-105, 2008. [http://dx.doi.org/10.1080/09640560701712275]

[24] I. Linkov, and T.P. Seager, "Coupling multi-criteria decision analysis, life-cycle assessment, and risk assessment for emerging threats", *Environ. Sci. Technol.,* vol. 45, no. 12, pp. 5068-5074, 2011. [http://dx.doi.org/10.1021/es100959q] [PMID: 21524065]

[25] F. Barekat, D.Y. Han, M. Dewan, and A. Qian, *An investigation into the use of wood vs. steel and concrete in construction of the new sub,* 2010.

[26] M. Salahuddin, M.R. Jeyagapal, M. Al-Tameemi, and V. Mogan, *An investigation into cistern for rainwater storage and sterilization for the new UBC Student Union Building,* 2011.

[27] C.H. Liaw, and Y.L. Tsai, "Optimum storage volume of rooftop rain water harvesting systems for domestic use", *J. Am. Water Resour. Assoc.,* vol. 40, no. 4, pp. 901-912, 2004. [http://dx.doi.org/10.1111/j.1752-1688.2004.tb01054.x]

[28] C. Studies, and L.C. Salt, *Weather,* November 2011.[Accessed Sep. 2014] [Online] Available: http://countrystudies.us/united-states/weather/utah/salt-lake-city.htm

[29] National Climate Data Center, *Salt Lake City monthly average precipitation, Western Regional Climate Center, Desert Research Institute,* 2011.[Accessed Sept. 2014] [Online] Available: http://www.wrcc.dri.edu/htmlfiles/ut/ ut.ppt.html

[30] J. Prusinski, "Life cycle inventory of slag cement concrete", *CTL Project No. 312012, CTL Construction Technology Laboratories, Inc,* 2003.[Accessed Oct. 2014] [Online] Available: http://www.slagcement.org/Sustainability/pdf/Slag%20Cement%20Concrete%20LCI%20Letter%20Report%20August%2011%202003.pdf

[31] K. Jones, *Density of Concrete, The Physics Factbook,* 1999.[Accessed Oct. 2014] [Online] Available: http://hypertextbook.com/facts/1999/KatrinaJones.shtml

[32] J.F. Schifo, and J.T. Radia, "Theoretical/Best Practice Energy Use In Metalcasting Operations", *KERAMIDA Environmental Inc.,* May 2004.[Accessed Oct. 2014] [Online] Available: http://www.afsinc.org/files/best%20practice%20energy-schifo-radia-may%202004.pdf http://www.afsinc.org/files/best%20practice%20energy-schifo-radia-may%202004.pdf [http://dx.doi.org/10.2172/1216246]

[33] M. Socks, *The promise of all-electric injection molding machines: a promise kept?*. Proceeding in ACEEE Summer Study on Energy Efficiency in Industry, 2005, pp. 155-166. [Online]

[34] *Stainless Steel Grade 304, UNS S30400, written by AZoM,* October 2001.[Accessed Oct. 2014] [Online] Available: http://www.azom.com/properties.aspx?ArticleID=965

[35] R. Brouwer, and R. van Ek, "Integrated ecological, economic and social impact assessment of alternative flood control policies in the Netherlands", *Ecol. Econ.,* vol. 50, no. 1-2, pp. 1-21, 2004. [http://dx.doi.org/10.1016/j.ecolecon.2004.01.020]

[36] J. Batchelor, L.S. Spang, T.T. Soo Hoo, C. Batchelor, M. Kalin, A. Borojerdi, and S. Fennessey, "Green roof planning study for the City of Boston", *Arrowstreet Report,* October 2009.[Accessed Nov. 2014] [Online] Available: http://www.mass.gov/eea/docs/doer/ green-communities/library/green-rof-boston-st2009.pdf

[37] L. Domènech, and D. Saurí, "A comparative appraisal of the use of rainwater harvesting in single and multi-family buildings of the Metropolitan Area of Barcelona (Spain): social experience, drinking water savings and economic costs", *J. Clean. Prod.,* vol. 19, no. 6, pp. 598-608, 2011. [http://dx.doi.org/10.1016/j.jclepro.2010.11.010]

[38] S. Ward, F.A. Memon, and D. Butler, "Performance of a large building rainwater harvesting system", *Water Res.,* vol. 46, no. 16, pp. 5127-5134, 2012. [http://dx.doi.org/10.1016/j.watres.2012.06.043] [PMID: 22840659]

[39] AnsiPro Pump Company Brochure, *Series AP96,* 2014.[Accessed Nov. 2014] [Online] Available: http://www.yorkfluid.com/pdf/ AnsiPro-AP96-Series-Pump-Brochure.pdf

CHAPTER 5

Technology Improvements or Influencing User Behaviour for Water Savings in Administrative and University Buildings: Which One Should Come First?

Asher Kiperstok* and **Alice Costa Kiperstok**

Teclim, Clean Technology Network, Department of Environmental Engineering, Polytechnics School, Federal University of Bahia, Salvador, State of Bahia, Brazil

Abstract: Technology upgrades or investment in changing user behaviour? This is a common dilemma when it comes to improving water savings in administrative buildings. The answer is obvious: both, but only after a management scheme is in place. Several real scale experiments have been carried out at the Federal University of Bahia, Brazil, and in the administrative buildings of the government of the State of Bahia over the last 17 years with significant results however, less than expected. This paper discusses the role played by so-called water saving devices and that played by maintenance activities and continuous calibration. A conceptual scheme to guide water savings in buildings as well as actions that must be considered in water saving programmes are presented. The conceptual guide considers the role of following-up water consumption on a daily basis by administrators but with data open to the public. It further presents an approach to understanding the reasons for water losses and water waste and the means to reduce them in institutions with clear technical and economic limitations. Consumption monitoring and control is the most important action to be taken and this has to precede further efforts or investments such as the acquisition of water saving devices or greywater use and rainwater catchment.

Keywords: Conservation, Institutional building, Rational use, University premises, Water saving, Water end-uses, Water consumption monitoring, Water awareness, Water demand management, Water soft path.

INTRODUCTION

This chapter discusses the implementation of water saving programmes in institutional buildings focusing 17 years of experience of Teclim/UFBA, a

* **Corresponding author Asher Kiperstok**: Department of Environmental Engineering, Polytechnic School, Federal University of Bahia, Salvador, State of Bahia, Brazil; Tel: +55 71 32839892; E-mail: asher@ufba.br.

research group at the Federal University of Bahia, Brazil. Besides having worked with large industrial sites, Teclim has constantly been taking part in water saving programmes in university buildings and governmental premises in Bahia. As explained in this chapter these programmes have led to a significant reduction in water waste and loss but they fall short of our expectations.

Perhaps our expectations were too high but they could not be different as we are aware of the hardship of constructing the utopia of sustainability. Moreover, we are fully aware that failure to reach minimum sustainability advances will mainly fall on the heads and shoulders of populations in developing regions. Society is failing to develop decent consumption practices especially at government and teaching institutions. Such places should lead by example and by failing to do so they jeopardize their main objectives.

We have reviewed several authors who suggest and describe programmes to rationalize water consumption in institutional buildings. As expected, their approaches have a lot in common. They all include technological and behavioural aspects that have to be brought together in a managerial context. Readers interested in putting forward organized action for saving water in an institution will find many examples in the technical literature. However, they should not expect it to be an easy task. It will require long-term persistence despite initial gains and economy. Long term durable solutions will require the building of a permanent awareness. That is, an institutional attitude capable of resisting the ever-changing characteristics of these organizations: A natural resources respecting culture.

Towards this we discuss the factors that define consumption patterns in these kinds of buildings and organize them in a conceptual model. The reader might agree that an effective way to promote savings of any resource is by charging for its use. This is very difficult to do in the type of edifications we are studying here and other mechanisms have to be used. Among these, we highlight the need for an Eco-team, an internal group responsible for conducting the water saving programme and implementing daily monitoring and control.

Furthermore, this work presents some basic tools to be included in a programme to promote the rational use of water in government administrative and university buildings which may also be useful for other types of premises for public use. These tools were used in the programmes developed with Teclim's participation, the results of which are described at the end of this chapter, just before our conclusion.

One word of warning however, although general action lines are useful to develop a water rationalization programme, each case is a particular case with its own wealth of details and, as the saying goes, "the devil is in the detail".

RATIONAL CONSUMPTION OF WATER IN THE CONTEXT OF THE ENVIRONMENTAL CRISIS

> *The problem with the concept of sustainability is that everyone seems to understand what is meant by it in its promiscuous usage, however, everybody understands it differently from all others* [1].
>
> *The world's current energy prospects are – put simply – unsustainable* [2].

Trivialization of the concept of sustainability [1, 3] makes it more difficult to implement the solutions for several aspects of the environmental crisis we are undergoing. Climate change, its most publicized face, have already shown that solutions will have to be very radical [4], much more radical than those acceptable to most authorities, voters and in particular, consumers. Market forces take advantage of this scenario by offering more goods to buy, aggravating the problem in two ways: Increasing consumption and forging the idea that we are the environmental heroes our kids expect us to be, just because we are now consuming "green" gadgets.

Climate change, however, is just one of the environmental sectors where the planetary boundaries have already been dangerously exceeded. The work of Rockström *et al.*, featured in Nature in September 2009, just before COP 15 [5] and Steffen *et al.* [6], allowed a broad discussion of nine, much interlaced sectors that must be considered when designing paths to deal with these environmental problems. The limits of three or four of these have already been dangerously surpassed and surprisingly water availability is not among them!

Imagine trying to explain that to people living in water stressed regions. If in California, Israel or Dubai we will get a clear, rapid, answer. Put more money into it! Put in more technology and the problem will be solved. However, if in arid or semi-arid poor or developing regions, response will be quite different.

This short discussion brings us into a more profound debate, as the one raised by Jaramillo and Destouni [7] and answered by Gerten *et al.* [8]. Global boundaries should not stop us from discussing local problems. This is even more so when solutions to environmental problems in one sector enlarge problems in other sectors.

Common in California, Israel and Dubai, seawater desalination is a solution to water scarcity. Making seawater available for urban use, is a typical supply side solution that, like other supply side options, is much more energy intensive than conventional solutions and emit greenhouse gases. It becomes much worse if the energy is supplied by shale gas, low quality coal or oil, as is the case in these places. However, less harm will be caused if these solutions are accompanied by water demand side alternatives.

The situation is aggravated by the lack of understanding of how the global environmental crisis is built. If we overload the atmosphere with more carbon gases in order to resolve regional or local water crisis, hydrological cycles will be affected at faster rates. This will demand further hydraulic work, which will then require additional energy, accelerating the vicious circle of environmental degradation.

But let's not forget, we have almighty clean energy to save humanity. However, as stated by the International Energy Agency: "*The change needed to achieve the aims of either long-term scenario is daunting, amounting to nothing less than an energy revolution*" [4]. For a problem of the magnitude that is being faced, the amount of cleaner energy available is what is measured not in megawatts but in "negawatts" or watts not spent [9]. That is, energy that is not wasted and therefore it is not necessary to produce. This means that solutions have to be found by reducing demand. To cut a long story short, demand side solutions cannot just be complementary to supply side ones. They ought to be the main path to be followed to face and overcome the water crisis embedded in the global environmental crisis.

Authors such as Wolff and Gleick [10] and Brooks and Holtz [11] used the term soft path for water, after the term soft path for energy coined by Amory Lovins in 1977 [12], to amplify demand side approaches to water supply. They included water policy and behavioural considerations to highlight solutions based more on human attitudes than resource intensiveness. These concepts have already been applied in regional planning processes enabling clarification of the differences between water demand management and water soft path [13].

With respect to water demand management approaches, these are far from being overcome and much work is needed to universalise them. A comprehensive IWA book on the subject reviews water demand actions including their impact on conventional sewage networks using methods that include life cycle assessment [14]. Water conservation methods and products, greywater treatment and use, rainwater catchment, economic and legal aspects, consumer reaction, decision support tools and other aspects that are analysed in this book have to be considered for designing evolution paths to converge from conventional sanitation to ecological or sustainable sanitation.

The United Nations World Water Development Report 2015, Water for a Sustainable World [15], envisions a desired situation to be achieved by means of hard work by 2050. The main attitudes for this refer to water management action including an increase of productivity in water use, diminishing water demand *per capita* and per unit of production, better levels for reuse or more adequate discharge standards, back to the environment. The document encourages more efficient technologies and equitable tariffs to improve demand for everyday use and production processes. It highlights the need to decouple economic growth from water use by stimulating sustainable production and consumption patterns. To achieve all these (utopic?) goals, much better supply and demand water management is needed.

WATER CONSERVATION IN INSTITUTIONAL BUILDINGS, HOW AWARE ARE THEY?

Many will relate the debate on water conservation to the construction of a new "water consciousness". We consider this as a quite loose concept that has also been trivialized. Great care has to be taken to avoid these concepts ending up in attitudes such as "telling others to do what you don't". Perhaps one must modestly attain to develop means to raise awareness related to water use that may, or not, support the achievement of higher levels of consciousness. In this sense, water demand management at home acquires a strategic stand. To foster and consolidate it, institutional and commercial premises may be useful. They are convenient practice spaces to promote learning through the example of water saving programmes, particularly the case of public premises. Schools and universities can be used as primary tools in this strategy where practices can be extended to reach students' homes. Accordingly, there is a need to develop more knowledge to support water savings in buildings of this kind. This, however, should not be motivated only by financial gains of reducing bills. It should support awareness raising about water issues as a whole.

There are several works on water use conservation in institutional premises in the technical literature, however they are much fewer in numbers than those related to household water conservation. Vickers' work, Handbook of Water Use and Conservation [16], contains a chapter dedicated to industrial, commercial and institutional premises. The author recognizes the difficulty of assessing water consumption in these kinds of buildings given their complexity and the large variety of ways the water is used.

Water saving in commercial and institutional premises requires a different approach from that used in residences, even when referring to domestic water end-uses. In this chapter we consider as domestic uses those occurring in toilets and kitchens in commercial and institutional premises. Other common end-uses of water in this sort of buildings are water replacement of cooling towers, large-scale floor cleaning, decorative devices such as fountains and water mirrors, irrigation of internal and external gardens and car washing.

Among the main reasons for the need to use a different approach are:

- Users do not directly pay for the water bill.
- They have a wide variety of habits and awareness levels to environmental issues.
- Maintenance is generally weak, mainly in public institutions
- Complexity of piping and installations make leaks more difficult to identify.
- A single metering device accounts for large and complex systems
- Sanitary devices are used more and, therefore, tend to be defective more frequently.

All these reasons contribute to the absence of appropriate indicators to guide water saving programmes in commercial and institutional buildings. In 2000 the American Water Works Association, conducted a survey in five categories of premises: schools, hotels, office buildings, restaurants and food stores selected from a list of 11 categories. The final AWWA report [17] reveals an opinion usually expressed by specialists: *"In general, water conservation programs that have been implemented are rarely well documented and evaluated. Many available documents lack direct information for generalizing water savings. There is a need for more information on program costs, implementation conditions and measurements of savings".*

Additionally to using numbers found in the literature as consumption reference, research conducted by Teclim/UFBA, indicate the importance of appropriate training of people involved in water saving programmes. The familiarization with their institution's water consumption behaviour is strongly recommended, in order to develop a sensibility to identify normal performances and when abnormal events are occurring. Moreover, daily monitoring and access to data of similar premises gradually support this knowledge building.

Water uses in non-residential buildings, used by a miscellaneous public, may vary significantly according to certain characteristics that are relevant in terms of water demands. The presence of large green areas, cooling towers or restaurants increase water consumption in absolute numbers and relative indicators. Buildings in tropical regions using air conditioning may use about half of its water demand to replenish water losses to atmosphere in their cooling towers. For this reason, water consumption indicators found in the literature must be carefully and critically used. Moreover, visible or invisible leaks, and the ability to identify and repair them may completely alter the consumption profile of a building. Therefore, monitoring systems and the capacity to react and repair the causes of wastes are crucial to define the amount of water consumed in buildings. Often just a broken seal on a tap can lead to significant waste if not promptly identified and repaired. However, the water use most common to all these buildings is in toilets, accounting for 40 to 90% of water consumption in buildings. These facilities should therefore be investigated when looking for water savings opportunities.

Familiarization with the buildings' water system and the behaviour of the user community is the first step towards raising awareness. But how can we encourage water awareness in institutional buildings? How can it be maintained and improved?

Steward *et al.* [18] examined the expansion of metering devices in a Web Based Knowledge Management System "*which integrates smart metering, end use water consumption data, wireless communication networks and information management systems in order to provide real-time information on how, when and where water is being consumed for the consumer and utility*". The authors propose this system for the future of water planning and management in Australia.

This interesting idea may be considered for realities where high standard operational and maintenance practices are in place and energy smart grids are already widespread. However, what can be used in more basic situations such as those found in most cities in the world? To help find solutions some fundamental things have to be taken into account. The most important approach has to foster water awareness in buildings. Another is to improve knowledge about end-uses of water and the factors that define water consumption profiles in buildings.

Governmental premises play an important role in developing more rational uses of natural resources as they are supposed to lead by example on the path to sustainability. They are also expected to comply with rational spending of precious public money and high levels of efficiency of resources. But, how frequently may we say that this really happens? Among public institutions, universities should lead the way. The same question arises: how frequently may we say that this really happens?

Teclim's experiences have been in these kinds of premises but also on industrial sites and few differences have been found in relation to the level of awareness towards water consumption in both. On industrial sites the main reason for a lack of commitment to water savings could be the very low price paid for water by the industries we have worked with. On the contrary, water tariffs paid by public buildings were up to tenths of times higher, but still it was difficult to develop full water saving programmes.

Work by several authors was consulted to understand the steps recommended for a Water Saving Programme in institutional premises including: academic thesis and papers, water associations, business organization and consultant offices' reports and universities' records and descriptions [16, 17, 19 - 35]. Several common steps were identified, as listed below.

Preliminary action and diagnosis: Definition of support from higher administration, historical data, water costs, identification of existing documents, plants and "as built" sketches, field surveys, profile of users and water uses, initial estimation of performance using indicators, possible measures and expected reduction scenarios, payback estimation.

Detailed water demand assessment or diagnosis - physical aspects: Buildings survey, hydraulic system, leak occurrence, special systems, existence of metering and sub-metering devices. Assessment of technical and economical viabilities including water losses and waste analysis in processes, apparels and hydraulic components. Identification of water pressure in piping and existence of pressure control devices, devices for water supply interruption by sectors. Probable causes for water quality loss.

Detailed water demand assessment or diagnosis - managerial aspects:Investigation of end-users proceedings, level of awareness, direct payment for the water and payment procedures, quality of maintenance practices and structure, calibration of main metering devices, installation of sub metering and temporary metering, frequency of metering, frequency of leak detection and repair, water flow tests.

Action plan preparation and implementation – conservation measures: Awareness campaigns, leak detection, water saving apparels, disclosure of waste reduction measures, specific educational campaigns. Implementation or improvement of a maintenance structure, personnel, equipment and procedures.

Action plan preparation and implementation – supply alternatives: Water reuse, rainwater catchment, local underground water extraction.

Implementation of water management system: Operational action, educational action, Institutional action, track and report progress reduction assessment, permanent monitoring, specific consumption indicators.

Several authors highlight some of the listed activities, which may be combined in different orders. Some works emphasize managerial action while others technical interventions. Most prioritize waste and loss reduction over supply alternatives, such as rainwater capture and use. The importance of combining physical, managerial and educational efforts is clear.

WHAT DEFINES WATER CONSUMPTION PATTERNS IN INSTITUTIONAL BUILDINGS?

To rationalize water consumption in commercial and institutional buildings requires recognizing not just water use from consumption but also wasted water from leaks and of course their causes. Research carried out in 2010 analysed the head office building of a public company located at the administrative centre of the government of the State of Bahia in Salvador. Two water meters were in use where the water entered the building. These meters were monitored for 106 weeks, before and during the different stages of the research. Results are shown in Fig. (**1**).

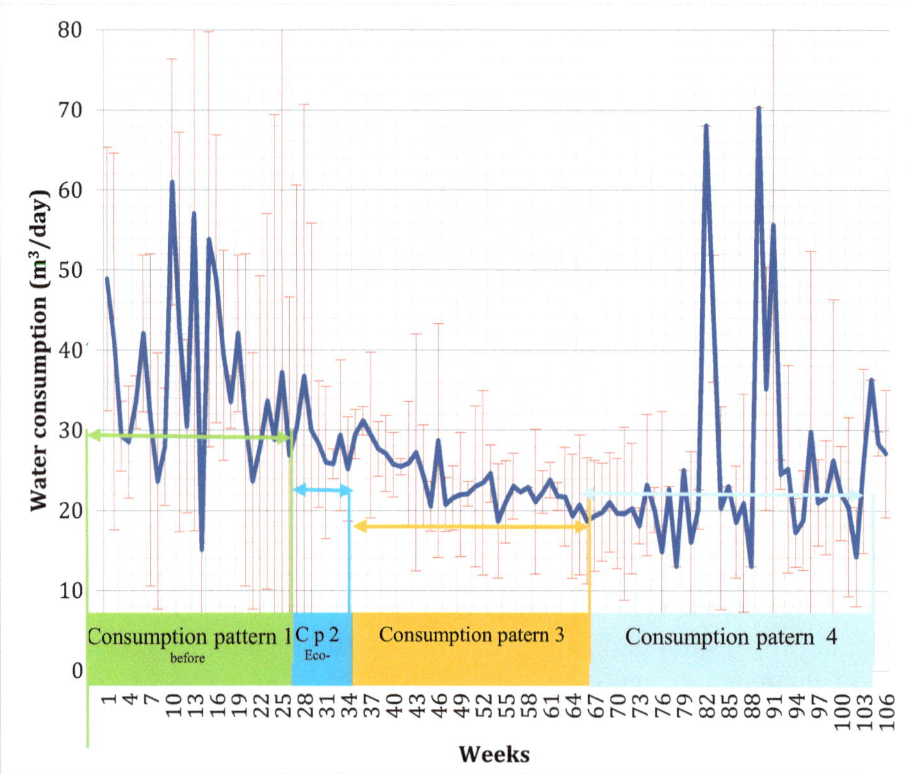

Fig. (1). Water consumption in an administrative building [36].

Consumption pattern 1 describes the situation before any intervention was implemented. It shows a significant variation in daily consumption. Each point of the graph indicates the mean value of the measurements registered in a certain week. Minimum and maximum daily values, within the weeks, are also shown. This phase lasted 30 weeks.

During the last part of this first stage, a water management team (or Eco-team), was created and trained to conduct water rationalization activities in the building, starting by daily following up of consumption in order to identify possible abnormalities. A clear modification of the consumption pattern can be identified in the second phase including a reduction in the weekly consumption average, from 35.5 to 27.7 m^3/day, smaller daily variations and closer values for the minimum and maximum daily consumptions within a week. During this period of approximately 10 weeks, daily water consumption was registered and publicized using "Aguapura Vianet" system, which is described further in this chapter. Small interventions to eliminate water losses were implemented, based on the observations made in this system. The plumbing system was checked to fix small problems and to regulate some devices.

During phase 3, which lasted for about 35 weeks, the water management team consolidated their work however, following orders from the company's higher administration, automatic faucets and low consumption water closets were installed to substitute existing ones. These substitutions were not recommended or requested by the research team but they came as a unidirectional initiative, from higher ranked officials, to collaborate with the water reduction effort. An additional reduction of the average weekly consumption, from 27.7 to 22.0 m^3/day was then obtained, accompanied by a more stable pattern as can be seen in Table **1**.

Table 1. Changes in water consumption patterns of an administrative building during intervention [36].

Water Consumption Pattern	Weekly Average (m^3/day)	Standard Deviation (m^3/day)	Coefficient of Variation (%)
1.Before intervention (30 weeks)	35.50	10.48	29.54
2.Water management team (10 weeks)	27.70	2.06	7.45
3.Substitution of hydraulic devices (35 weeks)	22.00	2.45	11.18
4.Changes in the water management team (10 weeks)	26.80	14.20	52.89

The fourth phase was also not planned by the research team. It was caused by the substitution of the coordinator of the Eco-team and its members. The new team did not receive adequate training and ceased to analyse the daily consumption data. The weekly average consumption returned to values similar to those obtained before the substitution of the old devices with the new saving water devices. Together with the increase in water consumption identified in this phase, its instability could also be observed. Nevertheless, consumption was daily registered by the researcher.

An analysis of the facts observed in this real scale experiment, while not very well controlled, enabled us to identify some trends. The most important is that a clear perception of water consumption by a managing team and also the interested consumers of the building positively influences the adoption of saving practices in a scale that appears to be more important that the simple substitution of hydraulic devices for water saving ones.

Another result from this experiment confirms a previous observation in industrial installations [37]. There is a positive correlation between water consumption and the stability of the consumption pattern. The more unstable consumption is, the greater the amount of water consumed. This seems to show that controlled systems tend to be more stable and therefore have fewer losses, leading to lower water consumption. If water consumption is controlled and publicized, consumer awareness may be raised leading to water savings. A high variability of daily consumption is, therefore, a good indicator that conservation action is needed and that significant savings can be achieved.

Building awareness for water conservation in non-residential buildings is a very important task for a conservation programme and its maintenance. Therefore, controlling water consumption cannot be directed just at the operators of the building, it has to spread across the community of users. The frequency of monitoring influences the performance of water conservation practices. Monthly checking of water consumption is indicated in California's Best Management Practices for Water Use in California State Government Facilities [38].

However, California is very different from most other places in the world and, even more different from the Brazilian northeast and other developing regions. Its *per capita* income is 15 times more and its technological capability is perhaps among the highest in the world. Therefore, a higher quality of water piping and devices and the maintenance systems can be expected to be in place there, directly influencing the occurrence of leaks. Meanwhile, monthly control clearly will not be enough in Bahia's reality. In our programmes we adopt a daily control, which has proved successful, allowing significant water savings even when other measures could not be implemented.

Daily water monitoring is used to foster awareness across managers and users to show how the water consumption of the building responds to different stimuli from regular events to leaks. Developing water awareness involves getting administrators to dedicate about a minute a day to this subject, to record and publicize water consumption, making the consumption data and control measures adopted transparent. It also includes a comparison performance, not just by administrators but also by others that may be interested.

To raise, develop and keep this water consumption awareness alive, it is recommended to empower a specific group. This is the main objective of what we have called the Eco-team. This team is also responsible for performing activities such as small repairs, inventorying the building's water system, communicating with peers and monitoring more complex maintenance. The Eco-team can also be the basic cell to foster a strategy that, starting from raising awareness on water consumption, may move towards building a more widespread and advanced environmental consciousness on the premises. Water saving is a concrete and observable problem to begin with, it can be easily linked to environmental and economic issues and appears to be a sympathetic cause.

The experience of our research group has found that consumption monitoring, on a daily basis, is not only a main factor, but is the first step to control and rationalization of water consumption. Other measures come after the system is controlled. Adopting them in a non-controlled building is a waste of money and time. Initial gains will rapidly recede. A controlled situation, at least in terms of being aware of how water is consumed, by itself helps to reduce water waste and losses.

CONCEPTUAL BASIS FOR WATER USE RATIONALIZATION IN NON RESIDENTIAL BUILDINGS

A conceptual basis for the water use rationalization programmes was developed to inform the user community and start to organise the work. It can be seen in Fig. (**2**) and includes: (a) Implementation of control measures (b) Understanding why and how water is consumed by users (effective consumption) (c) Perceiving the reasons behind water waste by users, and (d) Identifying what causes losses in piping and devices. Only after understanding all these aspects and considering action to face them, attention must be given to what we call the environmental quality of the building (e). In the latter we address the means to use alternative water sources which result in a more favourable energy/water relationship or lower environmental and social impact per unit of water used. However, action related to this last measure also requires the investment of financial resources. Since activities related to the first four topics do not require significant resources and will result in money savings, they can finance further work and the construction of segregated pipework, water reuse installations and rainwater catchment. Nevertheless, it is not always easy to do this in public institutions because they have to abide by strict and rigid budgetary rules to guarantee transparency rather than effectiveness and results.

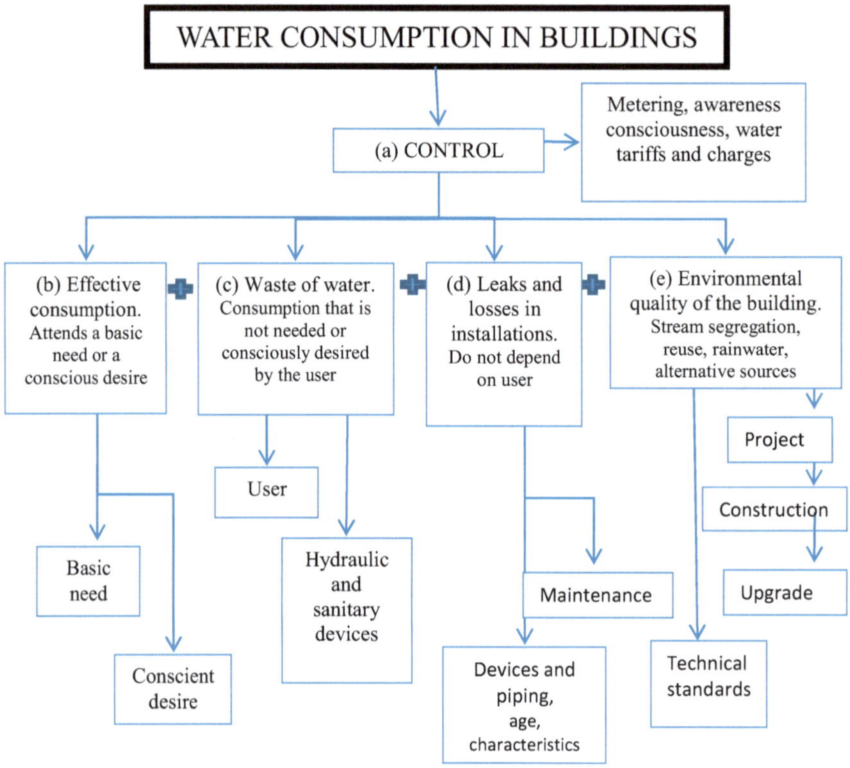

Fig. (2). Conceptual approach to water rationalization in buildings.

(a) Implementation of Control Measures

Controlling water use in premises open to a wide public requires a different approach from that usually applied in industrial premises, residences or even administrative offices. Control requires metering but also transmitting this information to users, to influence personal decisions on how much water he or she really needs or wants to use. If information about specific water consumption reaches users, decisions on how much water will be used depends either on their environmental consciousness or how much money they will have to pay for it.

Consciousness raising on environmental issues is widely covered in the literature. Our group's experience highlights the importance of a coherent behaviour on the institution side. Explicit examples of an institutional lack of commitment with water or energy savings will fail to encourage users to take greater care with more rational practices. Examples are water fountains spreading water around, unnecessary excessive lightning, etc. Explicit waste of natural resources is even sometimes used as a marketing strategy for certain luxury brands, however, to incur in these should be a matter of careful thought and decision taking by administrators of public premises and universities.

Charging for water in premises used by a wide public is very difficult. To charge for water use it has to be measured as close as possible to the user through a sub-metering system. Individual metering and charges in a university where there are research laboratories, restaurants, snacks bars, as well as car washing found fairly widespread in several Latin American countries can be implemented however, this is not always possible. Some of these services are outsourced and legislation may impose restrictions on charging for water consumption by institutions other than the municipal water supplier.

(b) Understanding Why and How Users (Effective Consumption) Consume Water

Another point considered in this conceptual approach is the effective consumption or proper use of water. It refers to water used to properly fulfil a clear objective. The objective can be to attend basic needs but also other uses consciously desired by the consumer. Using a regular faucet to clean teeth for example can use 100 ml of water, but a careless user may use more than a litre. The installation of automatic faucets that, if properly regulated, can release more than 200 ml of water for each time they are pressed can reduce water consumption by at least 400 ml. It is difficult to imagine that someone enjoys seeing water running from the tap as a conscious desire. The image and sound of running water may induce pleasing memories among some users but should the whole community pay for this? Are there other reasons to understand the use of decorative fountains?

The activity of taking a shower can better help us to understand this point. The time needed to take a shower to fulfil hygienic purposes may be 2 to 3 minutes. However, we all take a showers or baths to relax or for pleasure. Here there is a conscious desire, which can be considered a proper use of water. On the other hand, the shower tap may be left open while other activities are undertaken, *e.g.* soaping hair and water will be wasted needlessly.

Consider the use of either a regular toilet or a urinal for urinating. With a properly regulated device, using a water-flushing urinal a discharge can be made with 250 mL [39, 40]. If users choose to use a WC they will consume from 6 to 20 L per use. One may ask why a user will prefer a WC to urinate if he has a urinal at his disposal. Firstly urinals are not always available in public toilets. We should ask architects or even regulators why it is so. A survey run for several months at the

Federal University of Bahia showed that 16% of the toilets customers used a WC, instead of available urinals, for urinating. Further research performed with users in the International Airport of Salvador, confirmed these numbers and clarified that the main reason for this behaviour was the need for privacy [40]. For a number of users, privacy while urinating is a basic need. For others it is not but despite this they used the WC. One of the reasons behind this need for privacy is Paruresis. Also called Shy Bladder disorder, it is a social phobia that the International Paruresis Association claims to affect 6.6% of the population and may prevent people from using an open urinal or even public toilets [41]. Not many designers or engineers are aware of Paruresis but if public toilets are properly designed, users may be diverted from using WCs to use urinals. This can lead to very significant reduction in water consumption [40].

(c) Perceiving the Reasons behind Water Waste by Users

Aspect (c) of Fig. (**2**) refers to water being wasted directly by users and this includes all consumption that does not address basic needs or conscious desires, which are provoked by the users in contact with a sanitary device. These wastes may be caused either by the user lack of awareness as cited above, or induced by the characteristics of sanitary device and its maintenance conditions. Indeed, it is quite common in non-residential premises, to see water wastes provoked by improper adjustment of equipment. Results shown in Table **2** illustrate this. Two measurement campaigns were performed in July and November 2008 in the toilets of the International Airport of Salvador. Researchers observed a very significant range of discharge volumes in the toilets devices. For the urinals these varied from 50 mL to 2.5 L per discharge, with an average value of 0.9 L and 1.1 L per discharge respectively for each campaign. As manufacturers' indications and even technical rules usually indicate values between 350 mL and 1.2 L [42], inquiries were conducted with employees responsible for the maintenance and hygiene of the toilets to check if the urinals working with very low discharge volumes provoked any kind of annoyance to users, such as bad smell. No evidence of this kind was reported. Kammers and Ghisi [24] reported that according to manufacturers' information, urinals should discharge 7 L per use, which indeed appears to be more than necessary. On the other hand, average volumes released by self-closing faucets in hand washing sinks were 0.71 L. Similar numbers for the discharge volume for sinks were reported by Alves *et al.* [43]. Values indicated in technical norm ABNT NBR 13.713/96 vary from 0.2 to 1.0 L per discharge [42]. This shows that reasonable savings can be obtained just by regular adjustments.

These findings led the team to recommend and to establish routines to adjust hydraulic devices and to promote the use of urinals in the male toilet facilities.

The low volumes of water per urinal use may make redundant the need for waterless urinals using vegetal oils as sealing fluid. What indeed was found to be crucial in water savings in public toilets was diverting users from urinating in WCs to using well-adjusted urinals. In this sense, redesign of toilet facilities to favour the use of urinals instead of WCs is an effective measure in water use reduction.

Table 2. Water consumption in sanitary devices at the International Airport of Salvador, Bahia [40].

Parameters	1st Campaign	2nd Campaign
	10 & 18 July/08	21 & 28 Nov/08
Number of toilet rooms surveyed	10	24
N° urinals with optical sensor	22	54
N° lavatories with automatic (presure) faucets.	23	77
Average volume of water per discharge: urinals (L)	0.9	1.1
Average volume of water per discharge: lavatories (L)	0.71	0.71
Average time per discharge: urinals (s)	6	5
Average time per discharge: lavatories (s)	5	5
Average flow rate: urinals (L/s)	0.15	0.21
Average flow rate: lavatories (L/s)	0.14	0.14

(d) Identifying What Provokes Losses in Piping and Devices

The fourth aspect (d) considers the reasons that lead to water losses in piping and storage tanks. These losses depend on the characteristics of the building's water system and the maintenance thereof. The main characteristics that influence water losses are: the age and quality of the parts, the static pressure of the network and the quality and location of piping and tanks. In the cases studied by the author, there were several examples of underground tanks affected by the action of tree roots and piping running above ground subject to all kinds interference from traffic or animals for example. Many Brazilian water companies require their customers to build underground storage tanks to receive water from the mains. These are common and used even in sectors where water pressure is sufficient to take water directly to upper tanks.

As-built plans or other technical records are rarely available and this makes the identification of underground leaks more difficult. Often acoustic underground leak detectors and trained personnel need to be deployed and significant leakages occur for many days or months without being noticed. Follow-up of daily consumption by a trained observer may lead to the identification of new leakages.

However, especially if no trained observers exist, it is difficult to identify the existence of invisible leakages.

The quality of the maintenance structure directly influences the level of wasted water from toilet facilities and leaks in pipes. For a given hydraulic system wastes and losses related to equipment are depicted in Fig. (**3**). However, there is also a trade-off between acquisition and installation costs and maintenance costs. Better quality devices cost more but may require less frequent maintenance.

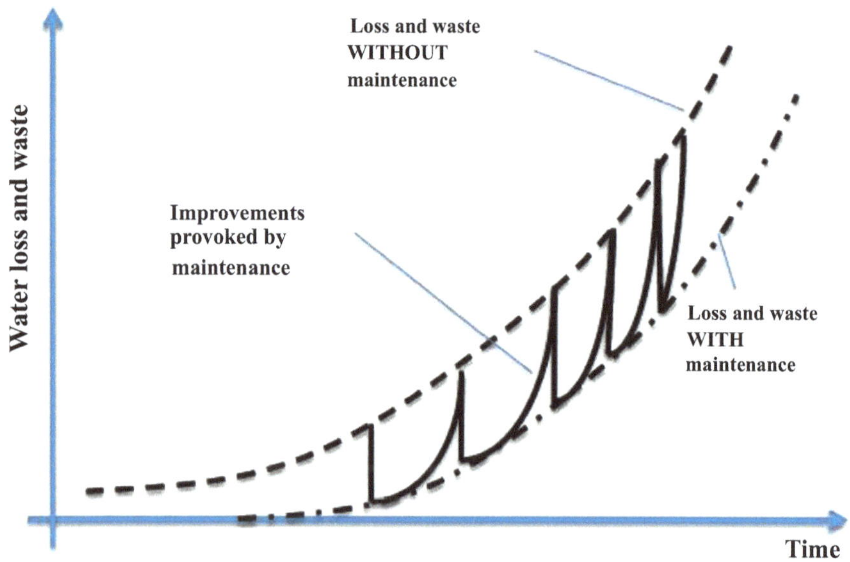

Fig. (3). Influence of maintenance in the reduction of water waste and loss (based on Lichentstein, 1985, *apud* [31]).

There are no indicators of what a proper consumption of a building should be and operators frequently confuse design parameters with operating indicators. For school installations, Ilha *et al.* [23] reviewed Brazilian and international technical literature and reported consumption indicators that vary from less than a litre to more than 500 L per person per day. However, in a survey with 12 municipal schools in the city of Campinas, São Paulo, they measured daily consumptions per student per day at between 5 and 50 litres. Numbers used for designing installations in schools are similar to those considered in universities where the number of students present in a building is difficult to assess. Experience shows that *per capita* consumption in this type of premises can easily be maintained below 20 or even 10 L per day.

Design parameters used as references for water consumption can mislead operators. Technical standards suggest considering a daily demand of 50 L per student. However, extra daily volumes can be added to this value, to attend specific demands. Designers have to insert additional safety factors to attend special or even unexpected situations. Furthermore, installations should be calculated to attend the expected conditions of the building once the project is finished and not reference values from the designing period.

Lack of adequate maintenance is also a problem, especially in public institutions. It mostly takes the form of corrective maintenance to eliminate visible leaks and is usually carried out by under-trained personnel and with almost no planning and after meaningful delays between identification of a leak and repair. Preventive and predictive maintenance rarely occur. Routine maintenance activities should include periodic regulation of hydraulic and sanitary devices and screening external and internal water networks to eliminate leaks. A complete diagnosis of the situation of hydraulic networks and a planned intervention programme is rarely planned and allocated a budget.

In their survey with state schools in São Paulo, Ilha *et al.* [23] identified the lack of maintenance and its inefficiency as reasons for a wasteful use of water, together with lack of awareness and tariffs not being directly paid by users. Vickers [16] analysed maintenance action in a chapter dedicated to industrial, commercial and institutional buildings and facilities. She illustrated the subject with a rich compilation of case studies addressing a wide range of institutions and emphasized the critical role of maintenance practices for the efficient use of water in old and even new facilities.

The ability to identify an abnormal event and to correct it depends on how the water system is monitored and by whom. This will depend on the importance the institution gives to the problem, its level of awareness and how much effort can or will to be invested.

(e) Environmental Quality of the Building

Actions to substitute water sources (aspect e) will only be fully effective if previous aspects (a to d) are considered. Substituting water from the mains with water from other alternative sources in a system without previous measures for water consumption reduction, leads to higher spending with poor results. However, it can offer significant financial and environmental savings if carefully studied and designed. Among these are: Direct capture of rainwater, exploitation of local aquifers and water reuse. In this case, responsibility for water treatment to assure drinking water standards will have to be assumed by the building operators. Institutions that perform low levels of operational and maintenance practices have to carefully analyse if they will be able to produce good quality water from alternative sources.

To allow higher levels of water reuse it is interesting to consider the segregation of water flows according to their characteristics. Pipework segregation in buildings includes both supply flow segregation into drinking and non-drinking water pipelines and greywater, brown water and yellow water for effluents. A deeper discussion on this topic can be found in the technical literature [44]. Lack of appropriate technical norms makes it more difficult to implement these options and they mostly tend to be considered when designing new buildings.

A PROGRAMME TO PROMOTE RATIONAL USE OF WATER IN PUBLIC ADMINISTRATIVE BUILDINGS: SOME USEFUL TOOLS

To design a policy to promote a more rational use of natural resources in public buildings several aspects have to be taken into consideration. It is often stated that it has to begin with the full commitment of the higher administration. However, administrations change and there is usually a huge gap between rhetoric and practice. Is the *top down* approach the only one to be considered? Are *bottom up* approaches effective? Should emphasis be given to technological advances or to behavioural changes? How do we choose the most adequate strategy to follow? We have discussed this for university buildings elsewhere [35, 45]. Decisions have to be made according to the characteristics and conditions available at every specific moment, each institution and building and the capabilities of the group who is directing the initiative. Sometimes, "genius is simply a fresh look at the obvious", as Amy Vickers quotes as an old saying [16].

Environmental issues involving a more rational use of natural resources, such as water, are rarely among the objectives of institutions, though it may be mentioned in official statements. When it comes to human or financial resource allocation, the central objectives of the institution always come first. So how can a programme to rationalise water consumption in university buildings and other public premises be designed and implemented? Can this get on the agenda at the level of the main objectives of the institution?

When working with universities and governmental institutions one must take into account that the environmental cause is an important part of their objectives. Almost every undergraduate or graduate course includes the environmental issue either as a subject or, at least, as a part of it. Environmental public policies and governmental environmental systems demand all state institutions to support sustainability action. If these commitments are taken seriously, these institutions will be asked to lead by example. Water and energy waste are sensitive subjects, and the institutions can be charged for a coherent behaviour in this respect.

A careful approach has to consider a step-by-step action involving awareness building (bottom up), higher administration commitment (top down), users' behaviour, technical responses and in particular good maintenance practices. All of these are expected to evolve and gradually achieve higher levels of performance that, in the longer term, could be able to respond to the challenges of the environmental crisis. In the case of poorly managed institutions it is very important to plan the involvement of intermediate managers and staff. A step-by-step process is needed to achieve results gradually and, when properly communicated, may foster further involvements from the parts.

The experiences presented here were built upon two programmes and one project developed and implemented with the partnership of Teclim, a research group of the School of Engineering of the Federal University of Bahia (UFBA) over the last 17 years. Some of our results are summarized in the following section and have been published elsewhere [27, 28, 35, 39, 40, 45, 46].

Aguapura started timidly in 1999 as an initiative of three lecturers at the school of engineering at UFBA. Initially, a small maintenance team specifically dedicated to combating leaks was set up. The group began analysing water bills, available from the university accounts department, and identified the consumers on the various campuses. As the main campus of UFBA is located in a central neighbourhood in Salvador, its buildings are connected to the local water company's mains and almost all have individual metering devices. Leaks, lasting for months, were identified particularly in underground tanks and pipes. Some leaks were so intense that they could be seen from the ground. The average monthly water consumption of the campus in Salvador, between 1999 and 2000 was 34,600 m^3 resulting in a *per capita* consumption of 43 litres per person per day (considering teaching staff, other university employees and registered students). In 2015 on the other hand, monthly consumption reached 21,400 m^3. Due to a federal expansion programme launched in 2009 the university experienced a rapid growth. As consumption kept its falling pattern a *per capita* consumption of 16 litres per person per day was achieved by 2015. We further comment on this results at the end of this chapter.

The programme, however, was not fully incorporated by the University and still relies on the research group [35]. It is clear that the results, although significant, could be much better. In addition to repairs to easily identifiable big leakages, a publicly accessible monitoring system was implemented. This system, Aguapura Vianet, has been fed by employees of each school and monitored by undergraduate students. Some devices have been substituted but maintenance has been poor over the years.

In 2007, as a consequence of results obtained by Aguapura in UFBA, the State Government of Bahia, through its Secretary for Management, SAEB, asked us to implement a similar programme in their administrative buildings. This programme ran between 2008 and 2015 with some intervals in between, leading also to significant results. Water savings of 50% were obtained producing financial savings of 7.7 million US dollars at 2014 exchange rates, for the whole period. For this programme a team of 4 professionals (engineers and an administrator) were contracted and some undergraduate students were given scholarships to work in it. With this workforce more efficient action could be taken including training civil servants, writing administrative procedures, and improving the monitoring system. Again, significant results were obtained despite the fact that not all the activities proposed were implemented. One of the results of this programme was more detailed planning that is presented later as a methodological tool for future works.

The third experience is a research project developed at the International Airport of Salvador between 2007 and 2009. After a thorough analysis of water consumption on the airport premises, several proposals were made, including a simulation of water savings that could be reached with their full implementation. The project identified that 40% of water consumed was from toilets and that a 77% reduction in consumption could be achieved by simple measures not involving significant investments. Proper design of toilets, for example, giving more privacy to urinals and including devices for women's use, may lead to substantial water savings.

Key Action: Daily Monitoring, Public Awareness and Rapid Responses, Making the Internal Public Aware of their Water Consumption

The reduction in *per capita* water consumption in UFBA form 40 to 18 L/per day, and more importantly sustaining it over the years, may be attributed to the very basic control of leaks and the implementation of a monitoring system based on daily data exposure. As simple and obvious as this seems, this is a fundamental step to any demand management programme.

Buildings should be aware of their water consumption and that is nothing more than managers + operators + user awareness. The usual approach in water saving programmes relies on "educational campaigns", consisting of a sequence of presentations with minimal results and that are discontinued after a short time due to the lack of financing resources. We understand that there are some cases of success, but this has not been our experience. Moreover, it is common to blame weak results to the public's low level of environmental education or consciousness.

Alternatively, communication technology based on automation and smart metering have been proposed and implemented. It is necessary to better assess the results of this choice and when and how it can be implemented in low-tech premises. In our programmes, we have been adopting a monitoring and communication system, called Aguapura vianet (*PURA* stands for programme for the rational use of water). It was developed by Teclim programmers and researchers for the web environment and has been available in Portuguese at www.teclim.ufba.br/aguapura. It includes public and internal interfaces integrating centralized and decentralized monitoring.

Aguapura Vianet includes the following functions:

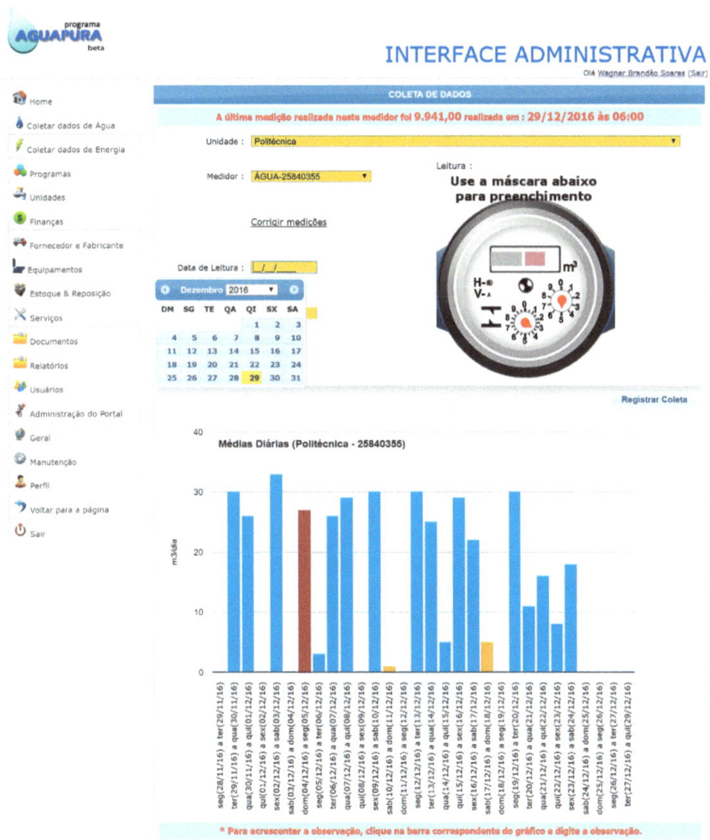

Fig. (4). Interface for daily insertion of data. The previous month's daily data histogram is shown to identify possible undesired events. Yellow bars identify weekends when consumption should be significantly lower. Failure to be so indicates a strong possibility of unidentified leaks.

Entering Consumption Data

This is the responsibility of each building or unit's leader. He or she selects an employee, usually a security guard, to read the metering device once a day, always at the same time. The number is put into the system by a person closer to the administrator and trained to correctly interpret any undesired events but also able to initiate action, if required. Automation of this process was considered and discarded for two reasons. The first, and more importantly, is that this manual action has been useful to gauge the level of awareness of the buildings administration and helping to develop a connection between the research group and each unit. The second is that often the maintenance culture is so low that more complex technology has to be carefully considered. Visual reading and manual insertion of data does not require additional work and takes less than 5 seconds per day to be done if properly implemented. Fig. (**4**) presents the user interface showing the places where data is entered and the immediate return screen that facilitates identification of possible anomalies.

Exposure of Consumption Data

Consumption data is shown on the site in a sequence of pages. The first page is shown in Fig. (**5**). It allows those interested in knowing the consumption of a specific building to access information and compare the performance of all units of the institution in terms of average consumption of the month and previous months. These two comparisons are the basis for forming an opinion about the pattern of consumption of the buildings. Clicking on the name of the unit takes the visitor to more detailed information about the unit he/she is interested in (Fig. **6**).

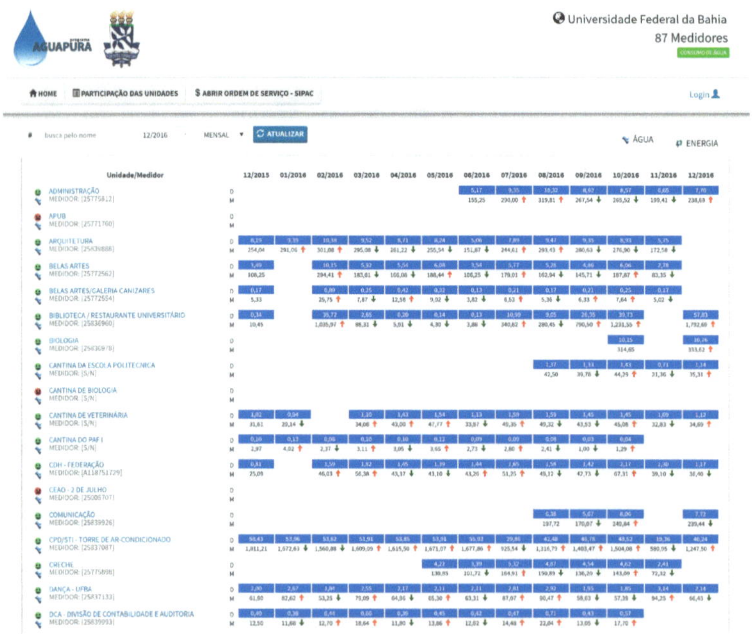

Fig. (5). Partial view o first page listing units of UFBA and general data for the last 12 months.

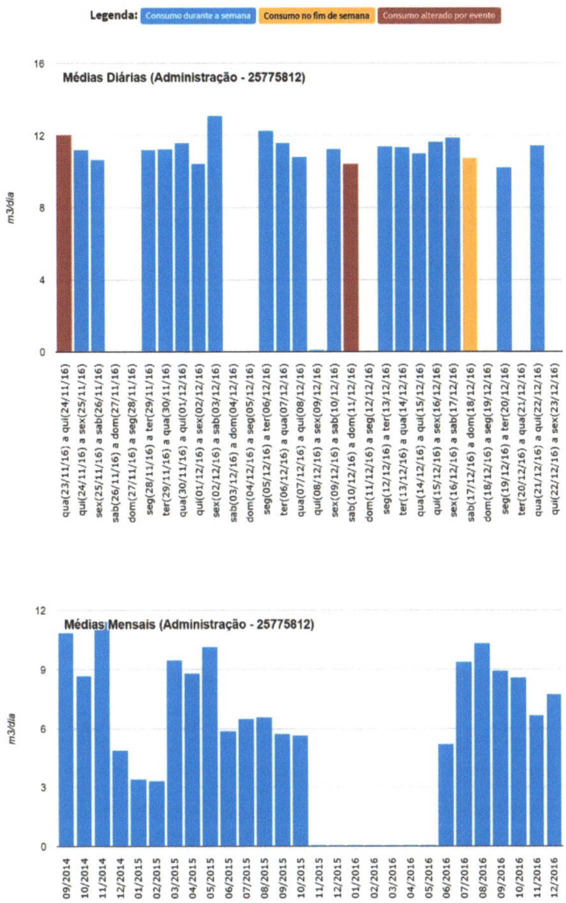

Fig. (6). Consumption shown on daily and monthly basis. Both graphs can be expanded to show the 24 last months.

At this point it is convenient to state that during the programme's life, it has not been possible to consolidate a team in the university administration to take over responsibility for the programme. It still depends on the research group. This reinforces the need to try to build widespread awareness on resource savings to sustain a more responsible attitude within the institution towards the environment.

Inserting a Comment to Explain Possible Causes of Significant Events

The interfaces shown in Fig. (**6**) are used to analyse consumption in the building and prompt the seeking of possible causes of wastes or losses. A significant alteration in the histogram should be accompanied by an observation of the likely causes, to be entered by the person responsible for the building. If an observation is inserted, the specific day bar will turn red. By passing the mouse over the bar, the comment will be made visible to the visitor. A log of registered observations promotes the creation of a culture of causes of water losses. This history will remain in the units and on Aguapura pages and be available to the public and other operators. It also provides useful material for training operators and for undergraduate classes.

Spreading Information

The system allows the user to easily communicate information about any events s/he considers relevant by email. The consumption histogram and observations will automatically be attached to the message. This tool allows communication not only to the employee responsible for taking action on a leak, for example, but also to higher ranked officials and the wider community. Any delays in action taking will be transparent to the community. This tool is mostly used by research students monitoring the behaviour of the unit, asking the teams to explain what could have possibly caused a specific consumption behaviour of a unit.

Registration of Consumption Devices and Plumbing

A database of all hydraulic devices, organized according to their location, is available on each units' page, with a code. In this database the characteristics of the plumbing, pipes, taps and sanitary equipment including their flow rates and adjusted times of operation are registered. Repair orders should also be inserted there. With this information maintenance can be more efficient as each device has a background history. This database can also inform replacement and buying procedures.

Other Functions

More than 20 other functions are available in the system, including finance (water bills), pictures describing the unit, its hydraulic installations, location of the metering device and how to read it and enter data to the system. A table with all the consumption data is also available and can be transferred to a spread sheet for further statistical analysis. Furthermore, maintenance services with a description of what was done, by whom and other relevant issues, reports, information about suppliers and manufacturers, are also functions of the program.

The system is also designed to be a repository of all relevant data related to the water and energy systems of the building and their behaviour.

Experience has been gained and spread motivating units to manage their water consumption using Aguapura Vianet. It is a strong tool used to underpin the water saving programmes and it has been used in our other programme and project presented below.

Planning a Programme for Water Saving in Buildings of Public Use

When invited to implement a full programme for water saving in the administrative buildings of the Government of Bahia, Teclim/UFBA initially proposed a pilot project, lasting 8 months, to verify how the administrative machine would react to the initiative. While analysing information related to water consumption, a monitoring scheme, using Aguapura Vianet, was started and operated, including training employees from 17 buildings in the Administrative Centre of Bahia. This pilot phase delivered significant physical and financial savings and provided the practical experience to propose the full programme, detailed below. Even though only part of the proposed activities were fully or partially implemented, the plan proposed was assessed and we found that it provides a strong basis for future developments of this type. A more complete report and analysis of action taken is available in Silva *et al.* [28]. An aerial image of the Administrative Centre of Bahia is shown in Fig. (**7**).

Fig. (7). Administrative Centre of Bahia, Salvador [Google maps].

The programme was made up of 5 components:

a. Definition of means for the programme management
b. Preparation of general water management procedures
c. General intervention
d. Local interventions
e. Follow up using Aguapura Vianet

Definition of Means for the Programme Management

As the programme was intended to train the Secretary of Administration to gradually assume full control, its management had to involve staff and authorities from the state secretaries. The initial proposal was submitted and adapted to take into consideration the opinions of the partners. During this stage, the technical team was recruited and gradually incorporated into the programme. The technical team followed a previously defined schedule of meetings with the technicians of the Secretary of Administration and other invited officials. An advisory Council was proposed to meet on a regular basis to formulate resource saving policies and directives, coordinate and articulate the programme's actions with the state's administrative machinery. It should have included strategically indicated officials representing the state secretaries. This council did not function, showing that involving higher ranking officials is not a trivial task, even if there is an order from higher ranks in place. During the programme 14 reports were delivered containing full documentation of each of the actions taken and documents produced.

Formulation of General Management Procedures

The second part of the programme dealt with managerial and technical propositions, to be included in the general information and procedures of the government, such as: Diagnosis of procedures related to maintenance, contracting services, public spending, general characteristics of the building; Designing the water management proposals; Drawing up the water control measures; Tools for publicizing information, awareness-building among the public and users. Adaptation, development and insertion of the water monitoring system (Aguapura Vianet) in the state's structure; Tools for the monitoring and assessment of administrative and technical advances achieved in each department and building, such as shown in Fig. (**8**).

General Intervention

Includes actions implemented or to be implemented for all secretaries and public buildings, such as: Organization of maintenance services including water saving initiatives; Identification of the building and user profiles; Drawing up of maintenance routines and procedures; Publicizing of the programme; Creation and training of the teams responsible for developing the programme and insertion of Aguapura Vianet procedures and routines in each department and building; Carrying out daily water consumption monitoring and supervision; Signalling of hydraulic and sanitary devices and rooms; Maintenance and savings; Definition of intervention priorities and blocks for action; Complementary projects such as studies for the installation of photovoltaic panels, rainwater harvesting, assessment of efficiency of hydraulic devices and identification of better alternatives to those used; General presentation events to spread knowledge about the programme. An important action that could not be implemented was contracting a company to assume maintenance services. Internal staff available in most buildings were not able to carry out adequate maintenance practices. Several regulatory barriers impeded the secretary for administration from accomplishing this.

Local Interventions

These refer to actions inside each department or building. Significant differences were found between units in terms of internal organization and interest in the programme, the operating conditions of sanitary devices, and the general maintenance of the building.

Local interventions were divided in three lines of action: Managerial aspects, physical aspects and behavioural aspects.

The managerial line of action included sensitizing leaders, authorities and the general public and consolidating an internal group to lead action towards water savings. This team was the so-called eco-team and one was created for each building. Advances in the programme in each unit were presented and shown using graphics. Fig. (**8**) illustrates the performance of the institutions in relation to: Level of internal institutionalization of the programme; Mobilization action; Training routines; Information System; Communication action; and Maintenance performance.

Other actions considered in the managerial line were: Definition and recommendation of specific norms and procedures; Definition of present performance levels and ones to be achieved, using water saving indicators and; Negotiating goals with representatives of the secretaries and defining commitments.

In relation to daily monitoring of local consumption, each unit should commit to: daily readings of metering devices, daily insertion of data in Aguapura Vianet, and interpretation of possible events responsible for the verified consumption. The researchers and students gave support to local officials of the eco-teams to interpret data and identify possible leaking events. A repository for harvesting ideas from local users was made available but did not have much success.

The physical aspects proposed could not be implemented because of the lack of specialized technicians in the buildings and failure to contract these services.

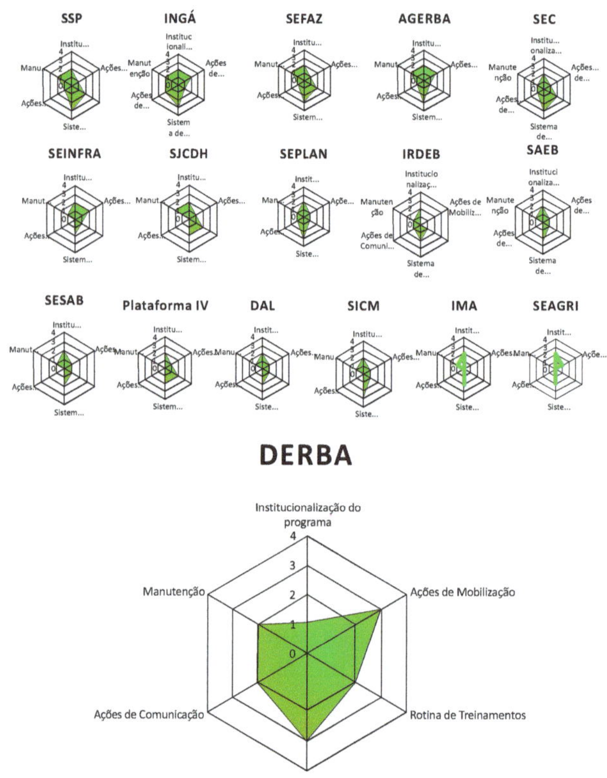

Fig. (8). Graph representation of progress of the program in in governmental buildings [47].

The physical aspects proposed could not be implemented because of the lack of specialized technicians in the buildings and failure to contract these services.

Programmed activities were:

1. General maintenance services of hydraulic and sanitary devices, including water tank piping.
2. Regular inspection and building surveys looking for invisible leaks and regulating sanitary devices.

3. Simulation of potential water savings according to the measures implemented. See next Item.
4. Organization of emergency action for significant leaks.
5. Planning and implementation of sectored sub-metering and daily insertion of data
6. Substitution of hydro-sanitary devices for water saving options
7. Reforming of buildings and installations
8. Conceptual design of solutions to use alternative water sources

The technical team developed detailed procedures for items (1) and (2), based on their former experience at UFBA. This information is available in Portuguese at www.teclim.ufba.br/aguapura. These procedures serve to guide maintenance teams during periodic inspections. An important action, when proceeding with regular inspections, is to adjust water flow rates of all devices to pre-defined values. In the cases of UFBA and the international Airport of Salvador, devices had a wide variation of flow rates as discussed previously.

After controlling water use, more expensive refurbishment can be done, such as promoting stream segregation leading to water reuse or rainwater harvesting. Most buildings in Brazil of over 40 years of age were built with very large underground water reservoirs. Operation irregularities, at that time, forced designers to include underground tanks because of the frequent occurrence of low water pressures in the distribution network having large volumes, because of usual interruption of supply. Nowadays supply conditions have improved making many of these tanks unnecessary or even undesirable because of the occurrence of leakages. However, if recovered and well maintained they can be used for rainwater storage as well as for grey water treatment and storage, significantly reducing the implementation costs of these solutions. Thinking further into the future, reforms may include yellow water segregation for phosphorus and nitrogen uptake.

The third line of action concerning local interventions refers to behavioural aspects. That is, action to grab the public's attention to the way they use water.

The following steps where planned:

1. A survey to understand user profiles, how much time they spent in the building, how they use water, what their understanding was in relation to the sources of water that supply the city, among others.
2. Designing a permanent communication plan.
3. Action to maintain permanent awareness and following up of the water consumption in the building.
4. Keeping the public informed about the Eco-teams activities.

5. Installation, maintenance and conservation of signalling of water devices, ambiences and equipment.

Consumption Monitoring using Aguapura Vianet

Aguapura Vianet monitoring system was used in the State administrative buildings after being adapted to reflect the hierarchy of units within each secretary. The idea was that buildings, departments, sectors and so on, could aggregate consumption data accordingly to each institution's organogram. This system was introduced not only to the buildings in the Administrative Centre of Bahia where the head of the secretaries are. Buildings where intermediary and final activities are developed, such as state schools and universities, were linked in each secretary page in Aguapura Vianet, even if further action was not intended in them at that stage of the programme.

Understanding How Users Behave in Public Spaces in Relation to Water Consumption

Surveys were conducted with passengers at the International Airport of Salvador and students, lecturers and employees at UFBA on various occasions to understand how the public uses hydraulic and sanitary devices on these kinds of premises (Fig. **9**).

In June 2008, 473 questionnaires where applied to passengers in the departing area of the airport, 44% to women and 56% to men. The experiment sample was defined for a 5% margin of error [39].

Fig. (9). Survey about habits related to water consumption in public toilets with 473 airport passengers [39].

Results showed that 69% of male passengers used the toilet during that occasion: 67% used them once, 29% twice and 4% three or more times. Men used the toilet to urinate and wash their hands in 84% of the cases, in 5% just to urinate, 3% to defecate and wash hands and 8% only to wash hands. If considering only excretion activities, 97% referred to urination!

Kammers and Ghisi [24] found that WCs and urinals represent 72.1% of end-uses in government buildings in Santa Catarina, Brazil. They indicated the possibility of using rainwater and treated greywater to supply these end-uses, as they did not need potable water to function.

In our work we were interested in the reasons why a significant number of users do not use urinals and preferred the closed WCs in public male toilets, especially if water consumption in these may be tens of times higher, depending on their characteristics and discharge adjustments. In 15% of the cases male users chose WCs. This meant that 80% of WC use was for urinating, indicating that urination in WCs was the main cause of water consumption in airport toilets such as the one surveyed.

As this requirement for privacy was the main reason for such a waste of water, we asked the public we interviewed, to identify what kind of urinals posts they would prefer to use. Fig. (**10**) was shown to assist them in making their choices. Only 16% of the public indicated that they preferred urinal type 3, which was the one in place at the airport. More private urinals (1 and 2) where preferred by 84% of those interviewed.

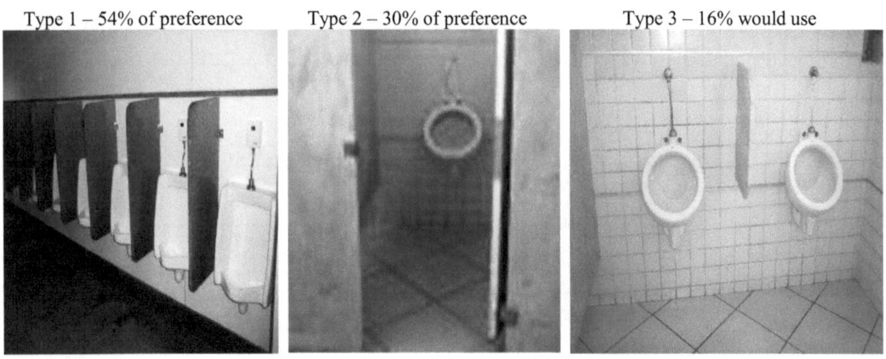

Fig. (10). Design preferences for using urinals [39].

Interviewing female passengers it was found that 68% of them used the toilet premises when departing from the airport. From these, 74% used them only once, 24% twice and 2% three or more times. Reason for using the toilets where, in 82% of the cases, for urinating and washing hands, 4% only urinating, 2% for defecation and washing hands and, 12% for only washing hands.

In relation to evacuation purposes, 98% of them referred to urinating and just 2% to defecation. These numbers indicate the importance of more serious thinking to the installation of female urinals. Women were asked if they would use female urinals if available in the premises. A picture of such a device was produced for them before answering because these devices are not known in Brazil. Women answered positively in 85% of the questionnaires.

Simulation of Results that can be Achieved with Specific Action, as Proposed at Salvador International Airport

When implementing a programme to reduce wastes and leaks in commercial and institutional premises, it is not easy to convince administrators put into action proposals made. To modify an administrative steady state requires significant effort and resistance should be expected from all levels of decision-making. To cope with these difficulties one strategy is to anticipate results that can be achieved and then propose a sequence of actions beginning with those that do not entail high costs.

At the International Airport of Salvador, after the diagnosis including reconciled estimates of water consumption by sectors, several propositions were made and their expected results were simulated to assess possible water savings [39].

Between 2006 and 2009, undertaking daily consumption control and improving maintenance practices slowly reduced consumption per passenger. Table **3** shows the changing consumption per passenger and also per user, including in this case companions bringing or receiving passengers as well as personnel working on the premises.

Table 3. Evolution of water consumption per passenger and per user at the Passenger Terminal of the International Airport of Salvador, 2006-2009 [39].

Period	Passengers/Day (a)	Passengers' Companions/Day (b)	Employees /Day (c)	All ** users/ Day (a)+(b)+(c)	Water Consumption (Litres/day)	Water Consumption per Passenger (Litres/day)	Water Consumption per User** (Litres/day)
Feb to Dec/2006	14,673	17,607	5,020	37,300	352,746	24.1	9.5
Jan to Dec/2007	16,269	19,523	5,667	41,459	376,692	23.3	9.1
Jan to Dec/2008	16,468	19,762	5,895	42,125	411,012	25.1	9.8

(Table 3) contd.....

Period	Passengers/Day (a)	Passengers' Companions/Day (b)	Employees /Day (c)	All ** users/ Day (a)+(b)+(c)	Water Consumption (Litres/day)	Water Consumption per Passenger (Litres/day)	Water Consumption per User** (Litres/day)
Jan to Dec/2009	19,309	23,171	6,586	49,067	429,353	22.3	8.8

* Average companions per passenger was considered to be 1.2 people. It was based in a survey conducted by the Brazilian Company for Airport Infrastructure, INFRAERO, in several airports in March, 2007. ** The term *Users* here refers to a simple sum of passengers, companions and personnel who worked on the premises.

The passenger terminal toilets represented 47.5% of the total consumption of the passenger terminal which itself represented 82% of the total consumption of the airport.

Actions proposed and considered in simulations were directed to the passenger terminal toilets and included substitution and adjustment of hydro-sanitary devices, refurbishing toilets to favour the use of urinals, reduction in the amount of water used to remove the urine and introduce female urinals; Water savings that could be obtained with these measures were calculated for the following scenarios:

Scenario 1: During the diagnosis phase, it was measured that self-closing device discharges in urinals vary from 50ml to 2.5 litres per action and from lavatory faucets from 200 mL to 2.5 L. We proposed adjusting both urinals and faucets to 250 mL per discharge. By doing so a 1.08 L and 0.71 L average reduction per discharge of urinals and faucets, respectively, could be achieved.

Scenario 2: As we saw above, urinating is by far the main reason for which toilet installations are used in airports. It was proposed to redesign the male toilets to make the use of urinals more attractive by increasing the number of urinals in relation to WCs, providing more privacy to their users and a better location to influence users' choice. It was considered that $4/5^{th}$ of users that formerly chose to urinate in the WCs, would be diverted to use urinals. This scenario results were added to results expected from Scenario 1.

Scenario 3: Accumulate former scenarios 1 and 2 and added the installation of dual discharges for WC's considering that $1/5^{th}$ of users would still prefer to use them for urinating.

Scenario 4: In this scenario possible savings related to urine removal means were extended to the women's toilets. It was simply considered to allocate to 50% of WCs the exclusive function of urine receiving and removal, by means of proper signalling and adjusting them to a 3 litres discharge.

Scenario 5: Considered further measures by extending urine removal to female premises by substituting 50% of WCs with female urinals adjusted to a 250 mL discharge.

Scenario 6: This repeats what was proposed for the third scenario and expanded this to women's toilets. That is, the remaining 50% of WCs, that were not substitute with urinals, were installed with dual discharges, for urine removal of female users WCs to use a 3L discharge. Table **4** summarizes the proposed scenarios.

Table 4. Proposed scenarios for water saving at the toilets in the passenger terminal at the Salvador International Airport (2009) [39].

Scenario Sc	Action Proposed
Original Situation	Represents the water consumption at the toilets of the passenger terminal before intervention, for 16,468 passengers per day (2008's average). Considers average water flows measured for urinals: 1,08 L/flow; taps: 0,71 L/flow. It was considered 2 activations of the tap for each use and 12 L per discharge for WCs.
Sc1	Reduction of urinal and tap flows to 250 mL per discharge.
Sc2	**Sc1** + displacement of 80% of uses of WCs for males urinating to urinals with 250 mL per use.
Sc3	**Sc2** + adjustment of half the WCs in male's toilets to 3 L per discharge, for urinating (with proper signalization).
Sc4	**Sc3** + adjustment of half the WCs in female's toilets to 250 mL per discharge, for urinating (with proper signalization).
Sc5	**Sc4** + installation of female urinals or adjustment of half the WCs in female's toilets to 250 mL per discharge, for urinating (with proper signalization).
Sc6	**Sc5** + replacement of half of the WCs in all toilets for dual discharge models considering 6 and 3 litres per discharge for defecation and urinating purposes, respectively.

The results obtained from the six simulations are shown in Table **5**. According to these simulations 77% of the water consumed in the airport toilets could be saved with very little investment if all proposed measure were implemented. These water savings could represent financial savings of US$ 300.000 per year at 2009 exchange rates. Measures could also be gradually implemented supporting further advances.

Use of Reconciliation Techniques to Estimate Consumption of Water by Sectors

Although the well-known quote "If You Can't Measure It, You Can't Manage It" may not be applicable to everything, it certainly does apply to water consumption savings. But how far should we go in implanting and operating metering systems in institutional and public buildings? It has been quite common to find one metering device per building but there are cases where individual metering does not exist and there is one single hydrometer for a whole campus or site. In some cases, sub-metering devices cover specific consumptions. This allows to understand how sectors participate in the overall consumption profile. Sector metering or sub-metering brings consumption data closer to managers and users empowering them to take more immediate action as well as to do better planning.

Table 5. Water savings that could be obtained at the toilets in the passenger terminal at the Salvador International Airport (2009) [39].

Consumption Reduction to be Obtained in the Toilets	Original Situation	Sc1	Sc2	Sc3	Sc4	Sc5	Sc6
Males (m^3/day)	67	42	22	20	20	20	17
Males, reduction (m^3/day)		25	45	47	47	47	51
Males, reduction (%)		37	68	71	71	71	75
Females (m^3/day)	148	137	137	137	89	58	33
Females, reduction (m^3/day)		11	11	11	59	90	115
Females, reduction (%)		7	7	7	40	61	77
All (m^3/day)	215	179	159	157	109	78	50
All, reduction (m^3/day)		36	56	58	106	137	165
All, reduction (%)		17	26	27	49	64	77
All, water bill reduction * (1000 US$/year)		66	102	106	194	250	301

* An exchange rate of R$ 1,958 per US$ was considered (average value of June 2009). Average tariff charged by the water company in June 2009 was US$ 5,0/m^3.

However, when sub-metering is not in place and there are no means to have it in a short time, the flow rates must be estimated to provide an initial basis for action and, if validated, guide sub-metering planning. To do so, it is necessary to assess the quality of these estimates to reduce errors. Reconciliation techniques can be useful in these cases.

Classically, data reconciliation is a technique used to distribute and reduce errors in a metering network. However, it has been successfully applied in situations where all available data is not necessarily obtained from metering devices and some of these are the best estimates one can get. It has been used in industrial water saving programmes providing essential support to water consumption knowledge and control action [40, 48, 49].

Data reconciliation is an optimization problem depicted by Equations (1) and (2) [40]:

$$\min_{V_R} \sum_{i=1}^{N} \frac{(V_{Ri} - V_{Mi})^2}{\sigma_i^2} \tag{1}$$

For the flows i = 1,..., N. Where: V_{Ri} = reconciled flow rates; V_{Mi} = measured flow rates; σ_i^2 = variance (related to the level of uncertainty of the metering device); in = incoming streams; out = outgoing streams; k = refers to control volumes; K = number of control volumes; l=refers to streams entering the control volume k; L = number of incoming streams for each control volume k; m= refers to streams leaving each control volume k; M = number of outgoing streams for each control volume k.

$$\text{s.t.} \sum_{i=1}^{L} (V_{Ri})_{in_k} - \sum_{m=1}^{M} (V_{Rm})_{out_k} = 0 \tag{2}$$

For k = 1....K control volumes or units.

Equation (1) is the objective function and looks for the minimum difference between measured and reconciled values. The second equation expresses restrictions that define the multidimensional space where minimum has to be found. Mass balances equations of flow rates, entering and leaving the water system and subsystems are the restrictions of the problem.

Redundancy in data is needed to run the model. When there is not sufficient data, instead of measured values we use mapped values, which can be either measured values or mapped ones. To each of these a quality of information value is assigned. These values qualify the mapped flow rates and are inversely proportional to the variance used in the above equations to describe uncertainty of the measurement devices or systems. Larger qualities of information are assigned to measured flow rates. As a result, equation (1) is re-written in the following way:

$$\min_{V_R} \sum_{i=1}^{N} \frac{QI_i^2}{V_{Mi}^2} (V_{Ri} - V_{Mi})^2 \tag{3}$$

Where Mi now represents the mapped flow rates and QIi the quality of information of each current. The values assigned to these coefficients are shown in Table **6**.

During the airport project 47 water balances were reconciled giving us a much better understanding of their flow rates. This information also guided the installation of new water meters in such a way that more cost effective measurement locations could be indicated for the improvement of the quality of information of the whole metering network.

Fig. (**11**) shows the water balance of the whole site of Salvador International Airport. The internal control volume refers to the passenger terminal. Ei are the streams entering the total site control volume and Si are the outgoing streams. QI is the quality of information assigned to each flow rate, either measured or mapped as indicated in Table **6**.

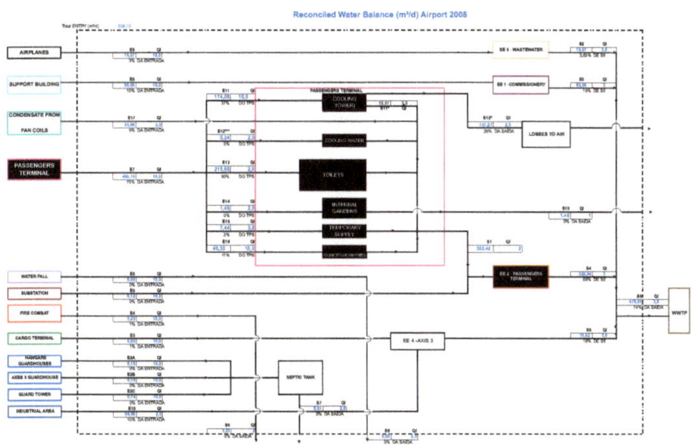

Fig. (11). Topology of the water balance at Salvador International Airport [39].

Table 6. Scale of values adopted for the quality of information [39].

QI (Quality of information)	Source of information
0.4 – Information with very low reliability	Rough estimative, without much consistency.
2.0 – Information with low reliability	Found in technical literature, old projects and simulations.
4.0 – Information with medium reliability	Reliable estimate obtained from field experiments, existing measurements and information obtained from experienced operators of the system.
10.0 – Information with high reliability	Data obtained from water meters. Was the best information available although obtained from devices that were not properly calibrated? Data was statistically analysed to remove spurious data

How to Define Water Consumption Indicators: the Equivalent Population Variable

Indicators of water consumption are important for a better understanding of a situation. They also help us to assign priorities for action to be taken and to enable comparison between buildings at a certain moment and to follow the evolution of the water conservation measures inside a building. A usual indicator in non-residential buildings is consumption in litres per user per day (L/p.d). In all three cases presented in this work the use of this indicator was not a straightforward task.

In the case of the Federal University of Bahia, none of the building registers the number of visitors entering the building per day. How long students stay in the buildings is very variable. Some stay the whole day as research students, some just go to a couple of lectures per day staying from two to four hours in the building.

To compare water consumption in university buildings one has to account for the different functions within the building such as the presence of laboratories, physical activities requiring showers, systems using cooling towers. For student's domestic demand, mainly related to water use in toilets, an indirect indicator has been developed. It defines an *equivalent consumer,* which is related to the number of hours of classes delivered in the building. The idea was to define the number of users equivalent to those staying for 8 hours in the building. Each 8 hours daily stay will then be considered a full user, comparable to employees' working day of 8 hours or a lecturer. To account for the presence of students the method considers the number of 2 hour classes delivered in the building per day and the number of students registered on the course.

For example, a course with 40 enrolled students, with four hours of class a day, will account for (40x2)/8 = 10 equivalent consumers. However, as normally students attend around 75% of all lectures, a correction factor of 0.75 is introduced making the number of equivalent persons of that class 7.5. That is, such a class is expected to provoke a water demand equivalent to 7.5 people working in the building in a normal working timetable. Courses delivered in the building, and the number of students registered in them, is easy information to access from academic databases. For the cases of research students, full time lecturers and administrative staff, each of them will count for one equivalent person. Sum of these 4 equivalent people will make up the denominator for the per-capita consumption indicator.

In the government buildings at the Administrative Centre of Bahia, as visitors do not represent a significant number compared to employees and civil servants, they alone provide an acceptable denominator for an indicator. Indicators such as consumption per area of the building did not deliver reasonable information.

In the airport project, consumption per number of passengers was found to be useful and is widely found in the technical literature. To enable comparison with other similar premises, such as shopping centres we added the number of companions to passengers, which is also a number airport's administration could usually provide.

Nevertheless, information obtained from these indicators can just be used as an initial approximation to the problem. To negotiate reduction goals with the buildings' administrators a better approach is to analyse the history of consumption especially if it is registered on a daily basis. Histograms of daily consumption, such as those obtained from Aguapura Vianet, enable the identification of possible causes of uses and wastes and therefore to arrange fast action.

We have not considered per-capita consumption factors used for the design for piping or hydro sanitary installations calculations, as a reference for conservation programmes. They usually indicate extremely high numbers, as their purpose is to define the size of the piping network and reservoirs. They usually include safety factors to guarantee the desired pressure of use in installations. At UFBA's School of Engineering, demand calculated using per-capita consumption from design manuals and texts, indicated a total consumption of the building, which is 4 to 5 times the one found today.

RESULTS OBTAINED BY TECLIM WATER SAVING PROGRAMMES AT A UNIVERSITY AND A GOVERNAMENTAL ADMINISTRATIVE CENTRE

Promoting water saving in buildings requires more than temporary programmes, it has to be a permanent activity. It is very rare that all the actions intended in the initial formulation of the programme are achieved. Nevertheless, partial results obtained constitute a solid foundation for further efforts.

The evolution of water consumption at UFBA between 1999 and 2015 is depicted in Fig. (**12**). It shows that there has been a 36% reduction in consumption over the period. It has to be taken into consideration however, that during the presidency of Luiz Inácio Lula da Silva, and his successor, Dilma Rousseff, the Brazilian Federal Government conducted a rapid expansion programme for professional and university education. UFBA's academic population including students, lecturers and administrative staff jumped from 35,000 in 2008 to 45,000 in 2015 [50][1]. To support this expansion several buildings were constructed on the campus between

2008 and 2014, most occurring between 2008 and 2011. This intense construction period generated several inefficiencies, among them an increase in water losses.

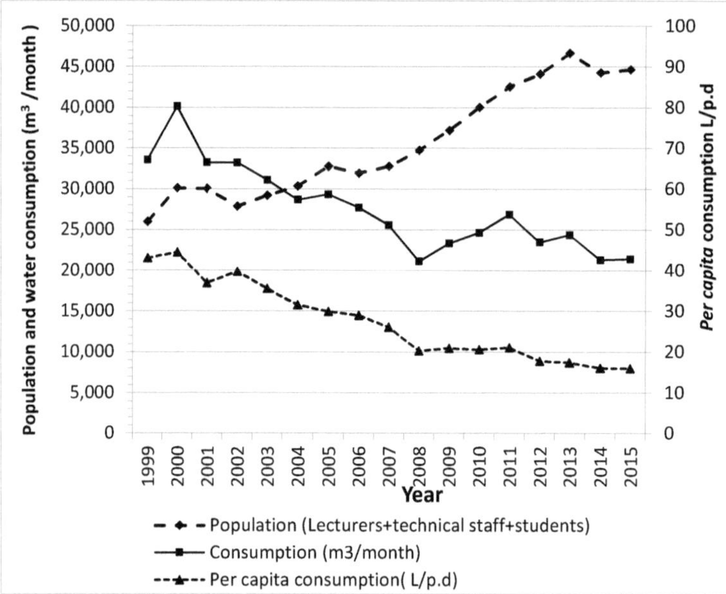

Fig. (12). Evolution of water consumption at the Federal University of Bahia 1999-2016 [www.teclim.ufba.br/aguapura. Federal University of Bahia, Brazil.].

It is interesting to observe that the consumption *per capita* indicator showed that a steady reduction was achieved. Over 14 years, *per capita* consumption was reduced by more than 63%. This result can be attributed to the open monitoring using the Aguapura Vianet system and basic maintenance work. However, other reasons can be added to explain this. The hydraulic installations of the new buildings are more efficient and present a lower rate of losses and the expansion rate was more significant in courses that did not demand higher water consuming labs. As previously mentioned, further savings can still be obtained if the proposed action is fully implemented. The programme's progress can be sustained by the economy obtained in early stages. The water rates, in Brazilian Reais, paid by the university in 2015 were 4.5 times the one paid in 2000. If UFBA were paying the monthly amount consumed in 2000 at today' water rates, it would be incurring additional monthly expenses of more than US$ 80,000 (based on September 2016 exchange rates).

With respect to the water saving programme with public buildings of the Government of Bahia, expressive results were also obtained in the 17 buildings where the programme was introduced at the initial stage, even if only partially implemented. Silva *et al.* [28] presents what actions could be implemented among those defined in the initial planning as described in a former section in this chapter. These mainly consist of creating and supporting Eco-teams, performing daily monitoring using Aguapura Vianet software, conducting basic preventive and corrective maintenance including regulation of sanitary devices flows and implementing some communication activities with users.

As shown in Fig. (**13**), the programme was developed in three stages. During the 12 months before the first stage the average monthly consumption was about 14.000 m^3. In the first stage the main characteristics of the buildings, the administrative structure, water consumption and user behaviour were studied. The objective of this stage was simply to identify the problem in order to select the units to be included and to plan future action. The only water saving action taken in those 6 months was to monitor the daily water consumption of each of the studied buildings using Aguapura Vianet programme. Although the programme was supposed to be continuous, administrative difficulties prevented this from happening and there was a break of almost 5 months between stages.

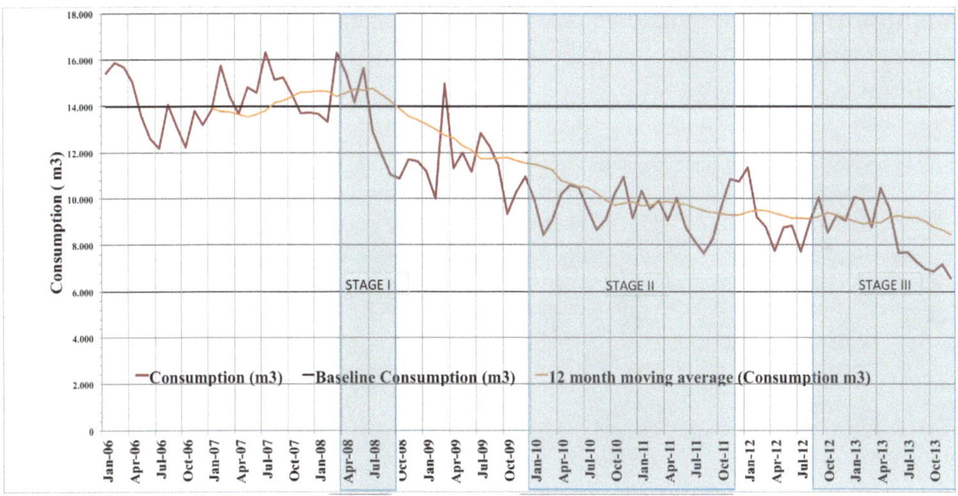

Fig. (13). Water consumption of 17 buildings at the Administrative Centre of Bahia (2006-2013) [28].

During the first stage water consumption declined from 14,400 to 11,600 m^3/month on average. During the 5-month interruption there was great fluctuation in consumption but the downward trend continued, reaching about 9,500 m^3/month by the end of the second stage. The third stage was initiated also after an unwanted interval of three months and, by the end of 2013, there was an average consumption of 7,000 m^3/month. A 50% reduction in water consumption was achieved in five years and a half considering 12 month rolling averages. As previously discussed, only daily monitoring, basic maintenance and some communication activities were responsible for this. Further significant savings could have been obtained if all the steps and actions planned initially had been put into action.

Other experiences in Brazil report similar or smaller savings obtained or they suggest that they could be obtained with more intensive conservation actions in maintenance and technology substitution including, greywater and rainwater use [24 - 26, 33, 44].

Reduction in water consumption was accompanied by financial savings. However, in absolute numbers, monthly water bills were only reduced in about 10% considering the 12 months average rolling values (Fig. **14**). These absolute numbers do not reflect the real savings that were obtained. Between 2008 and the end of 2013 water rates underwent five subsequent adjustments as indicated by the dashed line in the figure (baseline costs). To calculate real savings one has to consider the reality of inflationary economies. If one considers what the government would have paid for these water bills if they had maintained their consumption levels at the historic rates, an economy of 50% was achieved. These economies amounted to about US$ 64,000 a month (based on September 2016 exchange rates).

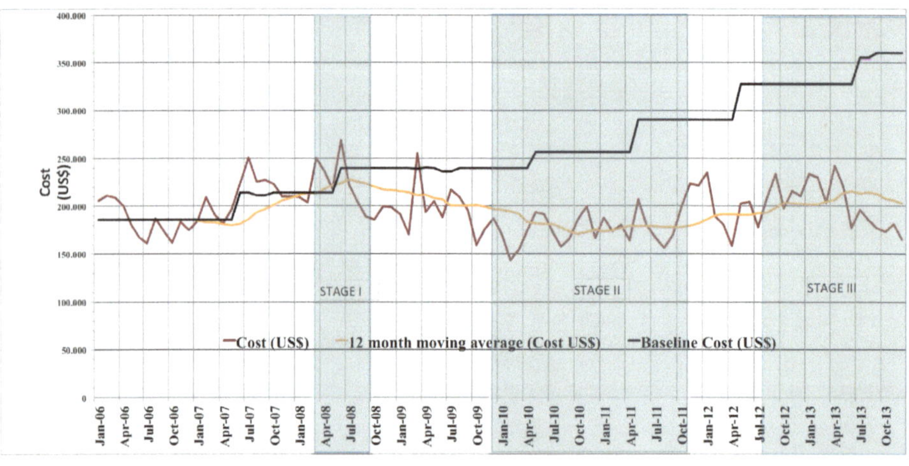

Fig. (14). Water cost and baseline cost for the 17 units at the Administrative Centre of Bahia (2006-2013) [28].

CONCLUSION

The magnitude of the environmental crisis requires more radical attitudes from society towards resource conservation. Urban water use is sustained using resources, which are extracted and transported from faraway places. Users are distant from watersheds that supply them. If you ask a city dweller where the water s/he uses comes from, the usual answer will be: from the tap.

The tap is seen as an endless source of water. Therefore, it is no surprise that water saving initiatives are so hard to introduce and sustain. When it comes to institutional or commercial buildings even water billing may not be enough to discourage water waste, as it does not directly affect the pockets of users. The cases reported here were university and governmental administrative premises. Making them more respectful to the environment, and to be more aware of their water consumption, involves a second social responsibility: There is a need for such institutions to lead by example and to promote more environmentally healthy attitudes. However, it has been observed that they can behave otherwise. Water waste and loss are constantly driven by weak maintenance schemes and the fact that water is free for the users of these premises.

Several authors have drawn attention to water saving programmes on these kinds of premises, highlighting the difficulties of implementing them. The solutions proposed have been mainly directed to the substitution of regular sanitation appliances for water saving ones, implementing water reuse alternatives and the capture of more environmental friendly sources, such as rainwater.

Water demand management strategies have to be reinforced and the "soft path for water" should enrich this discussion. The reasons for water loss and waste must be better investigated and either technological or behavioural solutions have to be more detailed. Most authors agree that raising water consciousness is fundamental to any water saving programme but they frequently just propose educational campaigns as the way to attain it.

Environmental consciousness is a broad and complex concept and has to be dealt with in a very ample and effective way. In our research projects we have emphasised raising water awareness by giving administrators and users the data about how and how much water is consumed in the building. Daily monitoring and exhibiting daily data in a graphic way on the web allows administrators and users to follow up their water consumption in real time. This helps administrators to take fast repair action in cases where consumption peaks are found. It may also provide users with a clearer understanding of the water issue. Public transparency fosters responsibility.

Water consumption depends on multiple variables; a conceptual approach has been presented to clarify how these variables interfere when developing long term water saving programmes. Several actions have been taken though it has not been possible to fully implement all the actions planned. Nevertheless, implementing continuous daily monitoring complemented with some basic maintenance has proved effective in bringing down consumption by more than 50% in the cases reported here.

One may affirm that for water saving programmes in institutional buildings, all actions usually proposed and implemented may be considered. However, their effectiveness will be seriously diminished if monitoring and control of water consumption are not in place and available to the public. Water saving has to be turned into a permanent concern, seeking not just for continuous but, for radical gains. The environmental crisis has turned chronic and it demands radical behavioural, technological and managerial innovation. The initial good results that can be obtained in water saving programmes cannot be a reason for accommodation. Water demand management efforts are to be brought to a stronger stand as the "soft path" may not be that soft…

NOTES

[1] During the period of expansion of the university new campuses were created outside of Salvador and other were turned independent from UFBA. Therefore the way to account for the university's population in official reports, underwent modifications. For this reason, the numbers presented in previous analyses [35] are somehow different from those here presented. Nevertheless, these differences do not alter the trends and overall conclusions obtained.

CONFLICT OF INTEREST

The authors confirm that they have no conflict of interest to declare for this publication.

ACKNOWLEDGEMENTS

The authors would like to acknowledge the Clean Technology Network (TECLIM), in particular the students and researchers involved in the projects discussed here. We would also like to thank the Federal University of Bahia (UFBA), the Secretary for Management of Bahia (SAEB), Brazilian Airport Infrastructure Company (INFRAERO), Projects and Study Financer (FINEP) and the National Council of Technological and Scientific Development (CNPq).

REFERENCES

[1] P. Rogers, *Water governance, water security and water sustainability.* Taylor & Francis: London, UK, 2006.
[http://dx.doi.org/10.1201/9781439834275.pt1]

[2] IEA, *Energy technology perspective. Scenario and strategies to 2050.* OECD/IEA: Paris, France, 2008.

[3] R. Engelman, "Beyond Sustainababble", In: E. Assadourian, T. Prugh, Eds., *State of the World 2013 - Is Sustainability Still Possible?.* Island Press: Washington, USA, 2013, pp. 3-16. [http://dx.doi.org/10.5822/978-1-61091-458-1_1]

[4] IEA, *Energy Technology Perspectives: Scenarios & Strategies To 2050.* OECD/IEA: Paris, France, 2010.

[5] J. Rockström, W. Steffen, K. Noone, A. Persson, F.S. Chapin III, E.F. Lambin, T.M. Lenton, M. Scheffer, C. Folke, H.J. Schellnhuber, B. Nykvist, C.A. de Wit, T. Hughes, S. van der Leeuw, H. Rodhe, S. Sörlin, P.K. Snyder, R. Costanza, U. Svedin, M. Falkenmark, L. Karlberg, R.W. Corell, V.J. Fabry, J. Hansen, B. Walker, D. Liverman, K. Richardson, P. Crutzen, and J.A. Foley, "A safe operating space for humanity", *Nature,* vol. 461, no. 7263, pp. 472-475, 2009. [http://dx.doi.org/10.1038/461472a] [PMID: 19779433]

[6] W. Steffen, K. Richardson, J. Rockström, S.E. Cornell, I. Fetzer, E.M. Bennett, R. Biggs, S.R. Carpenter, W. de Vries, C.A. de Wit, C. Folke, D. Gerten, J. Heinke, G.M. Mace, L.M. Persson, V. Ramanathan, B. Reyers, and S. Sörlin, "Sustainability. Planetary boundaries: guiding human development on a changing planet", *Science,* vol. 347, no. 6223, p. 1259855, 2015. [http://dx.doi.org/10.1126/science.1259855] [PMID: 25592418]

[7] F. Jaramillo, and G. Destouni, "SUSTAINABILITY. Comment on Planetary boundaries: Guiding human development on a changing planet", *Science,* vol. 348, no. 6240, p. 1217, 2015. [http://dx.doi.org/10.1126/science.aaa9629] [PMID: 26068843]

[8] D. Gerten, J. Rockström, J. Heinke, W. Steffen, K. Richardson, and S. Cornell, "SUSTAINABILITY. Response to Comment on Planetary boundaries: Guiding human development on a changing planet", *Science,* vol. 348, no. 6240, p. 1217, 2015. [http://dx.doi.org/10.1126/science.aab0031] [PMID: 26068844]

[9] A.B. Lovins, "The negawatt revolution", *Across Board (NY),* vol. 27, no. 9, pp. 18-23, 1990.

[10] G. Wolff, and P. Gleick, "The soft path for water", In: P. Gleick, Ed., *The World's Water 2002–2003..* Island Press: Washington, DC, 2002, pp. 1-32.

[11] D.B. Brooks, and S. Holtz, "Water soft path analysis: from principles to practice", *Water Int.,* vol. 34, no. 2, pp. 158-169, 2009. [http://dx.doi.org/10.1080/02508060902839940]

[12] A.B. Lovins, *Soft energy paths toward a durable peace.* Harper & Row: New York, NY, 1977.

[13] D.B. Brooks, C. Maas, O.M. Brandes, and L. Brandes, "Applying water soft path analysis in small urban areas: four Canadian case studies", *Int. J. Water Resour. Dev.,* pp. 37-41, 2015.

[14] D. Butler, and F.A. Memon, *Water demand management.* IWA Publishing: London, UK, 2006.

[15] WWAP (United Nations World Water Assessment Programme), *The United Nations World Water Development Report 2015: Water for a Sustainable World.* Paris, France, 2015.

[16] A. Vickers, "Industrial, commercial and institutional water use and efficiency measures", In: *Handbook of water use and conservation..* Waterplow press: Amherst, Massachusetts, USA, 2001, p. 446.

[17] AWWA Research Foundation, *Commercial and institutional end uses of water.* AWWA Research Foundation and the American Water Works Association: Denver, CO, USA, 2000.

[18] R.A. Stewart, R.M. Willis, D. Giurco, K. Panuwatwanich, and B. Capati, "Web-based knowledge management system: linking smart metering to the future of urban water planning", *Aust. Plan.,* vol. 47, no. 2, pp. 66-74, 2010. [http://dx.doi.org/10.1080/07293681003767769]

[19] L.H. De Oliveira, and O.M. Gonçalves, "Metodologia para a implantação do programa de uso racional de água em edifícios", *Tech Rep BT/PCC/247,* São Paulo, SP, Brazil, .

[20] S. da S. Nunes, Estudo da conservação de água em edifícios localizados no campus da Universidade Estadual de Campinas, University of Campinas: Brazil, 2000.

[21] "Programa de uso racional da água da Universidade de São Paulo", *"Diretrizes de uso racional da água - pura - USP,"*, 2011.[Accessed: 01-Jul-2016] [Online]. Available: http://www.pura.usp.br/ wp-content/uploads/sites/19/2015/02/PURA-USP_DiretURA_set11.pdf.

[22] L.A. Ywashima, Avaliação do uso de água em edifícios escolares públicos e análise de viabilidade econômica da instalação de tecnologias economizadoras nos pontos de consumo, University of Campinas: Brazil, 2005.

[23] M.S. de O. Ilha, L.P. Pedroso, and L.A. Ywashima, "Indicadores de consumo de água em escolas", In: *XII Encontro Nacional de Tecnologia do Ambiente Construído.* Fortaleza: Brazil, 2008.

[24] P.C. Kammers, and E. Ghisi, "“Usos finais de água em edifícios públicos localizados em Florianópolis, SC,” Ambient", *Construído,* vol. 6, no. 1, pp. 75-90, 2006.

[25] L.C. Proença, and E. Ghisi, "Water end-uses in Brazilian office buildings", *Resour. Conserv. Recycling,* vol. 54, pp. 489-500, 2010. [http://dx.doi.org/10.1016/j.resconrec.2009.10.005]

[26] L.C. Proença, and E. Ghisi, "Assessment of potable water savings in office buildings considering embodied energy", *Water Resour. Manage.,* vol. 27, pp. 581-599, 2013. [http://dx.doi.org/10.1007/s11269-012-0203-1]

[27] A.K. Nagawaka, Caracterização do consumo de água em prédios universitários: O caso da UFBA, Federal University of Bahia: Brazil, 2009.

[28] S.F. da Silva, V. Britto, C. Azevedo, and A. Kiperstok, "Rational consumption of water in administrative public buildings: The experience of the Bahia administrative center, Brazil", *Water,* vol. 6, no. 9, pp. 2552-2574, 2014. [http://dx.doi.org/10.3390/w6092552]

[29] A. A., "Quadros, R. A. Kalid, and A. Kiperstok, “Sistema de apoio gestão da demanda de água e uso racional da água em edificações públicas: AGUAPURA VIANET", *XXXII Congresso Interamericano de Engenharia Sanitária e Ambiental AIDIS,* 2010 Punta Cana: Dominican Republic

[30] R.F. Gonçalves, *Conservação de água e energia em sistemas prediais e públicos de abastecimento de água.* ABES: Rio de Janeiro, Brazil, 2009.

[31] L.P. Pedroso, Subsídios para a implementação de sistema de manutenção em campus universiário, com ênfase em conservação de água, University of Campinas: Brazil, 2002.

[32] L.P. Pedroso, Estudo das variáveis determinantes no consumo de agua em escolas: o caso das unidades municipais de Campinas, São Paulo, University of Campinas: Brazil, 2008.

[33] C.A. Sautchúk, Formulação de diretrizes para implantação de programas de conservação de água em edificações, University of São Paulo: Brazil, 2004.

[34] G.B. Fasola, and A.K. Marinoski, "Potencial de economia de água em duas escolas em Florianópolis, SC", In: *Ambient. construido,* vol. 11. 2011, no. 4, pp. 65-78.

[35] M. Marinho, M.D. Gonçalves, and A. Kiperstok, "Water conservation as a tool to support sustainable practices in a Brazilian public university", *J. Clean. Prod.,* vol. 62, pp. 98-106, 2014. [http://dx.doi.org/10.1016/j.jclepro.2013.06.053]

[36] L. Alcântara Santos, Gestão da água em edificações públicas: a experiência no prédio da Empresa Baiana de Águas e Saneamento S.A, Federal University of Bahia: Brazil, 2010.

[37] A. Kiperstok, K. Esquerre, R. Kalid, E. Sales, and G. Oliveira, "Rationalizing the use of water in industry—Part 1: Summary of the instruments developed by the Clean Technology Network in the State of Bahia and main results obtained", *J. Environ. Prot. (Irvine Calif.),* vol. 4, pp. 486-496, 2013. [http://dx.doi.org/10.4236/jep.2013.45057]

[38] D. of G. S. California DGS, *California best management practices for water use in California State government facilities,* 2016.[Accessed: 12-Aug-2016] [Online]. Available: http://www.dgs.ca.gov/dgs/Home/water.aspx.

[39] A. Kiperstok, M.T. Freire, and R.A. Kalid, "Tecnologias Poupadoras: Inovação, Aspectos Socioculturais e Manutenção", In: W. Cabral de Souza Júnior, E. Nolasco Ribeiro, Eds., *Uso eficiente de água em aeroportos.* Editora RIMA: São Carlos, Brazil, 2011, p. 318.

[40] M.T. Freire, O consumo racional de água no aeroporto internacional de Salvador, Bahia/Brasil, Federal University of Bahia: Brazil, 2011.

[41] International Paruresis Association (IPA), *FAQ,* 2016.[Accessed: 15-Jul-2016] [Online]. Available: http://paruresis.org/faq/.

[42] Associação Brasileira de Normas Técnicas (ABNT), *NBR 13.713/96: Aparelhos hidráulicos acionados manualmente e com ciclo de fechamento,* Rio de Janeiro, 1996.

[43] W.C. Alves, "A. Rocha, and R. F. Gonçalves, "Aparelhos sanitários economizadores", In: R.F. Gonçalves, Ed., *Uso racional de água em edificações..* ABES: Rio de Janeiro, Brazil, 2006, p. 352.

[44] R.F. Gonçalves, *Uso racional de água em edificações.* ABES: Rio de Janeiro, Brazil, 2006.

[45] M.B. Marinho, Universidades e sustentabilidade. Uma pesquisa em instituições de educação superior brasileiras, Federal University of Bahia: Brazil, 2014.

[46] V.R. Santiago, Barreiras para a implementação do programa de racionalização do consumo de água em prédios públicos administrativos do estado da Bahia, Federal University of Bahia: Brazil, 2016.

[47] Teclim/UFBA, "Racionalização do consumo de água e energia nos prédios públicos estaduais: Etapa 2 - Relatório final de acompanhamento das atividades desenvolvidas", *Tech Rep 071/09 SAEB-UFBA,* .

[48] D. Fontana, R. Kalid, A. Kiperstok, and M.A. Silva, "Methodology for wastewater minimication in industries in the petrochemical complex", In: *in EMPROMER.* Costa Verde-RJ: Brazil, 2005.

[49] M.A. Martins, C. Amaro, L. Souza, R. Kalid, and A. Kiperstok, "New objective function for data reconciliation in water balance from industrial process", *J. Clean. Prod.,* vol. 18, no. 12, pp. 1184-1189, 2010.
[http://dx.doi.org/10.1016/j.jclepro.2010.03.014]

[50] *Pró-reitoria de planejamento e orçamento, UFBA.* Salvador-BA, Brazil, 2016.

CHAPTER 6

Water Savings in Rainwater Harvesting Systems in Portugal: Influence of Weather and Type of Building

Cristina Matos Silva[1,*], **Vitor Sousa**[1] and **Inês Meireles**[2]

[1] *CERIS, Department of Civil Engineering, Architecture and Georesources, Instituto Superior Técnico, Universidade de Lisboa, Lisbon, Portugal*

[2] *RISCO, Department of Civil Engineering, Universidade de Aveiro, Aveiro, Portugal*

Abstract: Nowadays, water management is necessary to optimize the use of this key resource. As such, increasing water efficiency through the reduction of consumption and/or resorting to alternative water sources is a challenge of the near future. Rainwater harvesting is an accessible alternative source of water for non-potable and potable uses in many parts of the globe. Rainwater can be collected easily and, in many cases, its use may not require significant treatment, even for potable purposes. Yet, rainwater harvesting (RWH) systems are still not common in most of the water supply projects, namely in Mediterranean countries such as Portugal. The promotion of RWH systems involves viability analyses estimating the expected water savings for different climate and water consumption patterns. This chapter evaluates the performance of rainwater harvesting systems of several case studies in Portugal, covering different water use patterns and geographical locations. The water consumption pattern, both in time and end-use, influences the potential for rainwater use. Water consumption was monitored in residential buildings (a single family house and a building), a shopping centre and a university building. Simulations are presented for different locations in Portugal, in order to assess the influence of weather in the efficiency of RWH systems. In addition to the rainfall amount, its distribution in time and space also contributes for the rainwater availability. The operational data from an existing rainwater harvesting system is used to calibrate the model and evaluate its sensitivity to the main parameters. Tank optimization for each case study is discussed. The Mediterranean climate, characterized by distinct wet and dry seasons and significant variability of the rainfall events intensity and duration, shows substantial variation of the inter-annual water savings from a rainwater harvesting systems. The results show that rainwater harvesting systems in Portugal are a relevant alternative water source in different types of building.

[*] **Corresponding author Cristina Matos Silva:** DECivil, Instituto Superior Técnico, Universidade de Lisboa, Lisbon, Portugal; Tel: +351 218 418 356; E-mail: cristina.matos.silva@tecnico.ulisboa.pt.

Enedir Ghisi (Ed.)

Keywords: Commercial buildings, Portugal, Rainwater harvesting, Rainfall regime, Residential buildings, Simulation, University buildings, Validation, Water consumption pattern, Water savings.

INTRODUCTION

The XX century was characterized by population growth [1], water consumption increase and water sources pollution [2]. As a result, water became a key resource at risk. In 2000 its relevance was recognized internationally by the United Nations when defining the UN Millennium Development Goal of reducing by half the proportion of people without sustainable access to safe drinking water until 2015 [3]. Although 2.6 billion people have gained access to improved water sanitation since 1990, in 2015 the goal was far from being fulfilled in some parts of the globe [4] and universal access to clean water and sanitation became one of the 17 Global Goals on the 2030 Agenda for Sustainable Development [5].

Several parts of the globe are already facing problems related to water scarcity or will face them in a near future, most notably in Africa, with over 40 percent of the world's population affected in 2015. Projections point to an increased global water stress and scarcity in the future as a result of the forecasted climate change, with estimates that water related challenges will affect two thirds of the world's population by 2025 [2, 5 - 7]. This is more critical in developing countries, particularly in poor rural areas, where at least one-third of the population has little or no access to safe drinking water. As a consequence, major health problems from waterborne diseases add to the complex problem of stress and scarcity [4, 8]. In Europe, generally, the risks of water stress, and specially water scarcity, are smaller. However, public water supply systems imply a significant consumption of various other resources (*e.g.*, building, maintaining, operating and rehabilitating/replacing the supporting infrastructures [9, 10]) and represent a significant burden both at an economic and environmental levels. Considering the water-energy nexus alone will motivate the evaluation of alternatives for a more efficient water use, even in wealthy developed countries with water resources available higher than the water demand.

Improved water management is crucial in less developed countries, since several tend to locate in regions of the globe with higher water scarcity and stress and there are less resources and infrastructure available. In the more developed countries, water efficiency promotion by organizations with responsibilities in the water sector has been driven by sustainability. Resource optimization through the combined implementation of structural (*e.g.*, reduction of water losses) and non-structural (*e.g.*, education campaigns) measures [13] have proven to benefit the economy, environment and society [11, 12]. As a results, some countries where able to stabilize or even decrease the water use in various sectors (*e.g.*, residential; industry; agriculture) during the last decades.

Water management optimization solutions in buildings can be categorized into two main groups based on the underlying goal: i) water consumption reduction; and ii) alternative water sources use. The former includes promoting conscientious water consumption habits and lower water consumption devices development and implementation for a more efficient water use. Education campaigns, water tariffs varying with water consumption, low discharge fixtures and appliances are amongst the solutions to decrease water consumption. The latter involves the exploitation of alternative water sources to either replace or complement the public water supply. These alternative water sources include salt water, groundwater, greywater and rainwater.

The use of rainwater for both potable and non-potable applications is not a novelty, and people has resort to it for thousands of years throughout the world [14]. Nowadays, RWH in buildings is being used for different purposes depending on the country development. Coping with water shortages for potable and non-potable use [16] has been the main motivation underlying RWAH in developing countries such as Bangladesh, Botswana, China, India, Kenya, Mali, Malawi or Thailand [11, 15]. In more developed countries such as Belgium, France, Germany, Japan, New Zealand, Singapore or United States, RWH use is being promoted mainly to complement conventional systems for non-potable use, namely for toilet flushing, clothes washing, outside washes and irrigation [17 - 23], but also for potable uses (*e.g.*, Australia [24]). RWH is not limited to residential buildings and large scale systems are also found in collective housing and other types of buildings in countries such as Japan [25], the UK [26, 27], Germany [28] or Portugal [29].

Legal and/or regulation limitations and/or gaps, initial cost, social acceptance and treatment requirements have been some of the most significant obstacles to the implementation of RWH systems. In countries where water scarcity and stress is not yet a major concern, the most relevant stimuli for using RWH as an alternative water source are usually related to sustainability concerns, which include environmental issues (*e.g.*, climate changes, population growth, pollution) and the total costs of public water supply (investment, operation, maintenance and rehabilitation/replacement or disposal). This resulted in regional and/or national initiatives to promote RWH in countries such as Australia, Germany or the USA [30]. RWH has also been promoted to complement storm water management in urbanized areas (*e.g.*, Germany).

The main objective of this chapter is to evaluate the potential of using RWH systems in Portugal, a country which benefits from a Mediterranean climate, characterized by distinct wet and dry seasons and significant variability of the rainfall events intensity and duration. For that end, a detailed RWH simulation approach will be replicated in different regions and types of building, to take into account respectively the influence of weather and the influence of different end-use water consumption patterns on water savings from the use of a RWH system. The following sections present a review on RWH and the description of the methodology used, namely in terms of the RWH system simulation and rainfall regime and water consumption pattern characterizations. Next the case studies and results are detailed and discussed and the last section presents some final remarks.

RAINWATER HARVESTING

System Configuration

A RWH system involves the collection, storage, treatment and use of rainwater as either a principal or supplementary source of water. Despite the differences in sophistication of a RWH system components between developed and developing countries, it always encompasses three basic elements: i) the collection surface; ii) the conveyance system; and iii) the storage and distribution systems. In buildings, the roofs and terraces comprise the collection surface usually. The configuration (*e.g.*, area; slope) and material (*e.g.*, metal; clay) of the roofs and terraces will affect the rainwater quality and quantity. After collection, rainwater goes through the conveyance system to treatment, which normally includes a first flush device and a filtration device in modern RWH systems. The first flush device is installed with the aim of diverting the initial rainfall volume. This diversion aims at safeguarding water quality, since the first flow tends to be more contaminated due to washing of the pollutants deposited on the collection surface over the preceding dry weather period. In addition to the first flush, some water treatment may also be included. A common solution is the inclusion of a filtration providing some treatment for the physical characteristics of the rainwater. After this and other optional water quality controls, the rainwater is stored in a tank, from which it is conveyed to the end-use-point by a distribution system. A RWH system may have distinct storage and distribution tanks, or only one tank for both functions. Since the rainwater amount is highly variable, in most cases it is not reliable enough to meet the demand. In order to comply with the demand, another water source should be available. Usually, this is achieved by connecting the public water system to one of the tanks. The connection between the two systems must prevent the cross contamination. In Portugal, and many other countries in Europe, the regulations require an air gap when connecting the public water supply system with other water sources. With this requirement, the public water supply network pressure is lost. This may represent a waste of energy since, depending on the location of the tanks (elevated or underground) and of the end-use-point, a pumping system may be required to provide adequate water pressure at delivery.

Optionally, additional treatment stages can be included before or after the storage tank to assure water quality, but this is not generally required for non-potable uses. Alternative solutions also exist nowadays that don't require dispending the public supply water pressure to feed the RWH system (*e.g.* antipollution valves, special pumps), but these solutions are not allowed in many countries since the regulations require a physical separation between the alternative water sources and the public supply network for health safety reasons.

Rainwater Harvesting Design

The performance of a RWH system depends on a set of variables that include the: i) rainfall pattern; ii) water consumption pattern; iii) collection area, iv) water losses; and v) tank size. Usually, the main challenge when designing a RWH system is sizing the water tank correctly and provide an adequate storage capacity. Presently, there are various standards and guidelines on rainwater harvest systems design (*e.g.*, [31 - 36]). Some standards indicate methods for estimate the water tank size, including the Australian, the Brazilian, the English or the German practical methods. In Portugal, the German Practical Method is recommended in the guideline ETA 0701 [37]. The various design methods currently available can be organized into: i) simplified methods; and ii) simulation methods.

Amongst the simplified methods, the demand side and supply side approaches can be identified. In the first approach, the storage requirement defines the largest demand to be supplied. The demand side approach is best suited to areas with year-round rainfall and it comprises simply in multiplying the daily water demand by the average longest dry period, but variations of the approach have been developed (*e.g.*, [38]). The supply side approach is more appropriate for areas with low or uneven rainfall and is similar to the design of public water supply reservoirs. The storage capacity is defined has the largest difference between the cumulative demand and the cumulative rainfall collected. There are also simplified methods combining both of these approaches or complementing with user-defined rules (*e.g.*, [39]). Some simplified methods, namely graphical methods, disregard some of the rainfall spatial variability. Furthermore, these methods generally use a single average rainfall value to estimate the required size of a rainwater tank, completely disregarding the time variability of the rainfall. Therefore, simplified approaches should be regarded as approximated methods, useful for pre-design.

The simulation methods are deemed to be the most accurate for determining the optimal size thank for a RWH system since it can take into consideration rainfall and consumption patterns and the collection efficiency. The underlying principle of the simulation methods is to mimic numerically the performance of a real RWH system. Based on this logic, the simulation method allows for a variety of approaches. In addition to the number of variables considered in the model, the way those variables are quantified can have significant influence in the simulation results.

Rainfall is, probably, the most influential variable because it can be highly uncertain both in space and in time. As such, the design of rainwater tanks should be local specific and the length and resolution of the rainfall data series should be adequate for the purpose and context of the RWH system. Fewkes and Butler [40] evaluated the influence of the rainfall resolution (time-step of the simulations) on results using the ratio between the storage volume and the annual average inflow volume (Sf=Storage Volume / (Mean Annual Rainfall * Collection Area)). For UK rainfall and domestic water consumption patterns, they found that small ($Sf < 0.01$), medium ($0.01 < Sf < 0.125$) and large storage volumes ($Sf > 0.125$) require hourly, daily and monthly rainfall resolution, respectively. Mitchell [41] and Ghisi *et al.* [42] evaluated the influence of the length of the simulation on the results. For UK and Brazil the simulation results tend to stabilize for a rainfall data series 10 years. While most authors resort to historical rainfall data series, Basinger *et al.* [43] used a non-parametric stochastic rainfall generator from a 25 years rainfall data series while Guo and Baetz [44] used a statistical rainfall events model.

The collection efficiency relates the amount of rainwater precipitating over the collection area with the rainwater effectively stored in the tank. The difference between the two is usually modelled through the initial rainfall loss factor and the continuous rainfall loss ratio, that take into account the losses through surface retention and evapotranspiration, and the first flush, whenever a first flush device is installed. The latter will depend on the first flush device, where the most simple ones will divert a constant pre-defined volume of rainwater and the most sophisticated will divert the rainwater until a given water quality indicator threshold is not met. The runoff coefficient depends on the weather and roof characteristics. The initial and continuous rainfall losses are commonly accounted for using a runoff coefficient [45]. Farreny *et al.* [46] measured runoff coefficients for different roof types (varying slopes and coverings) in Spain, concluding that the rainfall intensity of the precipitation event also plays a role for some types of roofs.

Water demand in buildings has been thoroughly recorded and evaluated over the years because it is a key information for the design and operation of water supply systems. Since the second half of the XX century there have been various studies on water demand and its determinants. However, water consumption pattern depends on far more factors than just the cost. Factors related to biological, environmental and economic issues are known to influence water consumption. It depends also on social aspects, resulting in the existence of interaction between the effect of the factors at an individual level and at a group level. Considering the water consumption at a group level, namely a country, region or community, it depends on factors such as [47 - 49]: i) the development level (*e.g.*, developing or developed country; technical innovation); ii) the weather and climate (direct - temperature; indirect - rainfall); iii) the type (*e.g.*, rural; urban) and characteristics (*e.g.*, population; housing type) of the community, activity (*e.g.*, tourism; industry; services) and technology (*e.g.*, equipment; appliances); iv) water availability (*e.g.*, losses; pressure; service level; sources; treatment required); and v) policies (*e.g.*, pricing; regulations), amid others. The influence of some of the previous factors are questionable nowadays. For instance, the increase of water consumption with development level attested by Flörke and Alcamo [50] is only valid up to a certain threshold. For developed countries, where the water consumption *per capita* is larger, other factors are becoming more important. For instance, in Portugal, the water consumption (in cubic meter per person per day – m^3/ppd) has stabilized or decreased since 2000, despite the increase of the GDP until 2009. Another example is the influence of water pricing, which has been extensively studied for several decades [51 - 64]. Martins and Fortunato [65] analysed the residential water demand under block rates in Portugal and concluded that, consistently with most studies, the water price as a marginal negative effect on water consumption. However, the decreasing trend of the water consumption with the water price claimed by these studies and the OECD [66] report was challenged by recent research [67 - 69]. These studies demonstrate that, in most cases, residential water demand is price inelastic, taking into account its low relative cost when compared to other life essential goods. Therefore, it is reasonable to assume that, at least after a certain level of consumption that ensures comfort to the users, the water price influence is highly context specific. At an individual consumer level, which can be a single individual or a group that shares the same facility or activity, water consumption is further influenced by the dynamics of the individual or group (*e.g.*, family, colleagues) and the activity being performed (*e.g.*, domestic, sports, work). In addition to the occupants, the building characteristics and features also play a role in the water consumption. The number, age, education levels, attitudes, beliefs and behaviours of the occupants are some variables that help to explain the water consumption. The number, type and efficiency of water consuming devices (*i.e.*, washers, shower heads, tap fittings, dishwashers and flush toilets), the overall quality of the sanitary installations and the buildings' maintenance requirements are some of the buildings features affecting water consumption [50, 70 - 77]. In the Netherlands, the study on water consumption and the factors influencing it has been promoted by the Association of Dutch Water Companies (*Vereniging van waterbedrijven in Nederland* - VEWIN) since 1992. These studies revealed that water consumption varies with the location of the house, number and income of the individuals in the household, age and gender of the individuals and their ethnic group. The country, region or community level factors interrelate with individual factors. An important aspect in this affiliation is the connection between attitudes and beliefs concerning water and environment and their relationship with actual water consumption [78, 79].

Research Lines

Following the approach of Silva *et al.* [29], the research on RWH was organized into two groups of studies depending on their focus: i) water savings; ii) and water quality. The former can be further divided into: i) studies considering only a balance between rainfall and water consumption over a region or country (*e.g.*, in Jordan [80]; in Southern Brazil [81]; in Abeokuta, Nigeria [82]); ii) studies considering the simulation of RWH systems in a specific or typified building or set of buildings (*e.g.*, multi-unit building in Sydney, Newcastle and Wollongong [83]; single and multi-family buildings in Barcelona, Spain [84]; single family building in Almada and Porto, Portugal [85]); iii) studies considering the simulation of RWH systems at a neighbourhood (*e.g.*: Ringdansen neighbourhood in Norrköping, Sweden [86]; neighbourhood in Granollers, Spain [87]) or a city (Drancy and Colombes, France [88]; 62 cities in Santa Catarina, Brazil [89]; Morelia, Mexico [90]) scale. A fourth group could be identified, comprising studies evaluating RWH water savings potential for specific end-uses [91], non-residential buildings (*e.g.*, petrol stations [92]; airports [93]; commercial buildings [29]) or without considering a specific building configuration [94 - 99]. While the majority focus only in the technical viability of the RWH systems, some also evaluate its financial viability [86, 100 - 104]. These studies show a substantial variability of water savings potential from RWH depending on the specific context of the analysis and the adopted design method. The financial viability is also found to be context specific and not exclusively related to the water savings. It should be noticed that these financial viability assessments where made based on public water supply cost. However, water and sewer fees tend to represent a part of the costs directly supported by the water utilities. For instance, the European Union (EU) cohesion funds have been an important source of funding for water infrastructures that is not reflected on the fees. In the 1980s and 1990s, southern Europe benefitted from these funds with substantial impact on the water infrastructures. In Portugal, it supported the increase of the population ratio connected to public water supply from 61% in 1989 to 95% in 1999 [105, 106]. The improved access to potable water supply contributed to the water consumption increase, without the consumer bearing the cost for it directly. At a national level, tax funds are also used by most governments to finance public water utilities [107]. Most private water utilities also have received some form of public funds [108 - 110]. Furthermore, public water supply and sewer systems entail additional burdens that are not taken into account in the fees. For instance, the energy used to assure drinking water quality, deliver it to consumers and dispose of wastewater have an environmental burden that is not completely reflected in the fees [111 - 118]. The topic is starting the gain some interest in Portugal, with some studies performed on the technical and financial viability of RWH systems in residential [85, 119 - 121] and commercial buildings [29, 122], in various locations over mainland Portugal. Lúcio [123] also evaluated the technical performance of RWH at a city wide level in Portugal.

The harvested water can have chemical, microbiologic and physical contaminants. Amongst other factors, the types and degree of contamination depends on [124 - 163]: i) air that raindrops traverse before hitting the roof (called wet deposition), which relates to air pollution; ii) the rooftop and drainage pipes, which relates to the type of roof, the existing fauna, the antecedent dry weather period and the cleaning and maintenance strategies; and iii) the storage tank, which depends on the former and on the RWH system configuration and maintenance. Considering the results of Mendez *et al.* [164], it is possible to observe that the rainwater quality depends on the location. In the same location the water quality varies with time, even during a single rainfall event. Kus *et al.* [165] found that diverting the first 2 mm of rainfall assures compliance with the Australian Drinking Water Guidelines (ADWG) standards except for lead and turbidity, which required bypassing approximately the first 5 mm of rainfall. A first flush system improves the physicochemical quality of collected rainwater but it can't avoid microbial contamination of stored rainwater [166]. Despite the potential for contamination, in some regions of the globe rainwater has been identified as a major source also for drinking, cooking and sanitary purposes [167] since it does not present increased risk of gastrointestinal illness when compared with water from public supply systems [168, 169]. In developed countries, in general, the rainwater quality doesn't meet the water quality standards when compared to the public water supply sources [22, 46]. Still, Coombes *et al.* [100] and Muthukumaran *et al.* [170] concluded that rainwater quality could meet the parameters of the Australian Drinking Water Guideline if used to supply hot water systems operating at temperatures in the range 55°C and 63°C. This topic was recently reviewed by Abbasi and Abbasi [171]. Very little research has been carried on RWH quality in Portugal. Lança and Silva-Afonso [172] evaluated the quality of rainwater collected and stored in a buried tank without any treatment or filtration and the water quality analysis results complied with the requirements of the ETA 0701 [37].

One of the most relevant conclusions from the past research is that RWH systems performance can be highly variable. Local conditions and the system configuration govern water savings potential, economic viability and water quality. An accurate estimate of the water savings and economic viability of a RWH system requires an evaluation taking into consideration the specific water consumption and rainfall patterns, as well as the system design. If some degree of water savings is always possible, the economic viability is not assured. Regular water quality monitoring campaigns should be carried out for each location and installation, since this can vary significantly both in space and time. The low degree of control and the number of potential sources of pollution limit the use of rainwater for potable use without an adequate treatment scheme. Still, most research indicate that the required treatment is not extensive, especially when a suitable first flush is diverted. Furthermore, the rainwater suitability for non-potable use is supported by virtually all of the authors.

METHODOLOGY

Approach

The simulation method (or balance equations method) was adopted in the present research. Comparing to other author's implementations of the simulation method, the major distinctive features of the approach use are: i) the possibility to take into account the consumption variability throughout the year; and ii) the possibility to tune the amount of rain before and after consumption. The first feature was introduced because it was observed that the water consumption in Portugal may vary significantly throughout the year, particularly when comparing the summer months with the rest. The second feature was introduced as an alternative to the YAS (Yield After Spillage) and YBS (Yield Before Spill) algorithms, since probabilistically the rainfall may equally take place before, after or during consumption [122]. In order to explore several alternatives between consumption pattern, catchment area and tank size, the approach was implemented in Matlab following the flowchart depicted in Fig. (**1**). The approach was conceived to run in a daily scale, but it can be easily changed into a different time scale depending on the resolution of the data introduced.

Rainfall Regime

Portugal's rainfall regime is highly irregular both at the spatial scale and at the time scale [173 - 179]. Spatially, annual rainfall over mainland Portugal range from as low as 400 mm in the southeast (among the driest in Europe) to over 3000 mm in the northwest (among the wettest in Europe), with mean value of about 900 mm. This is consistent with the findings for the Mediterranean region, but the magnitude of the variability and length scale (800 km) are quite unique. In time, the rainfall variations exist at an annual and inter-annual scale. Regarding the former, there is a marked seasonal variability of the rainfall with a progressive transition between dry and wet periods. The disturbances associated with the polar front predominate during the 3 months' winter season (December to February) and, on average, are responsible for 42% of the annual rainfall. The Azores high pressure system moves poleward and becomes the prevailing system during the summertime, resulting in only 7% of the annual total rainfall occurring from June to August, on average. The interaction between both systems result in highly variable rainfall during the transition seasons, spring (March to May) and autumn (September to November). At an inter-annual scale, there is an interchange between dry and wet years or cycles of years [180, 181]. This variability can be explained, in part, by the North Atlantic Oscillation (NAO) [182 - 185].

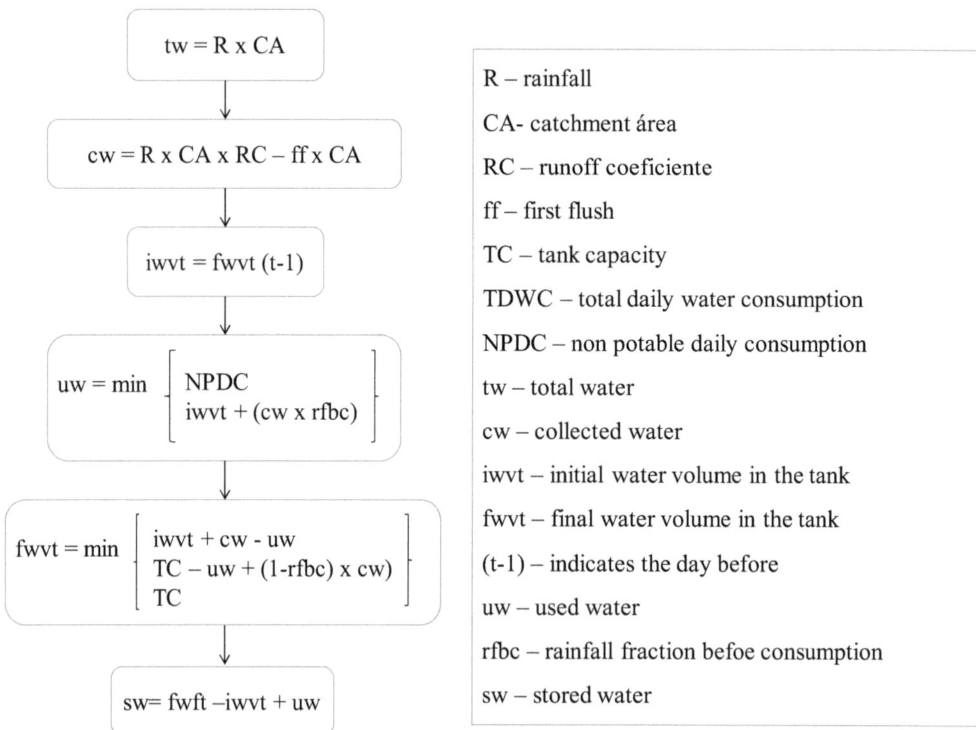

Fig. (1). Approach implemented.

Rainfall patterns were characterized for two locations in Portugal, Lisbon and Aveiro (Fig. **2**). The locations represent two situations of interest for RWH and were selected because: i) are located in the coast, where most of the population lives; ii) water consumption patterns studies are available; and iii) the two locations are only 240 km apart. The first two reasons for the locations selection are related with the relevance and accuracy of the study, while the last aims at evaluating the magnitude of the influence of the rainfall spatial variability on the performance of RWH systems performance in Portugal. Due to the length differences of the available measured rainfall time series in each location, the dataset of Belo-Pereira *et al*. [186] was used instead. The dataset is comprised of 53 years (1950 to 2002) of daily rainfall series in a 0.2° grid covering Portugal mainland. It was developed by interpolation of rainfall measurements from more than 800 stations spread through Portugal mainland, some of which with over 70 years of records.

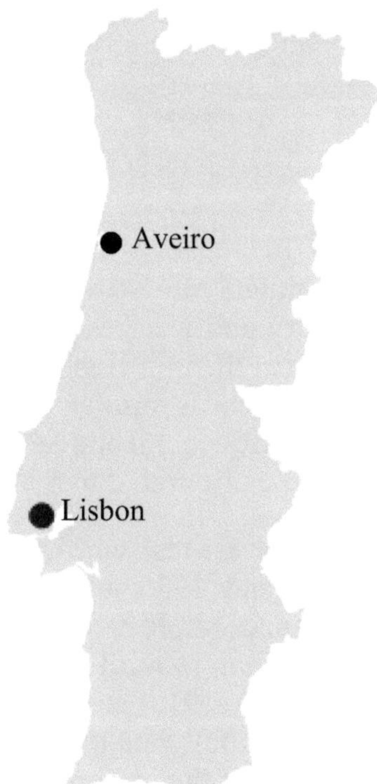

Fig. (2). Location of Lisbon and Aveiro in mainland Portugal.

The annual rainfall between 1950 and 2002 in Lisbon and Aveiro and the maximum number of consecutive days without rain are presented in Fig. (**3**).

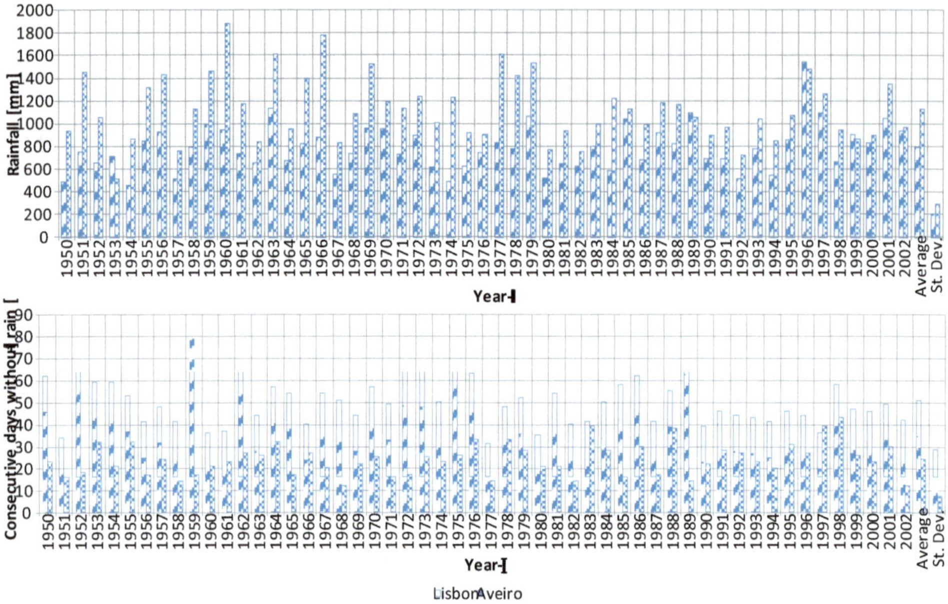

Fig. (3). Yearly rainfall series (top) and maximum number of consecutive days without rain (bottom) (1950-2002).

It can be seen a significant inter-annual rainfall variability in both locations, with different patterns of alternating dry and wet years. On average, there is roughly 300 mm more rainfall in Aveiro than in Lisbon. The average number of consecutive days without rain in a year is similar in Lisbon and Aveiro, but the average maximum and the variability in Lisbon is higher. The yearly median number of days without rainfall varies from to 2 to 6 in both locations.

Furthermore, an independent-samples t-test confirmed that the annual rainfall pattern in Aveiro between 1950 and 1979 and between 1980 and 2002 are different. Assuming unequal variances, there is statistically significant difference ($t(48)=2.57$, $p=0.013$) for the annual rainfall in Aveiro ($M=1207$, $SD=325$, between 1950 and 1979; $M=1022$, $SD=191$, between 1980 and 2002). This raises a series of questions regarding the rainfall data that should be used for simulating RWH systems that surpass the scope of the present chapter. To be consistent, and since the goal was not to evaluate the influence of the rainfall pattern change with time, the 53 years' rainfall series are used in both locations.

Consistent with the general characterization for Portugal, the rainfall in both locations is markedly seasonal (Fig. **4**). Officially, the hydrological year starts in October and the wet season lasts for 9 months (October to June) followed by 3 dry months (July, August and September). However, the transition between the two seasons is not abrupt, with the wet season showing a progressive evolution in the rainfall amount between the dry months and the wettest months (November, December, January and February)

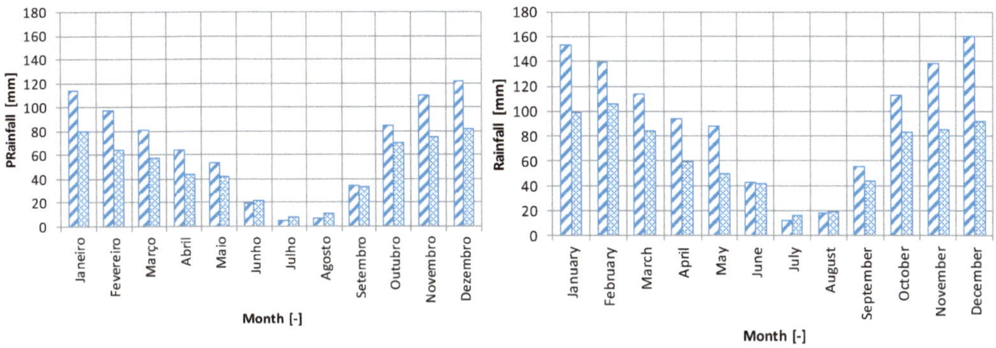

Fig. (4). Monthly rainfall statistics in Lisbon (left) and Aveiro (right) between 1950 and 2002.

At a daily scale it becomes evident that even during the wet season there are relatively short periods of rainfall, some of them of high rainfall intensity, followed by periods of several days or even weeks without rainfall. Fig. (**5**) presents the daily rainfall in 1958, which was selected because the rainfall during this year is the closest one to the average value of the period between 2001 and 2010.

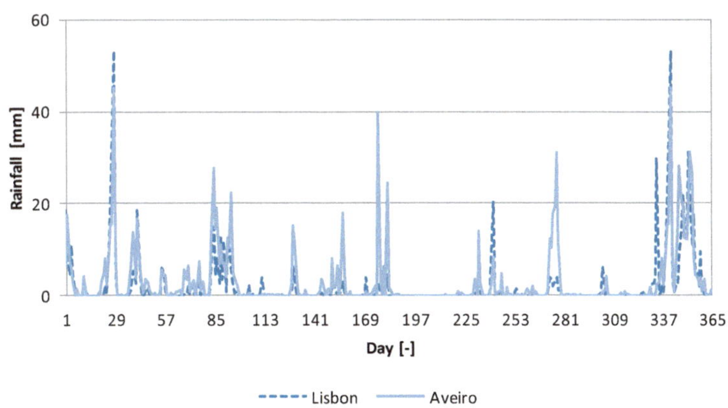

Fig. (5). Daily rainfall series (year of 1958).

The potential effect of climate change in the rainfall pattern is an increasingly important issue to take into account when evaluating RWH. Despite the differences in the quantification of the climate change effects in Europe, the studies available are consistent in predicting significant variation both in the amount and the time distribution of the rainfall. Invariably, it is estimated an increase of the rainfall amount in the Northern regions and in mountainous areas, a decrease in the Mediterranean region and relatively little change in the areas in between. The time distribution estimates show an almost general increase of the rainfall during the wet season and a decrease in the dry season [6, 187, 188].

An overall decrease in the rainfall amount is estimated, with more incidence in the spring and summer months for the Mediterranean region, and Portugal in particular. The estimates also indicate a decrease of the mean number of wet days, an increase in the periods without rainfall and an increase in the mean rainfall intensity on the wet days [175, 182, 189 - 191]. An increase of the rainfall variability in the wet days and of the annual rainfall amount is also estimated for Portugal, which is already among the highest in Europe [188 - 191].

If, in the one hand, the climate changes scenarios for Portugal are unfavourable for RWH when compared to wetter regions with rainfall throughout the year, on the other hand, they urge the need for an efficient use of water and for pursue of alternative water sources. In this context, RWH systems can be seen as additional water reserves that may be important for suppressing the needs during the dry season.

Water Consumption

Based on the critical review by Morrison and Friedler [192], it was devised that the methods used to measure water consumption and use patterns can be organized into three groups: a) direct methods; b) semi-direct methods; and c) indirect methods. Direct methods involve measuring the consumption in each fixture (direct metering). This approach was used by Edwards and Martin [193] and requires the installation of a meter dedicated to each fixture. In theory, this is the most accurate approach, but the overall system reliability, the meters' classes and the costs limit its use. The semi-direct methods are based in high frequency measurement to allow the disaggregation of the signal in order to identify the operation signature of each individual fixture. Larson *et al.* [194] used a pressure-base sensor (HydroSense) to record the pressure transients and tested two algorithms to identify each particular fixture or appliance pressure signature. This system show promising results, but it can be affected by pressure transients from other sources (*e.g.*, public network, other buildings or apartments) and there is always the issue of the pressure transient signature for partial openings [192]. The flow trace analysis is conceptually similar approach that uses a signal recognition technique to assign a specific fixture or appliance to each water-use event from high resolution flow data. This technology has been used successfully in several utility sponsored studies [195 - 203], as well as in some independent or academic research studies [204, 205], but it can't distinguish between similar fixtures or appliances [191] and loses accuracy when they are used concurrently [206]. The Identiflow system is another semi-direct method that identifies and classifies each water-use event of specific fixtures or appliances from flow data using a decision tree algorithm [207]. The system is only available through WRc consultancy services, which report high accuracy results, but the decision three will always fail in anomalous water-use events, it doesn't differentiate between similar fixtures [208] and it is prone to human error [192]. The last group, indirect methods, is composed by surveys, questionnaires, interviews or other forms of characterization of water consumption and end use from users. These approaches are the most used in practice for their simplicity and low-cost, having been used in several studies [209 - 214], but are dependent on the willingness of the participants or practical limitations [192]. Consequently, the results may be inaccurate or biased due to varying levels of participation of different types of participants, fluctuation of the level of participation with time, or possible behaviour change due to the awareness of being monitored, amongst other factors [215, 216].

According to the official statistics [217], the average daily residential water consumption in Portugal is 61 litres per capita. However, according to a study from the largest water supplier, EPAL (Empresa Pública das Águas Livres), the national average daily residential water consumption is 152 L *per capita* [218]. Other sources are consistent with the results from the EPAL study [85, 121, 123, 219, 220]. Within the present study, a mixed approach of questionnaires and metering was used to determine the water consumption pattern, being the details presented in each case study.

Validation

In 2011, Sonae Sierra installed a RWH system in Colombo Shopping Centre (CSC). It uses a 40,000 m^2 catchment area and it has a 150 m^3 tank. The rainwater is used in cooling towers, with an annual consumption of about 70,000 m^3 of water. Monitoring demonstrated an average 6,500 m^3 of rainwater used per year between 2012 and 2014, which is equivalent to about 9.4% of the cooling towers consumption in non-potable water savings.

The methodology was used to simulate the real system according to the consumption values recorded from 2012 to 2014. The rainfall records from the same period were retrieved from a weather station located in Cais do Sodré, Lisbon, about 12 km south of CSC. The analysis adopted the Yield After Spillage algorithm, a runoff coefficient of 0.8, a first flush of 1 mm, no annual continuity and no water in the tank at the beginning of the simulation. Other implementation details and results description may be consulted in Ponces [122].

Even without any parameter tuning, the results indicate that the existing RWH system used about 6,600m^3 of rainwater per year, which is in accordance with the 9,4% non-potable water savings measured on site.

The results of the parametric analyses are presented in terms of efficiency and non-potable water savings comparing the original simulation with alternative options for the input data. Five alternative scenarios were tested, as summarized in Table **1**. Each parameter was varied individually. Efficiency is defined in percentage from the relation between the stored rainwater and the collected rainwater. Non-potable water savings are computed dividing the used rainwater by the non-potable consumption.

Table 1. Parametric analyses data.

Parameter	**Original Simulation**	**Alternative**
Rainfall	Daily values	Average daily values
Consumption distribution	Monthly average	Annual average
Weather stations	Cais do Sodré	Caneças ; Point 174
First flush (ff)	1 mm	2 mm
Rainfall fraction before consumption (rfbc)	0	0.5 ; 1

Fig. (**6**) shows the efficiency of the system for different capacities of tank using both daily rainfall data and average daily data. Average daily rainfall values correspond to the average value of the rainfall in each day of the rainfall record series. Therefore, the average daily rainfall data shows more number of days with nonzero values and lower rainfall peaks. This results in a more uniform rainfall distribution throughout the year and, consequently, there is an unrealistic increase of the systems efficiency. The difference when compared to using the real daily rainfall data is about 20%.

Consideration of daily consumption values determined for each month was compared to daily consumption value determined as an annual average. However, Colombo Shopping Centre has a much higher consumption than rainwater availability, so this parameter does not show significant differences in this case.

Comparing the rainfall recorded at the Cais do Sodré station during the years of the RWH system operation with the historical 23 years rainfall series from the Caneças station (1983-2007), used in the RWH system design, and with the 52 years rainfall series (1951-2003) on point P174 from the study of Belo-Pereira *et al.* [186] it is possible to observe the importance of the rainfall data accuracy when analysing the efficiency of RWH system. The three sources of data yield significant differences in the final annual results, as illustrated in Fig. (**7**).

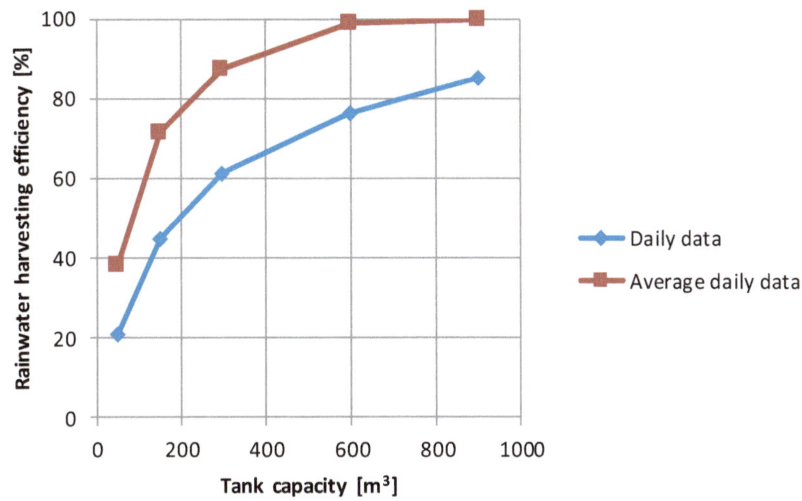

Fig. (6). RWH system efficiency depending on the rainfall data time scale.

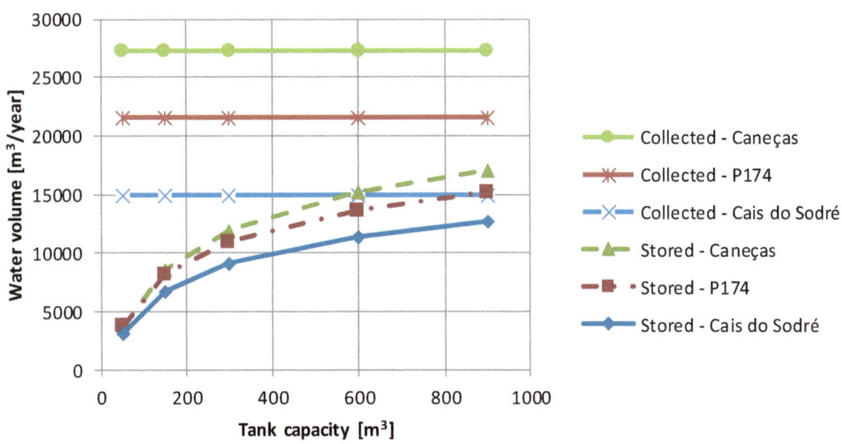

Fig. (7). Water volume variation for different weather stations.

First flush variation does not affect the efficiency of the RWH system since the collected rainwater, already excludes first flush. Fig. (**8**) presents the variation of non-potable savings for three different values of first flush (ff), indicating the higher first flush flows decrease the non-potable savings, especially for larger tanks.

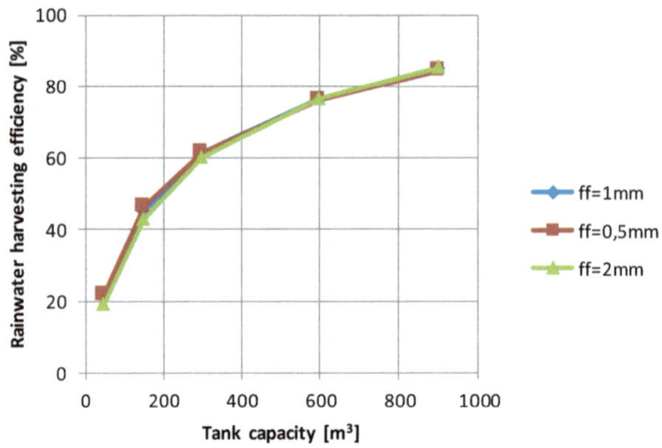

Fig. (8). Non-potable water savings for different values of first flush.

Assuming that the entire rainfall occurs either before or after the consumption does not exactly correspond to reality. Intermediate values represent a more distributed sequence of events, which is expected to increase the systems efficiency. However, for larger tanks the differences tend to vanish since there is more capacity regardless of when consumption and rainfall occur. This behaviour is captured in Fig. (**9**).

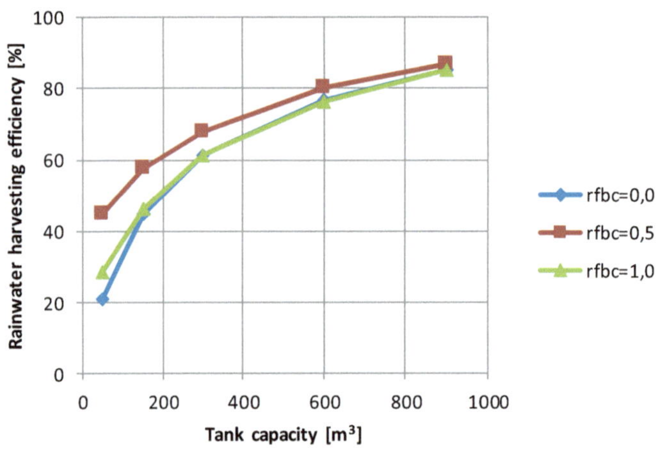

Fig. (9). Efficiency depending on rainfall fraction before consumption (pfbc).

CASE STUDIES AND RESULTS

Simulation Configuration

The four case studies were identified to test the influence of different water use patterns. Results of RWH efficiency and water savings (WS) efficiency are presented separately for each case study, for two geographical locations in

Portugal and for different collection areas and tank capacities. RWH efficiency is computed as the ratio between the harvested rainwater and the total rainwater while WS efficiency is defined as the ratio between the used rainwater and the non-potable water demand. Adequate seasonal variations of the water consumption were accounted for on a monthly basis depending on the case study. The Yield After Spillage algorithm, a runoff coefficient of 0.8 and a first flush of 1mm were adopted, following the most common international recommendations. No water in the tank at the beginning of the simulation and no annual continuity were assumed.

The results presented herein use the tank capacity in a continuous linear scale for visualization purposes only and discrete values may be identified by the points in the figures.

Single-Family House

A typical Portuguese single-family house with 2 bedrooms, 1 living room, 1 bathroom and 1 kitchen was chosen. Fig. (**10**) exemplifies typical houses as the one described.

The non-potable uses considered for rainwater in this case study include toilet flushing, clothes washing and outdoor irrigation and washes. Daily toilet flushing and clothes washing do not exhibit relevant fluctuations in Portugal, so it is realistic to assume they are uniformly distributed along the year. Following the results of the survey conducted in Silva *et al.* [29], this non-potable consumption corresponds to 38% of 154 litres per person per day (Memon e Butler).

Fig. (10). Typical single-family houses under study.

Five distinct water consumption patterns were tested. The first three assume constant total water consumption along the year and vary the expected number of permanent occupants of the single-family house from three to five. Non-potable water consumption pattern Q_1 (176 l/day) corresponds to three, Q_2 (234 l/day) to four and Q_3 (293 l/day) to five permanent residents in the house. The two other consumption patterns represent different water consumption patterns in the four residents house during summer months, due to holidays (Q_4) and higher irrigation needs (Q_5). Pattern Q_4 assumes only 2/3 of the occupants using the single family house during the months of July and August, and, consequently, smaller non-potable water consumption during this period (156 l/day *versus* 234 l/day). Pattern Q_5 is applicable whenever irrigation of the garden plays an important role and it includes an increment of the water needs from June to September in 114 l/day, corresponding to 6 mm/day irrigation of a 50 m^2 garden, which is a representative water demand for a Mediterranean climate such as Portugal [221].

Five distinct collection areas were identified: A-80 m^2; B-110 m^2; C-150 m^2; D-200 m^2 and E-5000 m^2. Surfaces A to D correspond to parts of the house roof and attached buildings, such as car garage or annexes for gardening tools. Collection area E is chosen to include the possibility of having an adjacent parcel of land next the house with a water cistern. The six different tank capacities tested follow similar hypotheses and vary from 200 l to 20 m^3.

The RWH efficiency for Lisbon and Aveiro is presented in Fig. (**11**) for consumption pattern Q_5 and for the different collection areas and tank capacities analysed. As expected, this indicator increases with the tank capacity and decreases with the collection area, due to the relation between the harvested rainwater and the non-potable water consumption. A bigger tank shows larger storage potential while a larger collection area increases the available rainwater, resulting in more spillage. Considering a tank with infinite capacity and the smallest collection area (80m^2), the RWH efficiency in Lisbon tends to an average maximum of 66% both in Lisbon and in Aveiro. The differences between both locations are more visible for larger collection areas. For instance, when adopting an intermediate value for the collection surface of 200 m^2 one obtains a maximum RWH efficiency of 50% in Lisbon and only of 42% in Aveiro, a difference of about 16%. This is mainly justified because the total rainwater in Aveiro is higher than the total rainwater in Lisbon, as described in the previous sections. Furthermore, depending on the rainfall series of each year, the RWH efficiency can vary widely, as concluded also, *e.g.*, in Silva *et al.* [29].

The water savings efficiencies shown in Fig. (**12**) increase with both the tank capacity and the collection area. As expected from the literature review, the influence of the collection area on the water savings becomes more relevant as the tank capacity increases, which is coherent with the results from other authors. The water savings efficiency in Lisbon tends to an average maximum of 67%, for collection area D, and 90%, for collection area E. This is due to the fact that for small water tanks the storage capacity is the limiting factor and with the increase in tank capacity the collection of rainwater becomes the limiting factor. In Aveiro, the performance of the system is more interesting since large collection areas (such as surface E) tend to 97% of average water savings efficiency and even surface D allows a maximum average water savings efficiency of 78%, a significant higher value when compared to Lisbon mainly due to the larger amount of rainfall expected for Aveiro. Again, the potential water saving varies significantly between years depending on the rainfall amount and temporal distribution.

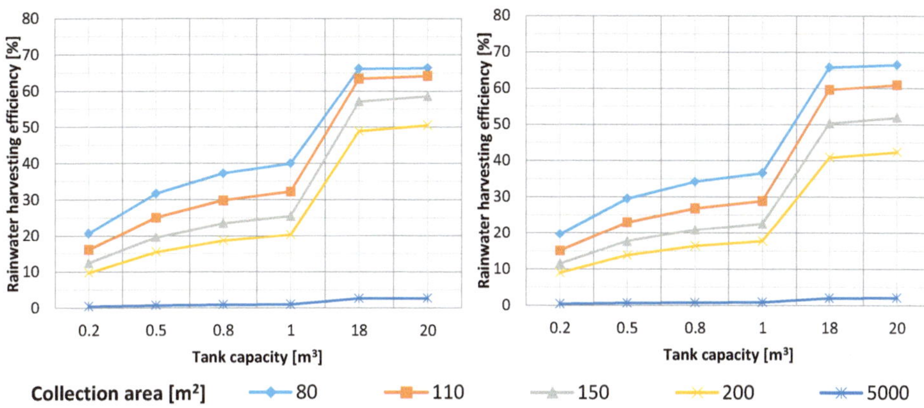

Fig. (11). RWH efficiency in Lisbon (left) and Aveiro (right) for different collection areas and water consumption pattern Q_5.

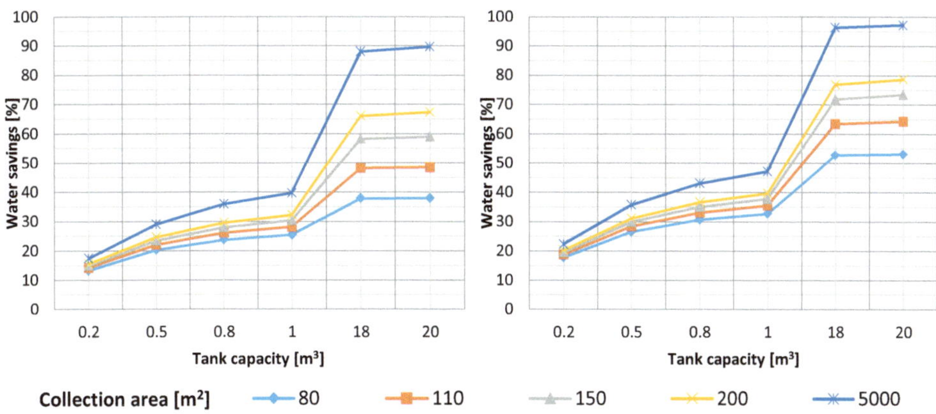

Fig. (12). Water savings efficiency in Lisbon (left) and Aveiro (right) for different collection areas and water consumption pattern Q_5.

The impact of different water consumption patterns in the single family house is illustrated in Fig. (**13**). More occupants of the house lead to more non-potable consumption and, consequently, smaller water savings efficiency (Q_3 *versus* Q_1). The water savings efficiency also changes if a non-potable consumption is expected to vary along the year. When comparing pattern Q_5, which is more demanding from June to September, with the uniform non-potable consumption of water pattern Q_2, it is noticeable that the water savings efficiency in case of water pattern Q_5 decreases (13.7%). This is easily explained because the consumption increase is coincident with the well-marked dry season of the rainfall regime in Mediterranean countries such as Portugal. The differences between Q_4 and Q_2 are not so perceptible (about 6%). However, since pattern Q_4 assumes smaller consumptions in July and August due to summer holidays and the rainfall is usually insignificant during this period, the performance of the RWH system is higher for pattern Q4 than pattern Q2. Similar results are obtained for Aveiro (Fig. **14**). In particular, the average maximum water savings efficiency in this case is 4.6% higher for consumption water pattern Q_5 and 13% smaller for consumption water pattern Q_4 when compared to the associated uniform water consumption pattern Q_2. The inversion on water savings efficiency due to consumption variation is explained by the fact that the rainfall in Aveiro is over the double of Lisbon during the dry season.

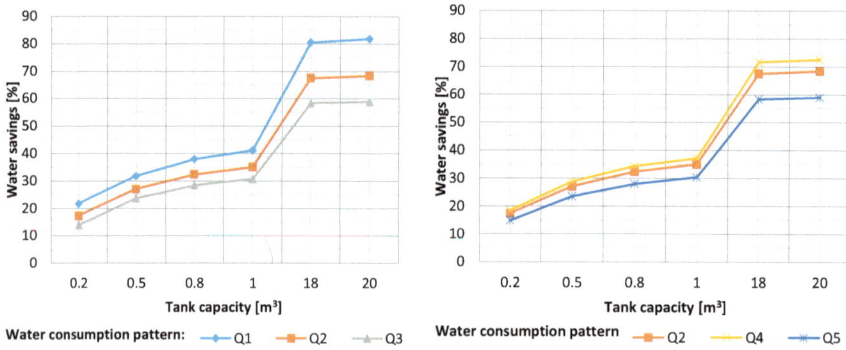

Fig. (13). Water savings efficiency in Lisbon for different water consumption patterns and collection area C – 150 m^2.

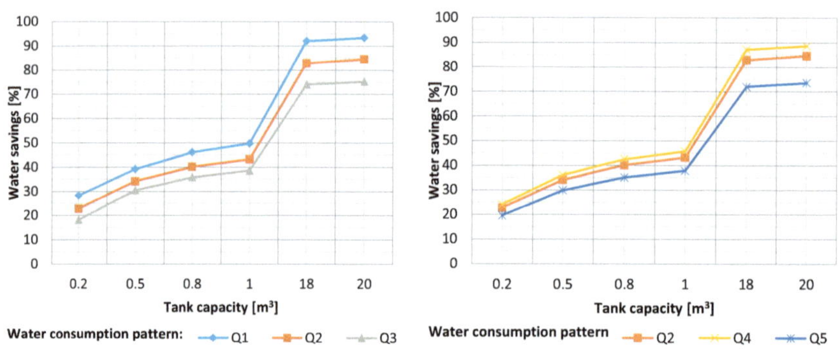

Fig. (14). Water savings efficiency in Aveiro for different water consumption patterns and collection area C – 150 m^2.

Residential Building

A four floors building with two independent apartments per floor was considered in this analysis. Fig. (**15**) illustrates a typical building with these conditions, quite common both in more cosmopolitan cities such as Lisbon and also in smaller towns such as Aveiro.

Fig. (15). Typical residential buildings under study.

The end-use water consumption of multi-family houses in Portugal is estimated from the survey provided in Lúcio [123]. The survey involved inquiries requesting the participants to measure the total water consumption and non-potable water consumption (clothes washing machines and toilet flushing) to identify the non-potable consumption susceptible of being replaced by rainwater. The household selection covered a wide range of occupancy typologies, including households of families with children of different ages, couples, single people, pensioners and students, and therefore it is considered representative for the present study. The results obtained are consistent with literature review, namely in terms of the total water consumption and toilet flushing. Therefore, the evaluation of the RWH savings potential in Lisbon computed in this work assumes a total daily water consumption of 146 l/ppd, of which roughly 28% (41.5 l/ppd) is used in non-potable end-uses.

In order to indirectly estimate the performance of higher residential buildings, as the one visible in the background of Fig. (**16**), different occupation rates of each apartment were investigated, leading to five distinct non-potable water consumptions patterns. As in the singly family house, constant non-potable water consumptions along the year are compared to non-uniform water patterns. Patterns Q_1 to Q_4 correspond to increasing number of occupants per apartment, from 1.5 residents/apartment to 3.5 residents/apartment. Lisbon average value of occupancy corresponds to 2.2 residents/apartment [123] which is in accordance with the adopted rates. These four water consumption patterns are assumed to be constant in all months. An alternative situation Q_5 is also included to simulate the influence of smaller occupancy rates during common summer holidays (July and August). The number of residents expected for each water consumption pattern scenario and the associated daily non-potable consumption of the building are shown in Table **2**.

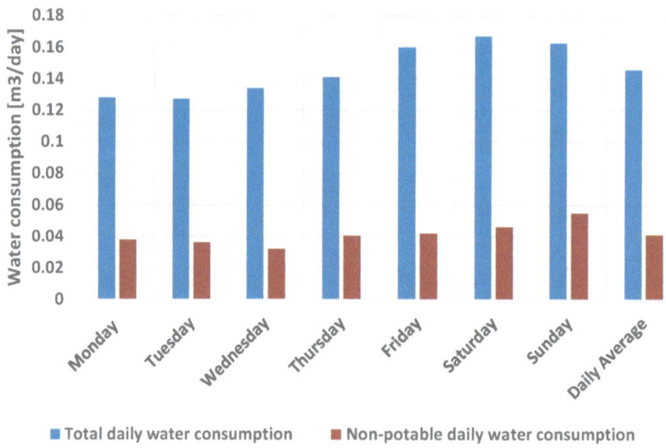

Fig. (16). Water consumption *per capita* during one week: potable and non-potable daily consumptions.

Table 2. Five different water consumption patterns in the residential building.

<table>
<tr><th rowspan="2"></th><th colspan="5">Water Consumption Patterns</th></tr>
<tr><th>Q1</th><th>Q2</th><th>Q3</th><th>Q4</th><th>Q5</th></tr>
<tr><td>Residents/apartment</td><td>1.5</td><td>2</td><td>2.5</td><td>3.5</td><td>2.5</td></tr>
<tr><td>Number of floors</td><td colspan="5">4</td></tr>
<tr><td>Number of apartments/floors</td><td colspan="5">2</td></tr>
<tr><td>Number of residents in the building</td><td>12</td><td>16</td><td>20</td><td>28</td><td>20</td></tr>
<tr><td rowspan="5">Daily water consumption (l/day)</td><td colspan="4">Jan-Dec</td><td>Jan-Jun and Sept-Dec</td></tr>
<tr><td rowspan="4">1752</td><td rowspan="4">2336</td><td rowspan="4">2920</td><td rowspan="4">4088</td><td>2920</td></tr>
<tr><td>Jul-Aug</td></tr>
<tr><td>1947</td></tr>
<tr></tr>
<tr><td rowspan="4">Daily non-potable water consumption (l/day)</td><td colspan="4">Jan-Dec</td><td>Jan-Jun and Sept-Dec</td></tr>
<tr><td rowspan="3">497.568</td><td rowspan="3">663.424</td><td rowspan="3">829.280</td><td rowspan="3">1160.992</td><td>829</td></tr>
<tr><td>Jul-Aug</td></tr>
<tr><td>553</td></tr>
</table>

The possibility of refining even more the water consumption pattern and choosing different non-potable consumptions values during the week was also analysed. However, according the Lúcio [123] survey (Fig. **16**), no relevant evolution is observed for the water consumption *per capita* during the week, so the average value was adopted as the daily consumption.

Seven different collection areas were tested (200, 250, 300, 350, 400, 450 and 500 m^2) covering usual values of the total area occupied by the building. Furthermore, the efficiency of the rainwater system in residential building was computed for seven tank capacities (2, 4, 8, 16, 24, 32 and 40 m^3). The maximum size of the tank was limited to the volume of roughly two parking lots, an option deemed viable for existing buildings. In a new building, the tank could easily be significantly larger.

When looking at the RWH efficiency results of the building (Fig. **17**), one identifies the same behaviour obtained for the single family house. No relevant differences are visible when comparing the performances in Lisbon and in Aveiro. Changes in the collection area have proportional changes in the RWH efficiency, with no significant influence of the tested tank capacities, especially in Aveiro In the case of the buildings under analysis, an average maximum is recognized only for the smallest collection area in Lisbon (66%), since the maximum tank capacity is still far away from an infinitive reservoir in the other tests.

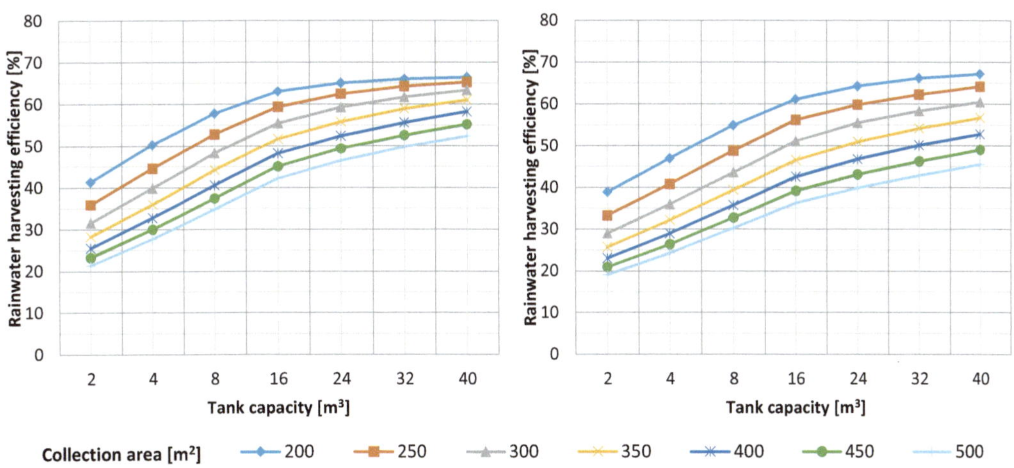

Fig. (17). RWH efficiency in Lisbon (left) and Aveiro (right) for different collection areas and water consumption pattern Q_3.

Fig. (**18**) illustrates the water savings efficiency in Lisbon and Aveiro for different collection areas and water consumption pattern Q_3. This parameter is quite different for Lisbon and Aveiro, due to the distinct rainfall regimes. Collection areas larger than 400-450 m^2 show similar performances, especially in Aveiro, even for bigger tanks. For Lisbon, the expected average maximum water savings in these cases is around 58% *versus* the 73% in Aveiro.

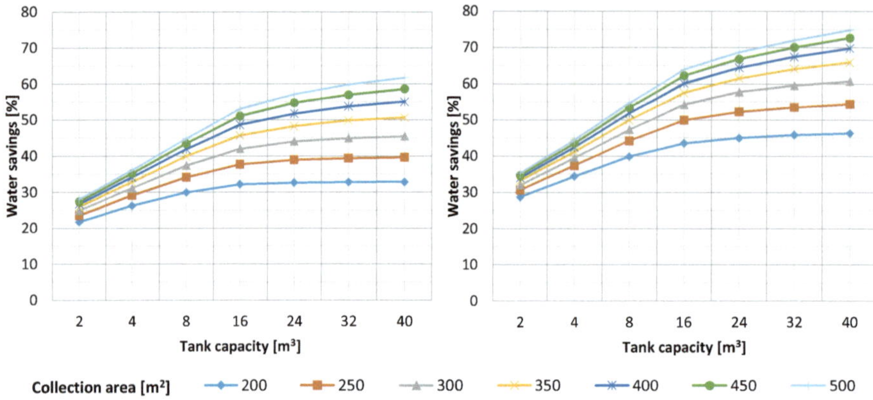

Fig. (18). Water savings efficiency in Lisbon (left) and Aveiro (right) for different collection areas and water consumption pattern Q_3.

The influence of the water consumption pattern is clarified in Fig. (**19**) for collection area C-350 m^2. As expected, water savings are higher in Aveiro for all water consumption patterns, with differences of about 25-35%. Water consumptions patterns with higher non-potable demands lead to smaller water savings of the RWH system (Q_1 to Q_4), but still reasonable values, always above 35% in Lisbon and 50% in Aveiro for medium size tanks. Pattern Q_5 presents a slightly better performance than pattern Q_3 (around 6% both in Lisbon and in Aveiro) due to its smaller consumption during July and August, when rainfall is typically scarce.

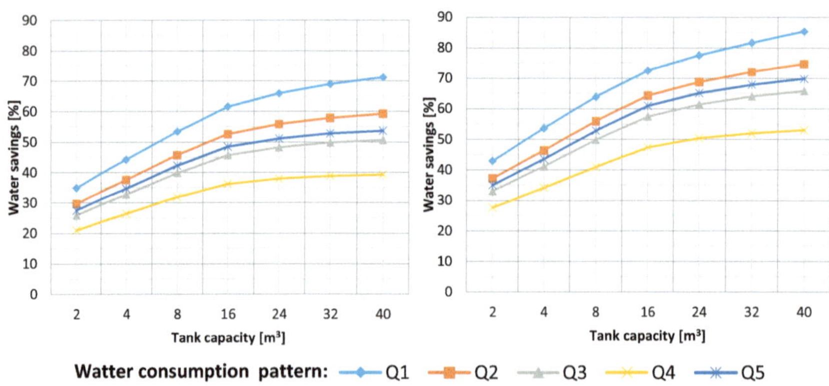

Fig. (19). Water savings efficiency in Lisbon (left) and Aveiro (right) for different water consumption patterns and collection area C – 350 m^2.

When comparing the performance of RWH system in buildings to similar systems in single-family houses, one obtains analogous behaviours but smaller water savings efficiencies for the building case, mostly because of the higher values of the non-potable water demand. Bigger tanks and collection surfaces could be obtained if the RWH was design over a certain block instead of isolated buildings.

University Campus Building

The Santiago campus of the University of Aveiro extends for an area of over 460,000 m^2 and is composed by 42 buildings. Each department has its separate building and there are also individual buildings for other services in the university (*e.g.*, library, sports, central administration, student's dormitory, and canteen).

The Civil Engineering Department (DECivil) building was selected to represent the geometry and architecture of a university campus building in Portugal. The small size of the population (roughly 300 people including students, researchers, professors and administrative and lab workers) and the dynamics of the building users allows for a high familiarity between staff member and students from all years, minimizing the variability from lack of participation or different behaviours between the various classes of occupants. In architectural terms the building can be characterized as modern with a rectangular layout and three floors (Fig. **20**). The ground floor is occupied by the laboratory and the administrative offices. The structures and hydraulic spaces of the laboratory extend up to the second floor, being the rest composed by classrooms and graduate students' rooms. The staff offices are located in the third floor, along with additional classrooms. The toilets in the second and third floor are aligned vertically and in the ground floor they are located close to the building main entrance.

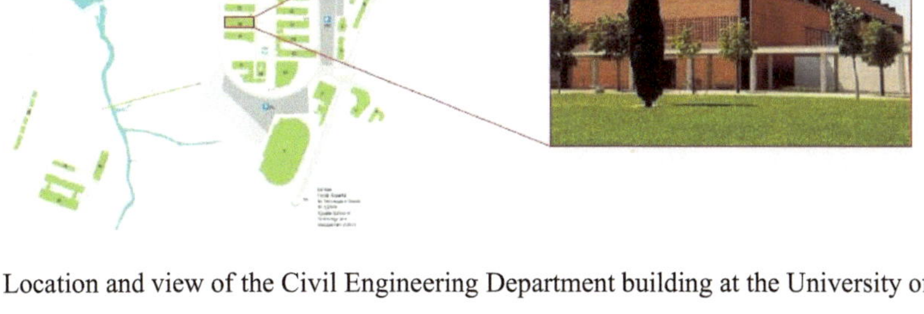

Fig. (20). Location and view of the Civil Engineering Department building at the University of Aveiro.

In the DECivil building water is required, mostly, to satisfy domestic/toilet demands, cleaning requirements and laboratory work (Table **3**). Toilets GT (Gentlemen toilets) and LT (Ladies toilets) 1, 2 & 3 are identified as having the most intensive use of water in the building, being responsible for roughly 70% of the building's water consumption. The methodology used to evaluate the end-use water consumption pattern encompassed two stages: i) identification and characterization of periods with different water consumption; and ii) determination of the water consumed in each end-use. The former was accomplished analysing the total water consumption in the building recorded and the latter through a monitoring campaign [222, 223].

Table 3. Characterization of the water devices in the DECivil building.

Toilet	Floor	Washbasin	Toilet		Urinal
Gentlemen					
GT1	1	2	2		4
GT2	2	3	2		3
GT3	2	2	1		3
Ladies					
LT1	1	3	2		-
LT2	2	2	2		-
LT3	2	2	2		-
Disabled					
DT1	1	1	1		-
DT2	2	1	1		
Laboratory					
LT1	0	2	1		2
LT2	0	2	2		-
Secretary					
ST	0	2	1		2
Laboratory	**Floor**	**Laboratory faucet**	**Sink faucet**	**Hose tap**	**Shower**
Laboratory 1	0	2	2	1	1
Laboratory 2	2	4	-	-	-

The total water consumption in the building is being recorded by a telemetry system at an hourly rate. Only the records from 2013 were analysed because there has been a gradual decrease on the total number of students, an increase of foreign students over the past years and it was out of the scope of the work evaluate potential consumption trends or variability. The complete dataset has 349 days of flawless daily records and 295 days of flawless hourly records. A preliminary analysis allowed the identification of possible distinct water consumption patterns between weekdays and weekends and between the various periods of the academic calendar. Regarding the latter, 7 distinct calendar periods were initially identified: i) breaks; ii) classes; iii) exams; iv) interruptions; v) thesis; vi) vacations; and vii) holidays. However, since except for one case the holidays coincided with weekends, breaks, interruptions or vacations, this period was discarded from the analysis. The distinction between breaks and interruptions is that the former occurs before a period of exams while the latter is used for various academic events. The thesis is a period before the summer vacations dedicated exclusively for thesis presentations. These determinants form a total of 12 groups of days with possible distinct water consumption patterns.

It is possible to observe that, with the exception of the breaks, exams and classes, the daily water consumption is highly irregular (Fig. **21**).

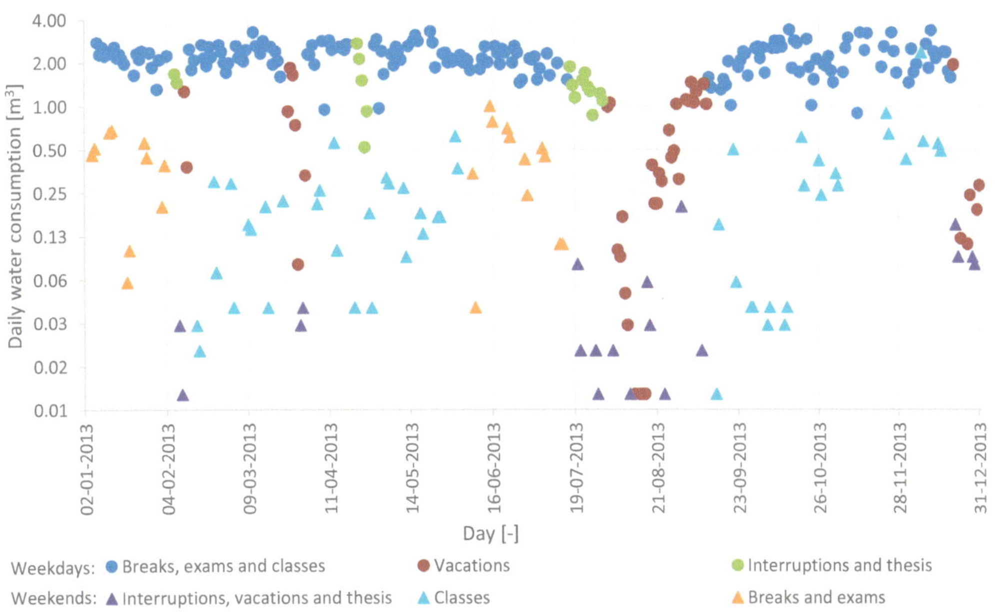

Fig. (21). Daily water consumption records in the year of 2013.

Even for the breaks, exams and classes the variability increases significantly with the resolution of the analyses, with the average daily water consumption pattern being characterized by two distinct periods of high (12:30-21:00, over 150 l/h) and low (22:00-11:00, bellow 50 l/h) consumption. Additionally, the water consumption in each hour can also be significant, has can be seen in the histogram of the water consumption at 14:00 (Fig. **22**).

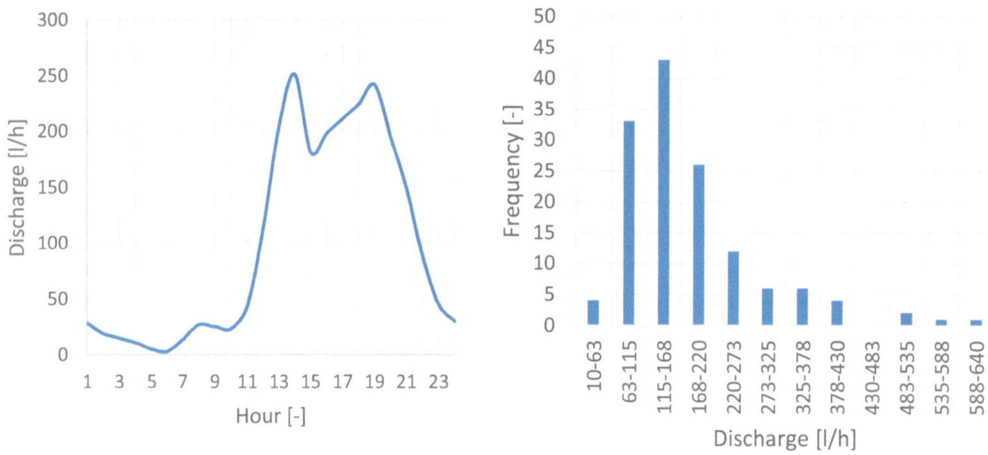

Fig. (22). Hourly discharge pattern during breaks, exams and classes weekdays: (left) average discharge during the day and (right) discharge histogram at 14:00.

The results of the previous stage were taken into account to define the monitoring campaign, which involved two steps. The first step involved the hydraulic characterization of all fixtures in the building toilets. To characterize the devices with storage (the flush toilets), four measurements of the volume of water used per flush in each device were performed. The relative differences between results obtained in different flushes in the same device were of up to 7%, which can be considered within the order of magnitude of the measurement error. However, for devices without water storage but with flush time setting (the urinals and the washbasin faucets), which are directly supplied by the building water system, discharge depends on the water pressure. Since water pressure varies with several factors, namely the total water discharge (and consequently the total head loss) in the building at the moment, the four measurements were taken in different periods. The volume of water consumed per use was also compiled for each measurement. For the urinals, although the discharge could vary significantly (up to 43%, although the second higher difference of 15% is significantly lower), the flushed volume of water was constant (with relative differences of up to 7%). On the other hand, for the washbasin faucets the discharge flow differences (up to 7%) were far less significant than the total discharge (relative differences of up to 23%), which was dictated by the discharge time regulation. The average values per type of device installed in the main toilets are presented in Table **4**. It should be noted, however, that significant differences were observed in the used volumes of water between devices within each type.

Table 4. Average volume/discharge water consumption per type of device.

Type of Device	Units	Quantity
Flush toilet	l	7.24
Urinal	l/min	14.72
	l	1.53
Washbasin	l/min	7.06
	l	0.76

The second step involved recording the number of uses of each fixture. Considering the methods for measuring water consumption and use patterns referenced in the literature, it was considered that most may not be suitable or viable for the case study. Direct metering was considered unviable due to the investment and complexity of installing such a system. The semi-direct methods based on analysing and disaggregating high resolution metering records were considered not suitable without prior validation since there will be several similar fixtures discharging simultaneously. Therefore, a diary survey was used in the monitoring campaign. Contrarily to residential buildings, the population in the case study building is significantly larger and it was not possible to ensure the complete participation of all users. In order to promote participation and obtain, to some extent, a partially accurate water consumption and end use pattern characterization, a modified diary survey was adopted involving a mixed approach entailing: i) semi-direct observation; and ii) indirect observation. The semi-direct observation required an observer permanently outside the toilets inquiring the users which fixture used and the number of discharges/pushes. This ensured the participation of all users on the toilets monitored. The indirect observation was based on the voluntary participation of the building users in filling the forms placed in each toilet. The form included the plant of the toilet with a coded identification of the fixtures and a question on the number of discharges/pushes on each fixture used and the hour it took place. In addition to the toilets, water is also consumed in the laboratory and for cleaning. The water consumption in the former was quantified with the aid from the laboratory manager and by checking the work program in each monitoring day. The latter was evaluated with the cleaning team and direct observation of the water meter records since it takes place before most of the users coming into the building. Complementary, the building water meter was checked before and after every monitoring period. Lastly, the number of users in the building was obtained by checking the staff/student presence and through the number of students in the classrooms recorded by the professors in the university online documentation system.

Water consumption was monitored every Tuesday and Friday, of selected 9 weeks, during the classes' period of the second semester of 2013/2014, between February and May, 2014, continuously, from 8:30 am to 6:30 pm. Since toilets GTi, LTi and DTi are grouped (where i = 1, 2 or 3), they were monitored simultaneously. When a group of toilets was monitored semi-directly, the others were monitored indirectly. All of the main toilets were semi-directly monitored 3 weeks and indirectly monitored 6 weeks, in an alternate order. All the other water devices presented in Table **1** were monitored indirectly 9 weeks.

Although was performed the characterization of all the water devices in the building, will only be presented results for the main toilets since they are the focuses of study.

The use of water in the main toilets corresponds to 70% of the consumed water in the building, on average. For all of the monitored days, this value ranged from 50% to 81%. Fig. (**23**) presents the example of the 25th of March, 2014, where 65% of the water was consumed in the main toilets. It shows that the value of the total hourly water consumption for the period between 9 am and 6 pm obtained by telemetry and from the monitoring campaign are not the same. The reason is that part of the users does not fill the questionnaire, when monitored indirectly, for which the acquired total hourly water consumption obtained from the monitoring campaign corresponds to a sample of the real total hourly water consumption in the monitored period. However, normalizing the water consumption by the respective average hourly water consumption it is showed that, for the 25th of March, 2014, the registered sample is representative of the total consumption of water in the building for the same period. In fact, the evolution of the normalized total hourly water consumption determined by telemetry and from the monitoring campaign is similar. In addition, it is also observed that the consumption of water in the main toilets follows the same behaviour of the total water consumption in the entire building.

Looking to the volume of water consumed per type of device in the main toilets, is observed that the flush toilets use more than 70% of the consumed water, followed by the washbasins and the urinals (Fig. **24**).

The efficiency of a RWH system in DECivil campus building was tested assuming two different catchment areas, one corresponding to the total area of the roof and other corresponding to half of it. Five different tank capacities were evaluated, varying from 5 to 60 m^3. A uniform total water consumption pattern (Q_1) of 1.57 m^3/day was adopted, considering a non-potable water consumption of 83% (1.30 m^3/day). Another alternative taken into account considers different water consumption patterns between the Holydays months and the rest of the year.

This second water consumption pattern (Q_2) adopts 1.80 m^3/day for all months except July and August that show only 0.50 m^3/day. Non-potable uses are again 83% of the total water consumption along all year. The performance of the building in its original location in Aveiro was compared to a similar building in Lisbon city. Fig. (**25**) shows the RWH efficiency and Fig. (**26**) shows the water savings obtained for both cases. The efficiency of the system is lower than the ones obtained for the single family house or residential building. This performance is also identified by the water savings efficiency of the RWH system of the university building.

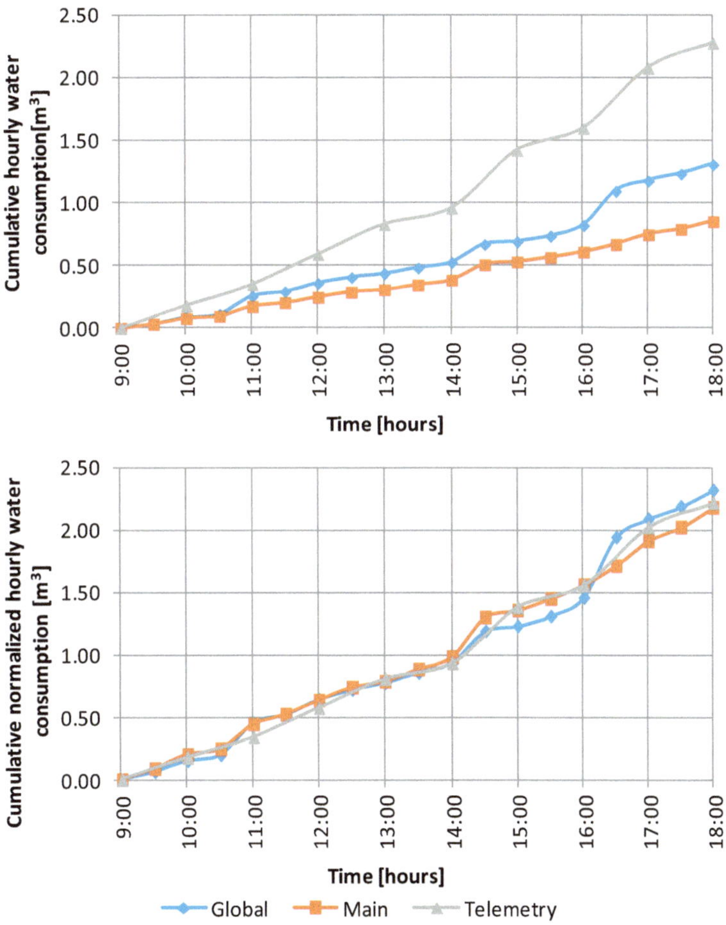

Fig. (23). Water consumption in the 25th of March, 2014 from 9 am to 6 pm: (left) total hourly water consumption in the indicated period; (right) total hourly water consumption in the indicated period, normalized by the average hourly water consumption in the same period ("Global" is the total water consumed in the building and "Main" the water consumed in the main toilets, both determined from the monitoring campaign, and "Telemetry" the total water consumed in the building determined by telemetry).

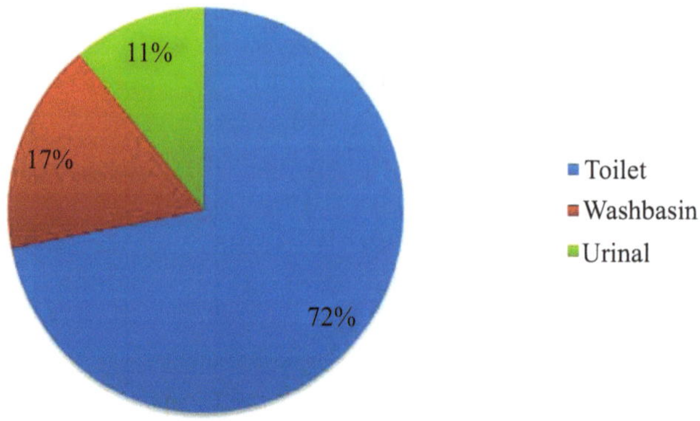

Fig. (24). Water consumption distribution per device.

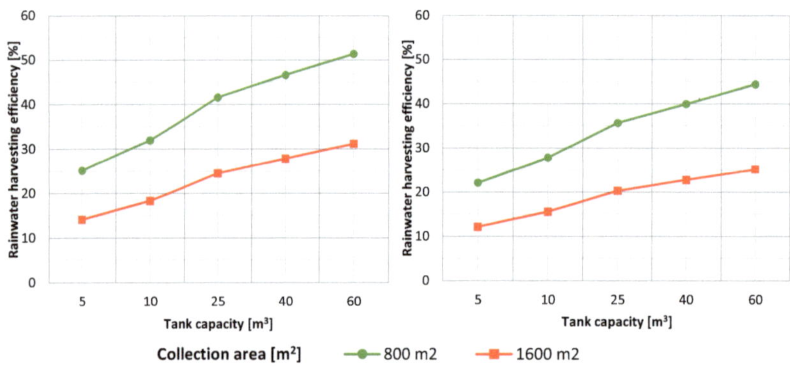

Fig. (25). RWH efficiency in Lisbon (left) and Aveiro (right) for the two tested collection areas and water consumption pattern Q_1.

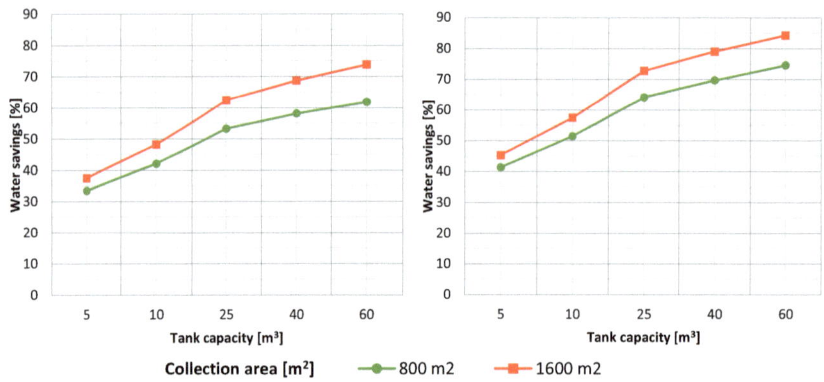

Fig. (26). Water savings efficiency in Lisbon (left) and Aveiro (right) for the two tested collection areas and water consumption pattern Q_1.

The results of the water savings efficiency for the two water consumption patterns are shown in Fig. (**27**). Again, the non-uniform water consumption pattern Q_2, that is less demanding during summer holidays, shows a better performance than uniformly distributed pattern Q_1. The differences in this case are also of about 6%.

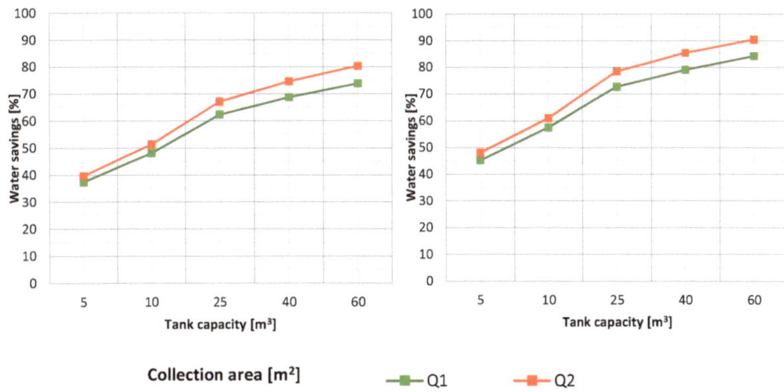

Fig. (27). Water savings efficiency in Lisbon (left) and Aveiro (right) for different water consumption patterns and collection area of 1600 m^2.

Commercial Building

Estação Viana Shopping (EVS) is located in Viana do Castelo Portugal and owned by Sonae Sierra. It was chosen as representative of a medium size commercial building in Portugal. Larger commercial buildings tend to concentrate only in the main cities (Lisbon and Porto) and smaller commercial buildings have been progressively closed. Fig. (**28**) presents the top view of the Shopping Centre.

Fig. (28). EVS: top view (Google Earth).

Fig. (**29**) indicates the water use in EVS during 2013, divided per end-use. Toilets make up the largest share.

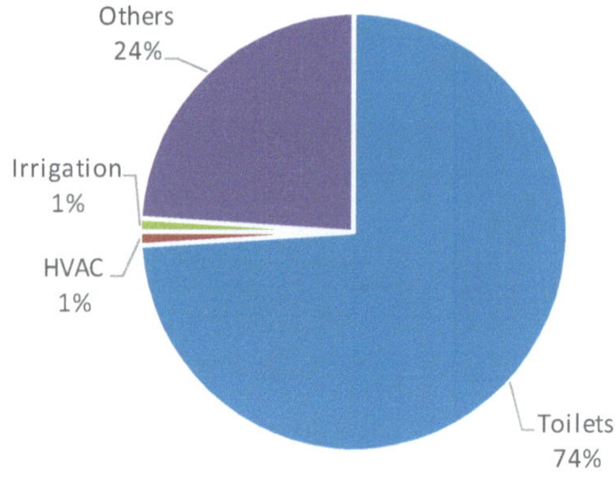

Fig. (29). EVS: water consumption in 2013.

The total non-potable water consumption was determined assuming that irrigation and Heating, Ventilation, and Air Conditioning (HVAC) are 100% non-potable uses and toilets and other uses are 90% non-potable uses. This is in accordance with Sonae Sierra monitoring results. Two water consumption patterns were adopted, one assuming the average monthly consumption measured by ESV management team and other with the average yearly value, as shown in Fig. (**30**).

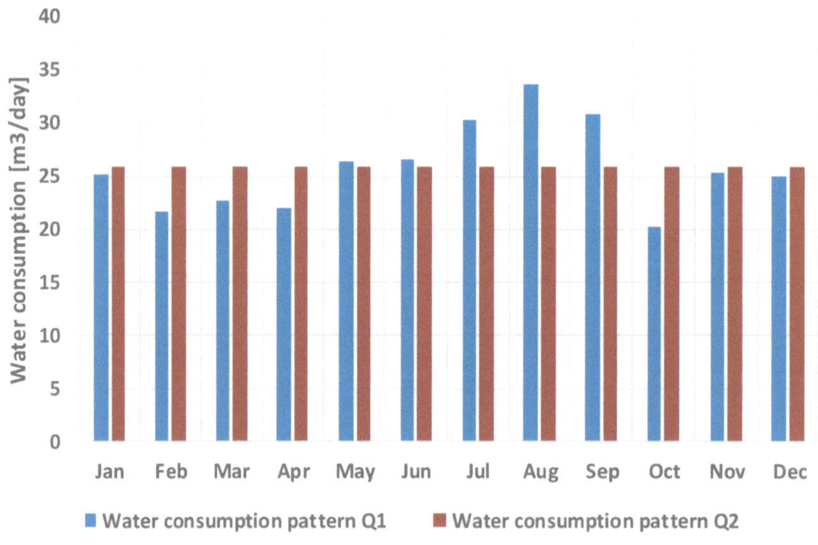

Fig. (30). EVS: two distinct water consumption patterns.

EVS architecture has three relevant catchment areas of A = 4280 m^2 and B = 9920 m^2 and c = 13470 m^2, as represented in Fig. (**31**). Tests were run for 10 different tank capacities, from 15 to 1000 m^3.

A B C

Fig. (31). EVS: Three alternative collection areas (Google Earth).

Fig. (**32**) and Fig. (**33**) illustrate the RWH system and water savings efficiencies for different tank capacities and for each one of collection areas, both in Lisbon and in Aveiro, respectively. It is possible to observe some stabilization around the 200 m^3 tank capacity, namely for the smallest collection area. Fig. (**34**) compares the water savings obtained for each water consumption pattern and collection area B. Although the differences are less noticeable in this type of building than in the other case studies, the same behaviour is recovered. The non-uniform water pattern consumption requires more water availability during summer months, leading to smaller efficiency values. The differences in this case are around 3% in Lisbon and 5% in Aveiro.

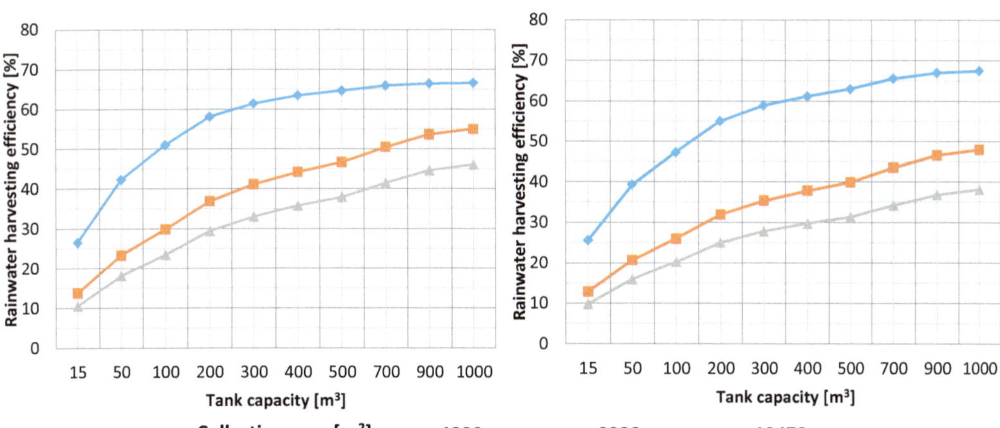

Fig. (32). RWH efficiency in Lisbon (left) and Aveiro (right) for different collection areas and water consumption pattern Q_1.

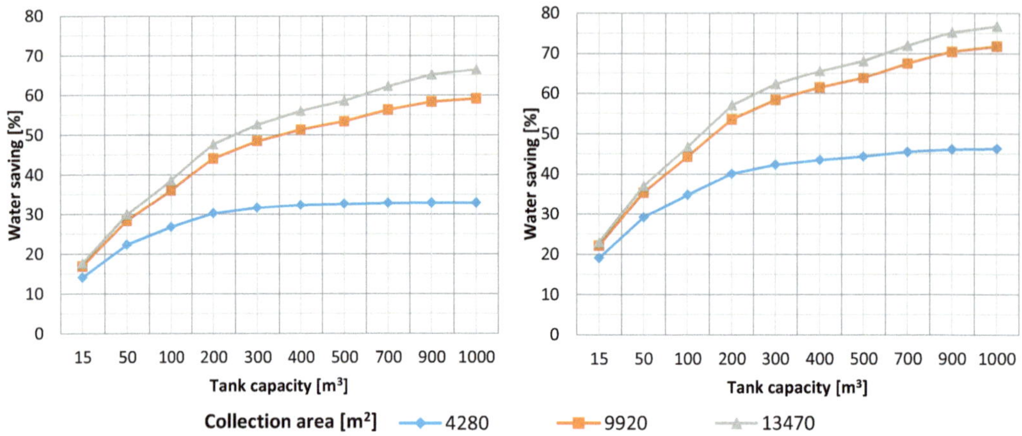

Fig. (33). Water savings efficiency in Lisbon (left) and Aveiro (right) for different collection areas and water consumption pattern Q_1.

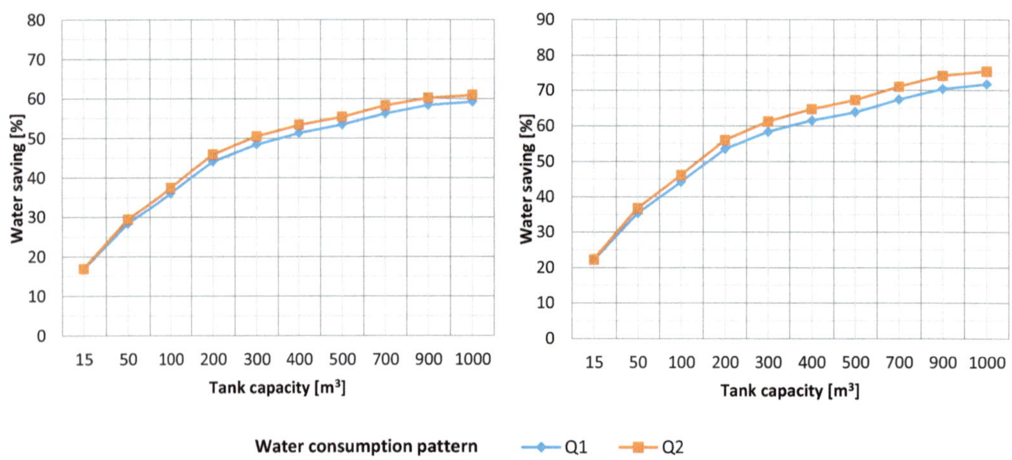

Fig. (34). Water savings efficiency in Lisbon (left) and Aveiro (right) for the two water consumption patterns and collection area B – 9920 m^2.

FINAL REMARKS

The performance of RWH systems in a Mediterranean climate is evaluated herein for two different geographical locations, four types of building and several water consumption patterns. The two chosen geographical locations cover the centre and north of mainland Portugal climate. The average annual rainfall in the location on the north of Portugal (Aveiro) is higher than the one in the centre of Portugal (Lisbon). Also the number of consecutive days without rain is different for both locations with direct implications in the performance of the RWH system. The efficiency of RWH systems of single family houses was related to the efficiency of residential buildings that show higher occupation rates. Other uses were also analysed, namely associated to university and commercial buildings. Moreover, uniform water consumption patterns were compared to non-uniform ones with higher demands in specific months of the year. This is essential do asses the influence of Mediterranean precipitation regime in the performance of the rainwater harvesting system, namely the influence of typical wet and dry seasons and significant variability of the rainfall events intensity and duration.

The RWH system efficiency shows similar values in all types of building. Since Aveiro expects more rainwater, it presents smaller values of this parameter. Differences become more relevant (about 15%) for larger collection areas. On the other hand, water savings in all building types and locations are mostly higher than 60%, proving that RWH systems are a relevant alternative water source in Mediterranean countries such as Portugal.

When comparing building types water savings efficiency, the single-family house and university building show the most competitive values, followed by the residential building and commercial shopping centre. However, water savings efficiency does not differ significantly, with average values of 75%-80% in Aveiro and 60%-70% in Lisbon, for all case studies and conventional values of the collection area and tank capacity in each case study. The single family house may achieve water savings of 95% if big collection areas are expected.

The use of non-uniform consumption patterns prove to be essential, especially when fluctuations of the non-potable demand are expected during summer period, when rainfall is scarce. Smaller non-potable consumptions during summer months increase the expected water savings of the system while higher non-potable demands in this period worsen the performance of the system. Differences of 13% are obtained in the first case and of about 6% in the second one, which may have direct impact in the viability analyses of the system and, therefore, should not be neglected.

In line with some of the most recent research on RWH (*e.g.*, [224 - 227].), the results obtained herein strengthened the need to consider the local climate (namely the rainfall), the water consumption pattern per end-use and the RWH system configuration (in particular the collection area and losses) when simulating RWH systems.

CONFLICT OF INTEREST

The authors confirm that they have no conflict of interest to declare for this publication.

AKNOWLEDGMENTS

The authors would like to acknowledge Sonae Sierra co-operation in this research.

DISCLOSURE

This chapter includes undergoing research on water efficiency developed by the authors. As a results, parts of the first two sections of this chapter have been previously published in Resources, Conservation and Recycling. Volume 94, January 2015, Pages 21-34 [http://dx.doi.org/10.1016/j.resconrec.2014.11.004].

REFERENCES

[1] United Nations, "World Population Prospects: The 2015 Revision, Key Findings and Advance Tables", In: *Working Paper No. ESA/P/WP.241, Department of Economic and Social Affairs, Population Division.* New York, 2015.

[2] United Nations Environment Programme, *Global Environment Outlook 3.* Earthscan Publications Ltd: United Kingdom and United States, 2002.

[3] United Nations, *Millenium Development Goals and Beyond 2015,* 2010.[Accessed 10th Aug. 2016] Available: http://www.un.org/millenniumgoals/

[4] United Nations, *The Millennium Development Goals, Report 2015.* United Nations: New York, 2015.

[5] United Nations, *Sustainable Development Goals,* 2016.[Accessed 10th Aug. 2016] Available: http://www.un.org/sustainabledevelopment/sustainable-development-goals/

[6] "Intergovernmental Panel on Climate Change", "Climate Change 2007: Impacts, Adaptation and Vulnerability", Tech. Report, Contribution of Working Group II to the Fourth Assessment Report of the Intergovernmental Panel on Climate Change

[7] B. Lehner, P. Döll, J. Alcamo, T. Henrichs, and F. Kaspar, "Estimating the impact of global change on flood and drought risks in Europe: A continental, integrated analysis", *Clim. Change,* vol. 75, no. 3, pp. 273-299, 2006.
[http://dx.doi.org/10.1007/s10584-006-6338-4]

[8] World Health Organization (WHO), *Report-reducing risks, promoting healthy life,* 2002.[Accessed 12th Aug. 2016] Available: www.who.int/whr/2002/en

[9] United States of America, U.S. Department of Energy, *Energy demands on water resources: Report to congress on the interdependency of energy and water,* 2007. http://www.sandia.gov/energy-water/docs/121-RptToCongress-EWwEIAcomments-FINAL.pdf

[10] A. Arpke, and N. Hutzler, "Domestic water use in the United States: A life-cycle approach", *J. Ind. Ecol.,* vol. 10, no. 1-2, pp. 169-184, 2006.
[http://dx.doi.org/10.1162/108819806775545312]

[11] UN-HABITAT, *Rainwater Harvesting and Utilisation, BlueDrops Series (3 books).* United Nations Habitat Programme: Nairobi, Kenya, 2005.

[12] S. White, A. Turner, S. Fane, and D. Giurco, "Urban water supply-demand planning: a worked example", *4th IWA Specialist Conference on Efficient Use and Management of Urban Water Supply,* 2007 Jeju: Korea

[13] T. Dworak, M. Berglund, C. Laaser, P. Strosser, J. Roussard, B. Grandmougin, M. Kossida, I. Kyriazopoulou, J. Berbel, S. Kolberg, J. A. Rodríguez-Díaz, and P. Montesinos, "*EU Water saving potential*", In: *Part 1 –Report ENV.D.2/ETU/2007/0001r, Ecologic - Institute for International and European Environmental Policy.* Berlin, Germany, 2007.

[14] A. Fewkes, "The technology, design and utility of rainwater catchment systems", In: D. Butler, F. A. Memon, Eds., *Water Demand Management.* IWA Publishing: London, U.K., 2006.

[15] TRCA, *Performance evaluation of rainwater harvesting systems, Sustainable Technologies Evaluation Program (STEP), Toronto and Region Conservation Authority.* TRCA: Toronto, Ontario, Canada, 2010.

[16] V. Meera, and M.M. Ahammed, "“Water quality of rooftop rainwater harvesting systems: a review”, J. Water Supply: Res. Technol. –", *AQUA,* vol. 55, pp. 257-268, 2006.

[17] T. Herrmann, and U. Schmida, "Rainwater utilisation in Germany: efficiency, dimensioning, hydraulic and environmental aspects", *Urban Water,* vol. 1, no. 4, pp. 307-316, 2000. [http://dx.doi.org/10.1016/S1462-0758(00)00024-8]

[18] CWWA, *“Report on rainwater harvesting and grey-water reuse for potable and non-potable uses”, Report to Canada Mortgage and Housing Corporation, Canadian Water and Wastewater Association.* CWWA: Ottawa, Ontario, Canada, 2002.

[19] B. de Gouvello, F. Derrien, and Y. Khouil, *The French experience in rainwater reuse in commercial buildings”, Plumbing Systems and Design.* Mar/Apr, 2005, pp. 12-17.

[20] O.M. Brandes, T. Maas, and E. Reynolds, *Thinking beyond pipes and pumps top 10 ways communities can save water and money”, Victoria, BC: The POLIS Project on Ecological Governance.* University of Victoria: Victoria, Canada, 2006.

[21] Government of France, *French Order of 21 August 2008 on Rainwater Harvesting and its Uses Inside and Outside Buildings.* Official Journal of the French Republic: Paris, France, 2008.

[22] F.M. Schets, R. Italiaander, H.H. van den Berg, and A.M. de Roda Husman, "Rainwater harvesting: quality assessment and utilization in The Netherlands", *J. Water Health,* vol. 8, no. 2, pp. 224-235, 2010. [http://dx.doi.org/10.2166/wh.2009.037] [PMID: 20154386]

[23] F. Golay, Rain water harvesting in urban New Zealand, MSc Thesis, École Polytechnique Fédérale de Lausanne/Massey University: Lausanne, Switzerland, 2011.

[24] MPMSAA, *“Rainwater Tank Design and Installation Handbook”, Australian Rainwater Industry Development Association, Master Plumbers and Mechanical Services Association of Australia.* MPMSAA: Australia, 2008.

[25] M. Zaizen, T. Urakawa, Y. Matsumoto, and H. Takai, "The collection of rainwater from dome stadiums in Japan", *Urban Water,* vol. 1, no. 4, pp. 335-359, 2000. [http://dx.doi.org/10.1016/S1462-0758(00)00028-5]

[26] J.C. Chilton, G.G. Mildment, T.D. Marriot, A. Francis, and G. Tobias, "Case study of a rainwater recovery system in commercial building with a large roof", *Urban Water,* vol. 1, no. 4, pp. 345-354, 2000. [http://dx.doi.org/10.1016/S1462-0758(00)00032-7]

[27] S. Hills, R. Birks, and B. Mckenzie, "The millennium dome ‘water-cycle’ experiment: to evaluate water efficiency and customer perception at a recycling scheme for 6 million visitors", *Proceedings of the IWA second world water congress,* 2001pp. 15-19 Berlin: Germany

[28] United Nations Environment Programme, *Rainwater Harvesting And Utilisation - An Environmentally Sound Approach for Sustainable Urban Water Management: An Introductory Guide for Decision-Makers,* 2002.[Accessed 15th Jun. 2016] Available: http://www.unep.or.jp/ietc/publications/urban/urbanenv-2/index.asp

[29] C.M. Silva, V. Sousa, and I. Ponces, "Technical-economic evaluation of rainwater harvesting systems in commercial buildings – Case studies from Sonae Sierra in Portugal and Brazil", *4th Annual Water Efficiency Conference, WatefCon 2016 - Water Frontiers: Strategies for 2020 and beyond,* 2016 Coventry: U.K.

[30] *International water-harvesting and related financial incentives.*[Accessed: 14th Aug. 2016] Available: http://www.rainwaterharvesting.org/policy/legislation_international.htm

[31] "Brazil, The Brazilian Association of Technical Standards (ABNT)", In: *ABNT NBR 15527: Rainwater - Catchment of roofs in urban areas for non-potable purposes – Requirements.* Brazil, 2007.

[32] United States of America, American Rainwater Catchment Systems Association (ARSCA), *Rainwater catchment design and installation standards.* Tempe, AZ, USA, 2009.

[33] United Kingdom, "Technical Committee CB/506", In: *BS 8515: Rainwater harvesting systems – Code of practice.* BSi British Standards: U.K., 2009.

[34] DIN Germany, *DIN 1989: Rainwater Harvesting Systems - Part 1: Planning, Installation, Operation and Maintenance.* Fachvereinigung Betriebs- und Regenwassernutzung e.V, fbr: Darmstadt, Germany, 2002.

[35] *Canada, Government of Ontario. "Ontario Guidelines for Residential Rainwater Harvesting Systems Handbook.* 1st ed Government of Ontario: Toronto, Ontario, Canada, 2010.

[36] H.J. Krishna, *The Texas manual on rainwater harvesting.* 3rd ed Texas Water Development Board, 2005.

[37] "Portuguese Association for the Quality and Efficiency in Building Services", In: *ETA 0701: Rainwater harvesting systems in buildings.* Technical Commitee 0701, ANQIP: Portugal, 2012.

[38] A. Fewkes, "Modelling the performance of rainwater collection systems: towards a generalised approach", *Urban Water,* vol. 1, no. 4, pp. 323-333, 2000. [http://dx.doi.org/10.1016/S1462-0758(00)00026-1]

[39] S. Ward, F. A. Memon, and D. Butler, "Rainwater harvesting: model-based design evaluation", *Water Science & Technology-WST,* vol. 61, no. 1, pp. 85-96, 2010. [http://dx.doi.org/10.2166/wst.2010.783]

[40] A. Fewkes, *Build. Serv. Eng. Res. Tech.,* vol. 21, no. 2, pp. 99-106, 2000. [http://dx.doi.org/10.1177/014362440002100204]

[41] V.G. Mitchell, D.T. McCarthy, A. Deletic, and T.D. Fletcher, "Urban stormwater harvesting - sensitivity of a storage behaviour model", *Environ. Model. Softw.,* vol. 23, no. 6, pp. 782-793, 2008. [http://dx.doi.org/10.1016/j.envsoft.2007.09.006]

[42] E. Ghisi, K.A. Cardoso, and R.F. Rupp, "Short-term versus long-term rainfall time series in the assessment of potable water savings by using rainwater in houses", *J. Environ. Manage.,* vol. 100, pp. 109-119, 2012. [http://dx.doi.org/10.1016/j.jenvman.2011.12.031] [PMID: 22377369]

[43] M. Basinger, F. Montalto, and U. Lall, "A rainwater harvesting system reliability model based on nonparametric stochastic rainfall generator", *J. Hydrol. (Amst.),* vol. 392, no. 3–4, pp. 105-118, 2010. [http://dx.doi.org/10.1016/j.jhydrol.2010.07.039]

[44] Y.P. Guo, and B.W. Baetz, "Sizing of rainwater storage units for green building applications", *J. Hydrol. Eng.,* vol. 12, no. 2, pp. 197-205, 2007. [http://dx.doi.org/10.1061/(ASCE)1084-0699(2007)12:2(197)]

[45] CMHC, *Guidelines for Residential Rainwater Harvesting Systems Handbook. Canada Mortgage and Housing Corporation.* CMHC, 2012.

[46] R. Farreny, T. Morales-Pinzón, A. Guisasola, C. Tayà, J. Rieradevall, and X. Gabarrell, "Roof selection for rainwater harvesting: quantity and quality assessments in Spain", *Water Res.,* vol. 45, no. 10, pp. 3245-3254, 2011. [http://dx.doi.org/10.1016/j.watres.2011.03.036] [PMID: 21492898]

[47] Organisation for Economic Co-operation and Development, *Household energy & water consumption and waste generation: trends, environmental impacts and policy responses.* ENV/EPOC/

WPNEP(2001)15/FINAL, Sector Case Studies Series, Working Party on National Environmental Policy, Environment Policy Committee, Organisation for Economic Co-operation and Development: Paris, 2001.

[48] *Australia, Queensland Government. "DNRM: Planning guidelines for water supply and sewerage.* Department of Natural Resources and Mines: Queensland, 2005.

[49] A. Turner, J. Willetts, S. Fane, D. Giurco, A. Kazaglis, and S. White, "Guide to demand management: The urban water planning framework", In: *WSAA Occasional Paper No.18.* Water Services Association of Australia (WSAA) and Institute for Sustainable Futures University Of Technology Sydney (UTS): Australia, 2008.

[50] M. Flörke, and J. Alcamo, *"European outlook on water use", Final Report.* Center for Environmental Systems Research, University of Kassel, European Environment Agency, 2004.

[51] C. Howe, and F. Linaweaver, "The Impact of Price on Residential Water Demand and its Relation to System Design and Price Structure", *Water Resour. Res.,* vol. 3, no. 1, pp. 13-32, 1967. [http://dx.doi.org/10.1029/WR003i001p00013]

[52] K. Gibbs, "Price Variable in Residential Water Demand Models", *Water Resour. Res.,* vol. 14, no. 1, pp. 15-18, 1978. [http://dx.doi.org/10.1029/WR014i001p00015]

[53] L.E. Danielson, "An Analysis of Residential Demand for Water Using Micro Time-Series Data", *Water Resour. Res.,* vol. 15, no. 4, pp. 763-767, 1979. [http://dx.doi.org/10.1029/WR015i004p00763]

[54] H. S. Foster, and B. R. Beattie, "Urban Residential Demand for Water in the United States", *Land Economics,* vol. 55, no. 1, pp. 43-58, 1979. [http://dx.doi.org/10.2307/3145957]

[55] R.B. Billings, "Specification of Block Rate Price Variables in Demand Models", *Land Economics,* vol. 58, no. 3, pp. 386-394, August 1982. [http://dx.doi.org/10.2307/3145946]

[56] J.E. Schefter, and L.E. David, "Estimating Residential Water Demand Under Multi- Part Tariffs Using Aggregate Data", *Land Econ.,* vol. 61, no. 3, pp. 272-280, 1985. [http://dx.doi.org/10.2307/3145842]

[57] D.L. Chicoine, and G. Ramamurthy, "Evidence on the Specification of Price in the Study of Domestic Water Demand", *Land Economics,* vol. 62, no. 1, pp. 26-32, February 1986. [http://dx.doi.org/10.2307/3146560]

[58] J.E. Moncur, "Urban Water Pricing and Drought Management", *Water Resour. Res.,* vol. 23, no. 3, pp. 393-398, 1987. [http://dx.doi.org/10.1029/WR023i003p00393]

[59] M. L. Niesdwiadomy, and D. J. Molina, "Comparing Residential Water Demand Estimates under Decreasing and Increasing Block Rates using Household Data", *Land Economics,* vol. 65, no. 3, pp. 281-289, August 1989.

[60] L.G. Hansen, "Water and Energy Price Impacts on Residential Water Demand in Copenhagen", *Land Econ.,* vol. 72, no. 1, pp. 66-79, 1996. [http://dx.doi.org/10.2307/3147158]

[61] M.E. Renwick, and S.O. Archibald, "Demand Side Management Policies for Residential Water Use: Who Bears the Conservation Burden?", *Land Econ.,* vol. 74, no. 3, pp. 343-359, 1998. [http://dx.doi.org/10.2307/3147117]

[62] M.E. Renwick, and R.D. Green, "Do residential Water Demand Side Management Policies Measure Up? An Analysis of Eight California Water Agencies", *J. Environ. Econ. Manage.,* vol. 40, pp. 37-55, 2000. [http://dx.doi.org/10.1006/jeem.1999.1102]

[63] S. Garcia, and A. Reynaud, "Estimating the benefits of efficient water pricing in France", *Resour. Energy Econ.*, vol. 26, pp. 1-25, 2004. [http://dx.doi.org/10.1016/j.reseneeco.2003.05.001]

[64] R. Martínez-Espiñeira, and C. Nauges, "Is all domestic water consumption sensitive to price control", *Appl. Econ.*, vol. 36, pp. 1697-1703, 2004. [http://dx.doi.org/10.1080/0003684042000218570]

[65] R. Martins, and A. Fortunato, *"Residential water demand under block rates – a Portuguese case study", Report GEMF, nº9-2005.* Faculty of Economics of the University of Coimbra: Coimbra, Portugal, 2005.

[66] Organisation for Economic Co-operation and Development (OECD), *"Household water pricing in OECD countries", ENV/EPOC/GEEI(98)12/FINAL, Working Party on Economic and Environmental Policy Integration, Environment Policy Committee, Organisation for Economic Co-operation and Development.*. OECD: Paris, 1999.

[67] G. Barret, "Water conservation: The role of price and regulation in residential water consumption", *Economic Papers,* vol. 23, no. 3, pp. 271-285, 2004. [http://dx.doi.org/10.1111/j.1759-3441.2004.tb00371.x]

[68] J. Schleich, and T. Hillenbrand, "Determinants of residential water demand in Germany", *Ecol. Econ.*, vol. 68, no. 6, pp. 1756-1769, 2009. [http://dx.doi.org/10.1016/j.ecolecon.2008.11.012]

[69] A.C. Worthington, and M. Hoffman, *A state of the art review of residential water demand modelling.* Accounting & Finance Working Paper 07/06, School of Accounting & Finance, University of Wollongong, 2007.

[70] M.L. Nieswaidomy, and D.J. Molina, "Comparing Residential Water Demand Estimates under Decreasing and Increasing Block Rates Using Household Data", *Land Econ.,* vol. 65, no. 3, pp. 280-289, 1989. [http://dx.doi.org/10.2307/3146672]

[71] M.A. Renwick, and S.O. Archibald, "Demand-side management policies for residential water use: who bares the conservation burden?", *Land Econ.,* vol. 74, pp. 343-359, 1998. [http://dx.doi.org/10.2307/3147117]

[72] P.W. Mayer, W.B. DeOreo, E.M. Opitz, J.C. Kiefer, W.Y. Davis, B. Dziegielewski, and J.O. Nelson, *"Residential end uses of water", Tech. Report.* AWWA Research Foundation: USA, 1999.

[73] M.A. Renwick, and R.D. Green, "Do residential water demand side management policies measure up? An analysis of eight California water agencies", *J. Environ. Econ. Manage.,* vol. 40, pp. 37-55, 2000. [http://dx.doi.org/10.1006/jeem.1999.1102]

[74] D. Inman, and P. Jeffrey, "A review of residential water conservation tool performance and influences on implementation effectiveness", *Urban Water J.,* vol. 3, no. 3, pp. 127-143, 2006. [http://dx.doi.org/10.1080/15730620600961288]

[75] J.M. Dalhuisen, M.R. Florax, H.L. de Groot, and P. Nijkamp, "Price and income elasticities of residential water demand: A meta-analysis", *Land Econ.,* vol. 79, no. 2, pp. 292-308, 2003. [http://dx.doi.org/10.2307/3146872]

[76] R.M. Willis, R.A. Stewart, K. Panuwatwanich, P.R. Williams, and A.L. Hollingsworth, "Quantifying the influence of environmental and water conservation attitudes on household end use water consumption", *J. Environ. Manage.,* vol. 92, no. 8, pp. 1996-2009, 2011. [http://dx.doi.org/10.1016/j.jenvman.2011.03.023] [PMID: 21486685]

[77] L. Höglund, "Household demand for water in Sweden with implications of a potential tax on water use", *Water Resour. Res.,* vol. 35, no. 12, pp. 3853-3863, 1999. [http://dx.doi.org/10.1029/1999WR900219]

[78] B.E. Nancarrow, L.M. Smith, and G.J. Syme, "The ways people think about water", *J. Environ. Syst.,* vol. 25, no. 1, pp. 15-27, 1996. [http://dx.doi.org/10.2190/5H45-YUXH-MPL0-PNT8]

[79] T. Hassell, and J. Cary, *"Promoting behavioural change in household water consumption: Literature review", Tech. Report.* Smart Water: Victoria, 2007.

[80] F.A. Abdulla, and A.W. Al-Shareef, "Roof rainwater harvesting systems for household water supply in Jordan", *Desalination,* vol. 243, no. 1–3, pp. 195-207, 2009. [http://dx.doi.org/10.1016/j.desal.2008.05.013]

[81] E. Ghisi, "Potential for potable water savings by using rainwater in the residential sector of Brazil", *Build. Environ.,* vol. 41, no. 11, pp. 1544-1550, 2006. [http://dx.doi.org/10.1016/j.buildenv.2005.03.018]

[82] O.O. Aladenola, and O.B. Adeboye, "Assessing the Potential for Rainwater Harvesting", *Water Resour. Manage.,* vol. 24, no. 10, pp. 2129-2137, 2010. [http://dx.doi.org/10.1007/s11269-009-9542-y]

[83] E. Eroksuz, and A. Rahman, "Rainwater tanks in multi-unit buildings: A case study for three Australian cities", *Resour. Conserv. Recycling,* vol. 54, no. 12, pp. 1449-1452, 2010. [http://dx.doi.org/10.1016/j.resconrec.2010.06.010]

[84] L. Domènech, and D. Saurí, "A comparative appraisal of the use of rainwater harvesting in single and multi-family buildings of the Metropolitan Area of Barcelona (Spain): social experience, drinking water savings and economic costs", *J. Clean. Prod.,* vol. 19, no. 6–7, pp. 598-608, 2011. [http://dx.doi.org/10.1016/j.jclepro.2010.11.010]

[85] C.M. Silva, V. Sousa, and N.V. Carvalho, "Evaluation of rainwater harvesting in Portugal: Application to single-family residences", *Resour. Conserv. Recycling,* vol. 94, pp. 21-34, 2015. [http://dx.doi.org/10.1016/j.resconrec.2014.11.004]

[86] E.L. Villarreal, and A. Dixon, "Analysis of a rainwater collection system for domestic water supply in Ringdansen, Norrköping, Sweden", *Build. Environ.,* vol. 40, no. 9, pp. 1174-1184, 2005. [http://dx.doi.org/10.1016/j.buildenv.2004.10.018]

[87] R. Farreny, X. Gabarrell, and J. Rieradevall, "Cost-efficiency of rainwater harvesting strategies in dense Mediterranean neighbourhoods", *Resour. Conserv. Recycling,* vol. 55, no. 7, pp. 686-694, 2011. [http://dx.doi.org/10.1016/j.resconrec.2011.01.008]

[88] A. Belmeziti, O. Coutard, and B. de Gouvello, "A new methodology for evaluating potential for potable water savings (PPWS) by using rainwater harvesting at the urban level: The case of the municipality of Colombes (Paris region)", *Water,* vol. 5, no. 1, pp. 312-326, 2013. [Switzerland]. [http://dx.doi.org/10.3390/w5010312]

[89] E. Ghisi, A. Montibeller, and R.W. Schmidt, "Potential for potable water savings by using rainwater: An analysis over 62 cities in southern Brazil", *Build. Environ.,* vol. 41, pp. 204-210, 2006. [http://dx.doi.org/10.1016/j.buildenv.2005.01.014]

[90] A. Bocanegra-Martínez, J.M. Ponce-Ortega, F. Nápoles-Rivera, M. Serna-González, A.J. Castro-Montoya, and M.M. El-Halwagi, "Optimal design of rainwater collecting systems for domestic use into a residential development", *Resour. Conserv. Recycling,* vol. 84, pp. 44-56, 2014. [http://dx.doi.org/10.1016/j.resconrec.2014.01.001]

[91] M. Basinger, F. Montalto, and L. Upmanu, "A rainwater harvesting system reliability model based on nonparametric stochastic rainfall generator", *J. Hydrol. (Amst.),* vol. 392, no. 3-4, pp. 105-118, 2010. [http://dx.doi.org/10.1016/j.jhydrol.2010.07.039]

[92] E. Ghisi, D.F. Tavares, and V.L. Rocha, "Rainwater harvesting in petrol stations in Brasília: Potential for potable water savings and investment feasibility analysis", *Resour. Conserv. Recycling,* vol. 54, no. 2, pp. 79-85, 2009. [http://dx.doi.org/10.1016/j.resconrec.2009.06.010]

[93] R.F.M. Neto, I.C. Carvalho, M.L. Calijuri, and A.F. Santiago, "Rainwater use in airports: A case study in Brazil", *Resour. Conserv. Recycl.,* vol. 68, pp. 36-43, 2012. [http://dx.doi.org/10.1016/j.resconrec.2012.08.005]

[94] A. Palla, I. Gnecco, and L.G. Lanza, "Non-dimensional design parameters and performance assessment of rainwater harvesting systems", *J. Hydrol. (Amst.),* vol. 401, no. 1–2, pp. 65-76, 2011. [http://dx.doi.org/10.1016/j.jhydrol.2011.02.009]

[95] A. Palla, I. Gnecco, L.G. Lanza, and P. La Barbera, "Performance analysis of domestic rainwater harvesting systems under various European climate zones", *Resour. Conserv. Recycling,* vol. 62, pp. 71-80, 2012. [http://dx.doi.org/10.1016/j.resconrec.2012.02.006]

[96] A. Campisano, and C. Modica, "Optimal sizing of storage tanks for domestic rainwater harvesting in Sicily", *Resour. Conserv. Recycling,* vol. 63, pp. 9-16, 2012. [http://dx.doi.org/10.1016/j.resconrec.2012.03.007]

[97] J.S. Mun, and M.Y. Han, "Design and operational parameters of a rooftop rainwater harvesting system: definition, sensitivity and verification", *J. Environ. Manage.,* vol. 93, no. 1, pp. 147-153, 2012. [http://dx.doi.org/10.1016/j.jenvman.2011.08.024] [PMID: 22054581]

[98] M.H. Rashidi Mehrabadi, B. Saghafian, and F. Haghighi Fashi, "Assessment of residential rainwater harvesting efficiency for meeting non-potable water demands in three climate conditions", *Resour. Conserv. Recycl.,* vol. 73, no. 2010, pp. 86-93, Apr. 2013. [http://dx.doi.org/10.1016/j.resconrec.2013.01.015]

[99] A. Rahman, J. Keane, and M.A. Imteaz, "Rainwater harvesting in Greater Sydney: Water savings, reliability and economic benefits", *Resour. Conserv. Recycling,* vol. 61, pp. 16-21, 2012. [http://dx.doi.org/10.1016/j.resconrec.2011.12.002]

[100] P.J. Coombes, "Argue; J. R.; Kuczera, G. "Figtree Place: a case study in water sensitive urban development (WSUD)", *Urban Water J.,* vol. 1, no. 4, pp. 335-343, 1999. [http://dx.doi.org/10.1016/S1462-0758(00)00027-3]

[101] C. Anand, and D.S. Apul, "Economic and environmental analysis of standard, high efficiency, rainwater flushed, and composting toilets", *J. Environ. Manage.,* vol. 92, no. 3, pp. 419-428, 2011. [http://dx.doi.org/10.1016/j.jenvman.2010.08.005] [PMID: 21056531]

[102] E. Ghisi, and D.F. Ferreira, "Potential for potable water savings by using rainwater and greywater in a multi-storey residential building in southern Brazil", *Build. Environ.,* vol. 42, pp. 2512-2522, 2007. [http://dx.doi.org/10.1016/j.buildenv.2006.07.019]

[103] E. Ghisi, and S.M. Oliveira, "Potential for potable water savings by combining the use of rainwater and greywater in houses in southern Brazil", *Build. Environ.,* vol. 42, pp. 1731-1742, 2007. [http://dx.doi.org/10.1016/j.buildenv.2006.02.001]

[104] E. Ghisi, and P.N. Schondermark, "Investment Feasibility Analysis of Rainwater Use in Residences", *Water Resour. Manage.,* vol. 27, pp. 2555-2576, 2013. [http://dx.doi.org/10.1007/s11269-013-0303-6]

[105] D. Hall, and E. Lobina, *"Water in Europe", Tech. Report.* Public Services International Research Unit, Business School, Univ. of Greenwich: London, 2008.

[106] D. Hall, and E. Lobina, *"The past, present and future of finance for investment in water systems", Tech. Report.* Public Services International Research Unit, Business School, Univ. of Greenwich: London, 2010.

[107] "*PEAASAR II: Plano estratégico de abastecimento de água e de saneamento de águas residuais 2007-2013*", In: *Ministério do Ambiente, do Ordenamento dos Território e do Desenvolvimento Regional (MAOTDR), Portugal.* Portuguese, 2007.

[108] *Financial Performance and Capital Investment of the Water Companies in England and Wales 1994-95.* Tech. Report, Water Services Regulatory Authority (OFWAT): Birmingham, 1995.

[109] W. Schönbäck, and G. Oppolzer, *R., Kraemer, W. Hansen, and N. Herbke, "International comparison of water sectors comparison of systems against a background of European and Economic Policy,* vol. Vol. 6. English Edition, Bundeskammer für Arbeiter und Angestellte: Vienna, 2004.

[110] *"The development of the water industry in England and Wales", Tech. Report, Office of Water Service (OFWAT), Dept. for Environment, Food and Rural Affairs.* DEFRA: UK, 2006.

[111] *Sustainability Indicators 2007/08.* Tech. Report, Water UK: London, 2008.

[112] *Sustainability Indicators 2007/08.* Tech. Report, Water UK: London, 2009.

[113] *Sustainability Indicators 2007/08.* Tech. Report, Water UK: London, 2010.

[114] *Sustainability Indicators 2007/08.* Tech. Report, Water UK: London, 2011.

[115] *Water & sustainability (Volume 3): U.S. Water Consumption for Power Production - The Next Half Century.* Report 1007862, Electrical Power Research Institute, 2002.

[116] *Ensuring a sustainable future: An energy management guidebook for wastewater and water utilities.* Tech. Report, Environmental Protection Agency (USEPA): USA, 2008.

[117] R.L. Lawson, J.R. Lyman, and B.J. Lyons, *"Impact of municipal, industrial, and commercial water needs on the energy water nexus: Challenges, solutions, and recommendations", Tech. Report.*. The Atlantic Council of the United States: USA, 2012.

[118] R. Lieste, J.P. Witte, A.C. de Nijs, C.J. Aggenbach, B.J. Pieters, J. Runhaar, and W. Verweij, *Beoordeling van de grondwatertoestand op basis van de Kaderrichtlijn Water.* Tech. Report 607300003, National Institute for Public Health and the Environment: Dutch, 2007.

[119] F.T. Oliveira, Rainwater harvesting in urban uses in mainland Portugal - viability assessment, Instituto Superior Tecnico, Universidade Tecnica de Lisboa, Portugal: portuguese, 2008.

[120] L.P. Barroso, Sustainable construction - Comparative solutions for efficient use of water in residential buildings, Faculdade de Ciências e Tecnologia, Universidade Nova de Lisboa, Portugal: portuguese, 2010.

[121] M.P. Amado, and L.M. Barroso, "Sustainable construction: Water use in residential buildings in Portugal", *Int. J. Sustain. Constr. Eng. Technol.*, vol. 4, no. 2, pp. 14-22, 2013.

[122] M.I. Ponces, *Technical-economic evaluation of rainwater harvesting systems in commercial buildings – Case studies from Sonae Sierra.* Instituto Superior Tecnico: Universidade Tecnica de Lisboa, Portugal, 2015. (in Portuguese)

[123] C.V. Lúcio, Potential reduction in potable water consumption by reusing rainwater in urban centers, Instituto Superior Tecnico, Universidade Tecnica de Lisboa, Portugal: Portuguese, 2015.

[124] I. Adeniyi, and I. Olabanji, "The physico-chemical and bacteriological quality of rainwater collected over different roofing materials in Ile-Ife, southwestern Nigeria", *Chem. Ecol.*, vol. 21, pp. 149-166, 2005.
[http://dx.doi.org/10.1080/02757540500117318]

[125] W. Ahmed, F. Huygens, A. Goonetilleke, and T. Gardner, "Real-time PCR detection of pathogenic microorganisms in roof-harvested rainwater in Southeast Queensland, Australia", *Appl. Environ. Microbiol.*, vol. 74, no. 17, pp. 5490-5496, 2008.
[http://dx.doi.org/10.1128/AEM.00331-08] [PMID: 18621865]

[126] W. Ahmed, A. Vieritz, T. Gardner, and A. Goonetilleke, "Microbial risks from rainwater tanks in southeast Queensland", *Water: Official Journal of the Australian Water and Wastewater Association,* vol. 36, no. 8, pp. 80-85, 2009.

[127] O.A. Al-Khashman, "Chemical characteristics of rainwater collected at a western site of Jordan", *Atmos. Res.,* vol. 91, pp. 53-61, 2009. [http://dx.doi.org/10.1016/j.atmosres.2008.05.007]

[128] H.J. Albrechtsen, "Microbiological investigations of rainwater and graywater collected for toilet flushing", *Water Sci. Technol.,* vol. 46, no. 6-7, pp. 311-316, 2002. [PMID: 12381006]

[129] A. Baez, R. Belmont, R. Garcia, H. Padilla, and M. Torres, "Chemical composition of rainwater collected at a southwest site of Mexico City, Mexico", *Atmos. Res.,* vol. 86, no. 1, pp. 61-75, 2007. [http://dx.doi.org/10.1016/j.atmosres.2007.03.005]

[130] R. Bannister, J. Westwood, and A. Mcneill, *"Investigation of microbiological and chemical water quality in rainwater tanks in Victoria", Tech. Report.* Department of Natural Resources and Environmental and Water: Victoria, 1997.

[131] P. Body, *"The contamination of rainwater tanks at Port Pirie", Tech. Report.* South Australian Department for Environment and Planning: Adelaide, 1986.

[132] H.A. Bridgman, "Evaluating rainwater contamination and sources in southeast Australia using factor analysis", *Atmos. Environ.,* vol. 26A, pp. 2401-2412, 1992. [http://dx.doi.org/10.1016/0960-1686(92)90370-Z]

[133] T.D. Bucheli, S. Muller, S. Heberle, and R. Schwarzenbach, "Occurrence and behavior of pesticides in rainwater, roof runoff, and artificial stormwater infiltration", *Environ. Sci. Technol.,* vol. 32, pp. 3457-3464, 1998. [http://dx.doi.org/10.1021/es980317n]

[134] T.D. Bucheli, S.R. Muller, A. Voegelin, and R.P. Schwarzenbach, "Bituminous roof sealing membranes as major sources of the herbicide (R,S)-Mecoprop in roof runoff waters: potential contamination of groundwater and surface water", *Environ. Sci. Technol.,* vol. 32, pp. 3465-3471, 1998. [http://dx.doi.org/10.1021/es980318f]

[135] Y-Z. Cao, S. Wang, G. Zhang, J. Luo, and S. Lu, "Chemical characteristics of wet precipitation at an urban site of Guangzhou, South China", *Atmos. Res.,* vol. 94, pp. 462-469, 2009. [http://dx.doi.org/10.1016/j.atmosres.2009.07.004]

[136] M. Chang, M.W. McBroom, and R. Scott Beasley, "Roofing as a source of nonpoint water pollution", *J. Environ. Manage.,* vol. 73, no. 4, pp. 307-315, 2004. [http://dx.doi.org/10.1016/j.jenvman.2004.06.014] [PMID: 15531389]

[137] P.J. Coombes, H. Dunstan, S. Anthony, C. Evans, and T. Harrison, "An overview of a decade of research into the quality of rainwater supplies collected from roofs", *12th International Rainwater Catchment Systems Conference,* 2005 New Delhi: India

[138] P.J. Coombes, H. Dunstan, A. Spinks, C. Evans, and T. Harrison, "Key messages from a decade of water quality research into roof collected rainwater supplies", *1st National HYDROPOLIS Conference,* 2006 Perth: Australia

[139] P.J. Coombes, G. Kuczera, and J. Kalma, "Rainwater quality from roofs, tanks and hot water systems at Fig Tree Place", *3rd International Hydrology and Water Resource Symposium: HYDRO 2000,* 2000 Perth: Australia

[140] T.A. Dillaha, and W.J. Zolan, "Rainwater catchment water quality in Micronesia", *Water Res.,* vol. 19, pp. 741-746, 1985. [http://dx.doi.org/10.1016/0043-1354(85)90121-6]

[141] C.A. Evans, P.J. Coombes, and R.H. Dunstan, "Wind, rain and bacteria: The effect of weather on the microbial composition of roof-harvested rainwater", *Water Res.,* vol. 40, no. 1, pp. 37-44, 2006. [http://dx.doi.org/10.1016/j.watres.2005.10.034] [PMID: 16343588]

[142] C.A. Evans, P.J. Coombes, R.H. Dunstan, and T. Harrison, "Identifying the major influences on the microbial composition of roof harvested rainwater and the implications for water quality", *Water Sci. Technol.,* vol. 55, no. 4, pp. 245-253, 2007. [http://dx.doi.org/10.2166/wst.2007.115] [PMID: 17425092]

[143] C.A. Evans, P.J. Coombes, R.H. Dunstan, and T. Harrison, "Extensive bacterial diversity indicates the potential operation of a dynamic micro-ecology within domestic rainwater storage systems", *Sci. Total Environ.,* vol. 407, no. 19, pp. 5206-5215, 2009. [http://dx.doi.org/10.1016/j.scitotenv.2009.06.009] [PMID: 19596393]

[144] L.J. Franklin, J.E. Fielding, J. Gregory, L. Gullan, D. Lightfoot, S.Y. Poznanski, and H. Vally, "An outbreak of Salmonella Typhimurium 9 at a school camp linked to contamination of rainwater tanks", *Epidemiol. Infect.,* vol. 137, no. 3, pp. 434-440, 2009. [http://dx.doi.org/10.1017/S095026880800109X] [PMID: 18687158]

[145] L. Gee, *Pilot survey of the microbiological and chemical aspects of water stored in domestic rainwater tanks.* Western Sector Public Health Unit, New South Wales Department of Health: Sydney, Australia, 1993.

[146] R.H. Haeber, and D.H. Waller, "Water quality of rain water collection systems in the Eastern Caribbean", *3rd International Rainwater Cistern Systems Conference,* 1987 Khon Kaen: Thailand

[147] L. Handia, "Comparative study of rainwater quality in urban Zambia", *J. Water Supply Res. T.,* vol. 54, no. 1, pp. 55-64, 2005.

[148] J.S. Heyworth, G. Glonek, E.J. Maynard, P.A. Baghurst, and J. Finlay-Jones, "Consumption of untreated tank rainwater and gastroenteritis among young children in South Australia", *Int. J. Epidemiol.,* vol. 35, no. 4, pp. 1051-1058, 2006. [http://dx.doi.org/10.1093/ije/dyl105] [PMID: 16723368]

[149] R. Huston, Y.C. Chan, T. Gardner, G. Shaw, and H. Chapman, "Characterisation of atmospheric deposition as a source of contaminants in urban rainwater tanks", *Water Res.,* vol. 43, no. 6, pp. 1630-1640, 2009. [http://dx.doi.org/10.1016/j.watres.2008.12.045] [PMID: 19171360]

[150] R.H. Kim, S. Lee, Y.M. Kim, J.H. Lee, S.K. Kim, and S.G. Kim, "Pollutants in rainwater runoff in Korea: their impacts on rainwater utilization", *Environ. Technol.,* vol. 26, no. 4, pp. 411-420, 2005. [http://dx.doi.org/10.1080/09593332608618546] [PMID: 15906493]

[151] J.P. Koplan, R.D. Deen, W.H. Swanston, and B. Tota, "Contaminated roof-collected rainwater as a possible cause of an outbreak of salmonellosis", *J. Hyg. (Lond.),* vol. 81, no. 2, pp. 303-309, 1978. [http://dx.doi.org/10.1017/S0022172400025146] [PMID: 701791]

[152] P. Mantovan, A. Pastore, L. Szpyrkowicz, and F. Zilio-Grandi, "Characterization of rainwater quality from the Venice region network using multiway data analysis", *Sci. Total Environ.,* vol. 164, no. 1, pp. 27-43, 1995. [http://dx.doi.org/10.1016/0048-9697(95)04432-Z]

[153] S. May, and R. Prado, "Experimental evaluation of rainwater quality for non potable applications in the city of São Paulo, Brazil", *Urban Water J.,* vol. 3, no. 3, pp. 145-151, 2006. [http://dx.doi.org/10.1080/15730620600961213]

[154] V. Meera, and M.M. Ahammed, "Water quality of rooftop rainwater harvesting systems: A review", *J. Water Supply Res. T.,* vol. 55, no. 4, pp. 1-13, 2006.

[155] J. Radaideh, K. Al-Zboon, A. Al-Harahsheh, and R. Al-Adamat, "Quality assessment of harvested rainwater for domestic uses", *Jordan J. Earth. Environ. Sci.,* vol. 2, no. 1, pp. 26-31, 2009.

[156] E. Sazakli, A. Alexopoulos, and M. Leotsinidis, "Rainwater harvesting, quality assessment and utilization in Kefalonia Island, Greece", *Water Res.,* vol. 41, no. 9, pp. 2039-2047, 2007. [http://dx.doi.org/10.1016/j.watres.2007.01.037] [PMID: 17363028]

[157] G. Simmons, V. Hope, G. Lewis, J. Whitmore, and W. Gao, "Contamination of potable roof-collected rainwater in Auckland, New Zealand", *Water Res.,* vol. 35, no. 6, pp. 1518-1524, 2001. [http://dx.doi.org/10.1016/S0043-1354(00)00420-6] [PMID: 11317899]

[158] J. Spinks, S. Phillips, P. Robinson, and P. Van Buynder, "Bushfires and tank rainwater quality: a cause for concern?", *J. Water Health,* vol. 4, no. 1, pp. 21-28, 2006. [PMID: 16604835]

[159] R. Thurman, "Evaluation of rainwater stored in collection tanks", *Australian Microbiologist,* pp. 20-22, March 1995.

[160] B. Uba, and O. Aghogho, "Rainwater quality from different roof catchments in the Port Harcourt district, River State, Nigeria", *Water Supply Res. T.,* vol. 49, no. 5, pp. 281-287, 2000.

[161] P. Vasudevan, and M. Tandon, "Microbial quality of rainwater from roof surfaces", *J. Sci. Ind. Res. (India),* vol. 67, pp. 432-435, 2008.

[162] A. Vázquez, M. Costoya, R.M. Peña, S. García, and C. Herrero, "A rainwater quality monitoring network: a preliminary study of the composition of rainwater in Galicia (NW Spain)", *Chemosphere,* vol. 51, no. 5, pp. 375-386, 2003. [http://dx.doi.org/10.1016/S0045-6535(02)00805-6] [PMID: 12598003]

[163] J.D. Willey, R.I. Bennett, J.M. Williams, R.K. Denne, C.R. Kornegay, M.S. Perlotto, and B.M. Moore, "Effect of storm type on rainwater composition in southeastern North Carolina", *Environ. Sci. Technol.,* vol. 22, no. 1, pp. 41-46, 1988. [http://dx.doi.org/10.1021/es00166a003] [PMID: 22195508]

[164] C.B. Mendez, J.B. Klenzendorf, B.R. Afshar, M.T. Simmons, M.E. Barrett, K.A. Kinney, and M.J. Kirisits, "The effect of roofing material on the quality of harvested rainwater", *Water Res.,* vol. 45, no. 5, pp. 2049-2059, 2011. [http://dx.doi.org/10.1016/j.watres.2010.12.015] [PMID: 21232781]

[165] B. Kus, J. Kandasamy, S. Vigneswaran, and H.K. Shon, "Analysis of first flush to improve the water quality in rainwater tanks", *Water Sci. Technol.,* vol. 61, no. 2, pp. 421-428, 2010. [http://dx.doi.org/10.2166/wst.2010.823] [PMID: 20107269]

[166] D. Georgios, G.D. Gikas, and V.A. Tsihrintzis, "Assessment of water quality of first-flush roof runoff and harvested rainwater", *J. Hydrol. (Amst.),* vol. 466–467, pp. 115-126, 2012.

[167] L.C. Duncker, *Hygiene awareness for rural water supply and sanitation projects,* Report no. 819/1/00, Water Research Commission: Pretoria, RSA, .

[168] J. Heyworth, "A diary study of gastroenteritis and tank rainwater consumption in young children in South Australia", *Proc. 10th Int'l Rainwater Catchment Systems Conf.,* 2001pp. 141-148

[169] L. Handia, J.M. Tembo, and C. Mwiindwa, "Potential of rainwater harvesting in urban Zambia", In: *Phys. Chem. Earth,* vol. 28. 2003, no. 20-27, pp. 893-896.

[170] S. Muthukumaran, K. Baskaran, and N. Sexton, "Quantification of potable water savings by residential water conservation and reuse – A case study", *Resour. Conserv. Recycling,* vol. 55, no. 11, pp. 945-952, 2011. [http://dx.doi.org/10.1016/j.resconrec.2011.04.013]

[171] T. Abbasi, and S.A. Abbasi, "Sources of Pollution in Rooftop Rainwater Harvesting Systems and Their Control", *Crit. Rev. Environ. Sci. Technol.,* vol. 41, no. 23, pp. 2097-2167, 2011. [http://dx.doi.org/10.1080/10643389.2010.497438]

[172] I. Lança, and A. Silva-Afonso, "Rainwater storage and reuse. A safety case study in a groundwater storage installation", *Symposium CIB W062,* 2011 Aveiro: Portugal

[173] J. Corte-Real, B. Qian, and H. Xu, "Regional climate change in Portugal: precipitation variability associated with large-scale atmospheric circulation", *Int. J. Climatol.,* vol. 18, no. 6, pp. 619-635, 1998.

[http://dx.doi.org/10.1002/(SICI)1097-0088(199805)18:6<619::AID-JOC271>3.0.CO;2-T]

[174] R.M. Trigo, and C.C. DaCamara, "Circulation weather types and their influence on the precipitation regime in Portugal", *Int. J. Climatol.,* vol. 20, no. 13, pp. 1559-1581, 2000. [http://dx.doi.org/10.1002/1097-0088(20001115)20:13<1559::AID-JOC555>3.0.CO;2-5]

[175] C.M. Goodess, and P.D. Jones, "Links between circulation and changes in the characteristics of Iberian rainfall", *Int. J. Climatol.,* vol. 22, no. 13, pp. 1593-1615, 2002. [http://dx.doi.org/10.1002/joc.810]

[176] P.M. Miranda, F.E. Coelho, A.R. Tomé, M.A. Valente, A. Carvalho, C. Pires, H.O. Pires, V.C. Pires, and C. Ramalho, "20th century Portuguese climate and climate scenarios", In: F.D. Santos, K. Forbes, R. Moita, Eds., *Climate Change in Portugal: Scenarios, Impacts and Adaptation Measures (SIAM Project).* Gradiva, 2002, pp. 23-83.

[177] J.A. Santos, J. Corte-Real, and S.M. Leite, "Weather Regimes and their Connection to the Winter Rainfall in Portugal", *Int. J. Climatol.,* vol. 25, no. 1, pp. 33-50, 2005. [http://dx.doi.org/10.1002/joc.1101]

[178] A.C. Costa, and A. Soares, "Trends in extreme precipitation indices derived from a daily rainfall database for the South of Portugal", *Int. J. Climatol.,* vol. 29, no. 13, pp. 1956-1975, 2009. [http://dx.doi.org/10.1002/joc.1834]

[179] APA and Ecoprogresso, *Fifth national communication to the United Nations framework convention on climate change. Second National Communication in the context of the Kyoto Protocol.* Portuguese Environment Agency (APA) with the cooperation of Ecoprogresso – Environment and Development Consultants, SA: Amadora, 2010.

[180] M.R. Haylock, and C.M. Goodess, "Interannual variability of European extreme winter rainfall and links with mean large-scale circulation", *Int. J. Climatol.,* vol. 24, no. 6, pp. 759-776, 2004. [http://dx.doi.org/10.1002/joc.1033]

[181] S. Mourato, M. Moreira, and J. Corte-Real, "Interannual variability of precipitation distribution patterns in Southern Portugal", *Int. J. Climatol.,* vol. 30, no. 12, pp. 1784-1794, 2010.

[182] U. Ulbrich, M. Christoph, J.G. Pinto, and J. Corte-Real, "Dependence of winter precipitation over Portugal on NAO and baroclinic wave activity", *Int. J. Climatol.,* vol. 19, no. 4, pp. 379-390, 1999. [http://dx.doi.org/10.1002/(SICI)1097-0088(19990330)19:4<379::AID-JOC357>3.0.CO;2-8]

[183] A. Rocha, "Low frequency variability of seasonal rainfall over the Iberian Peninsula and ENSO", *Int. J. Climatol.,* vol. 19, no. 8, pp. 889-901, 1999. [http://dx.doi.org/10.1002/(SICI)1097-0088(19990630)19:8<889::AID-JOC404>3.0.CO;2-P]

[184] S. Antunes, O. Pires, and A. Rocha, "Detecting spatio-temporal precipitation variability in Portugal using multichannel singular spectral analysis", *Int. J. Climatol.,* vol. 26, no. 15, pp. 2199-2212, 2006. [http://dx.doi.org/10.1002/joc.1358]

[185] M. Fragoso, and P. Tildes Gomes, "Classification of daily abundant rainfall patterns and associated large-scale atmospheric circulation types in Southern Portugal", *Int. J. Climatol.,* vol. 28, no. 4, pp. 537-544, 2008. [http://dx.doi.org/10.1002/joc.1564]

[186] M. Belo-Pereira, E. Dutra, and P. Viterbo, "Evaluation of global precipitation data sets over the Iberian Peninsula", *J. Geophys. Res.,* vol. 116, no. 20, pp. 1-16, 2011.

[187] J.H. Christensen, and O.B. Christensen, "A summary of the PRUDENCE model projections of changes in European climate by the end of this century", *Clim. Change,* vol. 81, no. 1, pp. 7-30, 2007. [http://dx.doi.org/10.1007/s10584-006-9210-7]

[188] R. Dankers, and R. Hiederer, *"Extreme temperatures and precipitation in Europe: Analysis of a high-resolution climate change scenario", JRC Scientific and Technical reports, Institute for Environment and Sustainability, Joint Research Centre (JRC).* European Commission: Luxembourg, 2008.

[189] F.S. Rodrigo, and R.M. Trigo, "Trends in daily rainfall in the Iberian Peninsula from 1951 to 2002", *Int. J. Climatol.,* vol. 27, no. 4, pp. 513-529, 2007. [http://dx.doi.org/10.1002/joc.1409]

[190] R. Durão, M.J. Pereira, A.C. Costa, J.M. Corte-Real, and A. Soares, "Indices of precipitation extremes in Southern Portugal – a geostatistical approach", *Nat. Hazards Earth Syst. Sci.,* vol. 9, no. 1, pp. 241-250, 2009. [http://dx.doi.org/10.5194/nhess-9-241-2009]

[191] M.I. de Lima, F. Espírito Santo, J.L. de Lima, and A.M. Ramos, "Recent precipitation variability over 67 years in mainland Portugal", *Die Bodenkultur,* vol. 64, no. 3-4, pp. 21-26, 2013.

[192] J. Morrison, and E. Friedler, "A critical review of methods used to obtain flow patterns and volumes of individual domestic water using appliances", *Urban Water J.,* vol. 12, no. 4, pp. 328-343, 2015. [http://dx.doi.org/10.1080/1573062X.2014.900090]

[193] K. Edwards, and L. Martin, "A methodology for surveying domestic water consumption", *Water Environ. J.,* vol. 9, no. 5, pp. 477-488, 1995. [http://dx.doi.org/10.1111/j.1747-6593.1995.tb01486.x]

[194] E. Larson, J. Froehlich, T. Campbell, C. Haggerty, L. Atlas, J. Fogarty, and S. Patel, "Disaggregated water sensing from a single, pressure-based sensor: An extended analysis of HydroSense using staged experiments", *Pervasive Mobile Comput.,* vol. 8, no. 1, pp. 82-102, 2012. [http://dx.doi.org/10.1016/j.pmcj.2010.08.008]

[195] W.B. DeOreo, J.P. Heaney, and P.W. Mayer, "Flow Trace analysis to assess water use", *J. Am. Water Works Assoc.,* vol. 88, no. 1, pp. 79-90, 1996.

[196] P.W. Mayer, D. Bennett, W.B. DeOreo, R. Harris, and D. Muir, "Great expectations – actual water savings with the latest high-efficiency residential fixtures and appliances", In: *Water sources conf.*. EUA: Las Vegas, 2002.

[197] P.W. Mayer, W.B. DeOreo, E.M. Opitz, J.C. Kiefer, W.Y. Davis, B. Dziegielewski, and J.O. Nelson, *Residential end uses of water.* AWWA Res. Foundation, 1999.

[198] P.W. Mayer, W.B. DeOreo, E. Towler, and D.M. Lewis, *Residential indoor water conservation study: evaluation of high efficiency indoor plumbing fixture retrofits in single family homes in the east bay municipal utility district service area.* Prepared by Aquacraft inc. Boulder Colorado, for EBMUD and the US EPA, 2003.

[199] P.W. Mayer, W.B. DeOreo, E. Towler, L. Martien, and D.M. Lewis, *Tampa water department residential water conservation study: the impacts of high efficiency plumbing fixture retrofits in single-family homes.* Prepared by Aquacraft inc. Boulder Colorado, for Tampa Water Department and the US EPA, 2004.

[200] P. Roberts, *Residential end use measurement study.* Yarra Valley Water: Melbourne, 2005.

[201] N. Mead, "V. and Aravinthan, “Investigation of household water consumption using a smart metering system", *Desalination Water Treat.,* vol. 11, no. 1-3, pp. 1-9, 2009. [http://dx.doi.org/10.5004/dwt.2009.850]

[202] R.M. Willis, R.A. Stewart, K. Panuwatwanich, S. Jones, and A. Kyriakides, "Alarming visual display monitors affecting shower end use water and energy conservation in Australian residential households", *Resour. Conserv. Recycling,* vol. 54, no. 12, pp. 1117-1127, 2010. [http://dx.doi.org/10.1016/j.resconrec.2010.03.004]

[203] R.M. Willis, R.A. Stewart, P. Williams, C. Hacker, S. Emmonds, and G. Capati, "Residential potable and recycled water end uses in a dual reticulated supply system", *Desalination,* vol. 272, no. 1–3, pp. 201-211, 2011. [http://dx.doi.org/10.1016/j.desal.2011.01.022]

[204] N. Mead, Investigation of domestic end use, Thesis (Honours), University of Southern Queensland: Australia, 2008.

[205] M. Heinrich, *"Water end use and efficiency project – final report", Study Report 159.* BRANZ: Judgeford, New Zealand, 2007.

[206] C. Wilkes, A. Mason, L. Niang, K. Jensen, and S. Hern, *Evaluation of the Meter-Master data logger and the Trace Wizard analysis software. Special appendix to quantification of exposure-related water uses for various U.S. subpopulations.* Prepared for the US EPA: Las Vegas, NV, USA, 2005.

[207] M. Kowalski, and D. Marshallsay, "A system for improved assessment of domestic water use components", *Efficient 2003 – 2nd International Conference on Efficient Use and Management of Urban Water Supply,* 2003 Tenerife, Canary Islands: Spain

[208] A. Clarke, N. Grant, and J. Thornton, *Quantifying the energy and carbon effects of water saving.* 2009.

[209] M.C. Almeida, D. Butler, and E. Friedler, "At-source domestic wastewater quality", *Urban Water J.,* vol. 1, no. 1, pp. 49-55, 1999. [http://dx.doi.org/10.1016/S1462-0758(99)00008-4]

[210] D. Butler, "A small-scale study of wastewater discharges from domestic appliances", *Water Environ. J.,* vol. 5, no. 2, pp. 178-185, 1991. [http://dx.doi.org/10.1111/j.1747-6593.1991.tb00605.x]

[211] D. Butler, "The influence of dwelling occupancy and day of the week on domestic appliance wastewater discharges", *Build. Environ.,* vol. 28, no. 1, pp. 73-79, 1993. [http://dx.doi.org/10.1016/0360-1323(93)90008-Q]

[212] E. Friedler, and D. Butler, "Quantifying the inherent uncertainty in the quantity and quality of domestic wastewater", *Water Sci. Technol.,* vol. 33, no. 2, pp. 13-24, 1996. [http://dx.doi.org/10.1016/0273-1223(96)00190-4]

[213] E. Friedler, D. Brown, and D. Butler, "A study of WC derived sewer solids", *Water Sci. Technol.,* vol. 33, no. 9, pp. 17-24, 1996. [http://dx.doi.org/10.1016/0273-1223(96)00365-4]

[214] E. Friedler, D. Butler, and D. Brown, "Domestic WC usage patterns", *Build. Environ.,* vol. 31, no. 4, pp. 385-392, 1996. [http://dx.doi.org/10.1016/0360-1323(96)00008-X]

[215] P. Levallois, N. Guevin, S. Gingras, B. Levesque, J.P. Weber, and R. Letarte, "New patterns of drinking-water consumption: results of a pilot study", *Sci. Total. Environ.,* vol. 209, no. 2, pp. 233-241, 1998.

[216] J.M. Parker, and R.L. Wilby, "Quantifying household water demand: a review of theory and practice in the UK", *Water Resour. Manage.,* vol. 27, no. 4, pp. 981-1011, 2013. [http://dx.doi.org/10.1007/s11269-012-0190-2]

[217] *Per capita consumption of the water supplied by the public system in Portugal,* 2015.[Accessed: 28-Oct-2015] [Online]. Available: http://www.pordata.pt/Home.

[218] Associação Portuguesa de Distribuição e Drenagem de Águas (APDA), *Crisis reduces water consumption in Lisbon.*[Accessed: 28-Oct-2015] [Online]. Available: http://www.apda.pt/pt/noticia/1123.

[219] P. Vieira, M.C. Almeida, J.M. Baptista, and R. Ribeiro, "Household water use: a Portuguese field study", *Water Sci. Technol.,* vol. 7, no. 5–6, pp. 193-202, 2007. [http://dx.doi.org/10.2166/ws.2007.098]

[220] C. Matos, C.A. Teixeira, A.A.L.S. Duarte, and I. Bentes, "Domestic water uses: Characterization of daily cycles in the north region of Portugal", *Sci. Total Environ.,* vol. 458–460, pp. 444-450, Aug 2013. [http://dx.doi.org/10.1016/j.scitotenv.2013.04.018]

[221] G. Cudell, *Manual for installation of irrigation systems,* 2000.[Accessed: 25-March-2016] [Online]. Available: http://cudell.pt/sites/cudell.pt/files/files/OS/Manual_Instalador/Manual_de_Instalacao_de_Rega.pdf.

[222] I. Meireles, P. Gonçalves, V. Sousa, and A. Silva-Afonso, "Water efficiency potential in university buildings: The DECivil building of the University of Aveiro", *40th IAHS World Congress,* 2014 Funchal: Portugal

[223] I. Meireles, V. Sousa, and A. Silva-Afonso, "Statistical analysis of water consumption pattern in university buildings", *41st IAHS World Congress,* 2016 Albufeira: Portugal

[224] M.A. Imteaz, A. Ahsan, J. Naser, and A. Rahman, "Reliability analysis of rainwater tanks using daily water balance model: Variations within a large city", *Resour. Conserv. Recycling,* vol. 56, no. 1, pp. 80-86, 2013. [http://dx.doi.org/10.1016/j.resconrec.2011.09.008]

[225] M.A. Imteaz, U. Paudel, A. Ahsanb, and C. Santos, "Climatic and spatial variability of potential rainwater savings for a large coastal city", *Resour. Conserv. Recycling,* vol. 105, pp. 143-147, 2015. [http://dx.doi.org/10.1016/j.resconrec.2015.10.023]

[226] A. Campisano, and C. Modica, "Selecting Time Scale Resolution to Evaluate Water Saving and Retention Potential of Rainwater Harvesting Tanks", *Procedia Eng.,* vol. 70, pp. 218-227, 2014. [http://dx.doi.org/10.1016/j.proeng.2014.02.025]

[227] A.S. Silva, and E. Ghisi, "Uncertainty analysis of daily potable water demand on the performance evaluation of rainwater harvesting systems in residential buildings", *J. Environ. Manage.,* vol. 180, pp. 82-93, 2016. [http://dx.doi.org/10.1016/j.jenvman.2016.05.028] [PMID: 27208997]

CHAPTER 7

Health Issues and Security of Water Saving Systems

Cristina Santos[1,*], **Cristina Matos**[2] and **Armando Silva-Afonso**[3]

[1] *Department of Civil Engineering, Faculdade de Engenharia da Universidade do Porto, Porto, Portugal*

[2] *C-MADE- Centre of Materials and Building Technologies; University of Beira Interior, Departamento de Engenharias, Escola de Ciências e Tecnologia, Universidade de Trás-os-Montes e Alto Douro, Vila Real, Portugal*

[3] *Department of Civil Engineering, University of Aveiro, Board of ANQIP (National Association for Quality in Building Services), Aveiro, Portugal*

Abstract: This chapter focuses on the health issues related to the application of Water Saving Systems (WSS) in buildings. It stars by showing the environmental and economic advantages of these systems and then presents the main obstacles to their large-scale implementation: concerns about water quality, the security of the users and the implications on the public infrastructures. For these reasons, it is important to carefully study this question, list all health issues related to the use of WSS and show how they can be minimized and avoided. The base for this chapter is several feasibility studies, made in different types of buildings, and population surveys showing concerns about the user's security. Studies about water quality in existing WSS are also presented, as well as important security measures presented in international legislation and standards. The conclusions allow us to understand if this question is being properly discussed (if all the main problems are being covered), which measures are being applied and if others are needed.

Keywords: water efficiency, rainwater, greywater, advantages, obstacles, public perception, health issues, contamination, security, maintenance.

INTRODUCTION

Water Saving Systems (WSS)

Currently used in an unsustainably way, not only because of the exponential growth of the population on the planet, but mainly due to the global model of

* **Corresponding author Cristina Santos**: Department of Civil Engineering, Faculdade de Engenharia, Universidade do Porto, Porto, Portugal; Tel: 54 19 508 22 +351; Fax: 91 88 880 93 +351; Email: csantos@fe.up.pt

Enedir Ghisi (Ed.)

economic growth that over-consumes the resources, fresh water is becoming a limited resource. As climate change will worsen the situation in many areas of the planet, the adoption of sustainable practices in water use has become urgent and indispensable.

The first step to sustainability is the efficiency. In buildings, the best way to achieve an efficient use of water can be summarized as a principle which is known as "the 5R principle" (Fig. **1**) [1].

- *Reduce consumption*
- *Reduce losses and waste*
- *Reuse water*
- *Recycle water*
- *Resort to alternative sources*

} *- WATER EFFICIENCY IN BULDINGS*

Fig. (1). The 5Rs principle for water efficiency in buildings.

The first R (reduce consumption) includes the adoption of efficient products, devices and procedures (such as smart landscaping), without prejudice to other nontechnical aspects of economic or sociological nature. In this scope, the labelling of water efficiency of products is an essential measure to provide information to consumers.

The second R (reduce losses and waste), may involve interventions such as the monitoring of losses in buildings (flushing cisterns, watering gardens, *etc.*) or the installation of circulation and return circuits of sanitary hot water.

The "reuse" or "recycling" of water differs from a traditional use by the re-introduction of water at the start of the circuit (after treatment). There is the possibility of using treated wastewater for some purposes, such as watering gardens, and also using only greywater (wastewater from washing basins, baths or laundries) which has a recognized viability in some situations and has already several applications known in many countries [2 - 4]. In fact, domestic wastewater can be divided into black water, dark greywater and light greywater. The black water is generated in the toilet, dark greywater is generated in the dishwasher, in the kitchen sink and in the washing machine, and light greywater is produced in the washing basin and bathtub. Total greywater (dark + light) represent a significant amount of domestic wastewater and, according to some studies, contribute to an increased concentration of some pollutants and contaminants in wastewater [5], which can be harmful to public health and the environment.

Finally, the resort to alternative sources may involve the use of rainwater, groundwater and even salt water. Rainwater harvesting has a great potential to be included in urban water cycle, even in seemingly adverse climates such as the

Mediterranean (where the hot summers are typically dry) and presents additional advantages, such as the reduction of the urban runoff, reducing the flood peaks in urban areas and the pollutant loads. If applied in a large scale, rainwater harvesting actually reduces a city's external water demand and relieves water stress by promoting significant potable water savings [6].

Rainwater harvesting (RWH) is a widely known and used technique in Asia, Africa, and has a great technological development in some European countries. Worldwide, regulations and incentives to foster the use of rainwater have been reported [7]. A RWH system generally consists of a catchment area (usually the roof area), a filter, a storage tank, a supply network, pipes and an overflow unit [8]. The feasibility of these systems depends greatly on rainfall magnitude and intensity. In regions characterized by low precipitation rates, rainwater harvesting alone is not enough to meet the water supply demands of rural and urban population [9].

On the other hand, the interest in greywater reuse (GWR) is rapidly growing in cities located in arid and semiarid zones of growing economies because of fresh water scarcity, the increasing water demand associated with the increase of population, and rapid industrialization. In these regions, greywater reuse may contribute to alleviate water shortage and stress on existing water sources, exacerbated by the climate change. Greywater treatment and reuse is an alternative to provide non-potable water to households and reduce water usage per person by up to 50%, or even more [7, 10 - 12].

All these measures can be easily considered in new buildings or even in refurbishment projects. For existing buildings, water efficiency audits are the appropriate procedure to find the best options, in the technical-economic point of view, to improve the efficiency.

Environmental and Economic Advantages of WSS

The various measures described for an efficient use of water, summarized by "the 5R principle", may be addressed by its environmental and economic advantages point of view. In fact, the environmental advantages are strictly connected with the economical ones: a reduction of potable water consumption leads to significant reduction on water bills and the reduction on hot water consumptions (by introducing efficient showers, for example) leads also to a reduction on the energy spend in the heating process.

Dwellings generally have a high concentration of energy demand and consequently are responsible for a high amount of greenhouse gas (GHG) emissions, including CO_2, which induces important environmental impacts.

Globally they are responsible for approximately 40% of total world energy consumption. Most of this energy is for the provision of lighting, water heating, cooking and air-conditioning [13]. However, more than 50% of the energy consumption can be reduced by efficiency measures [14].

To meet these high values, in 2005 was established in Europe the EPBD - Energy Performance of Buildings Directive (Directive no. 2002/91/EC) which requires that buildings and their heating, cooling and ventilation systems are designed and executed so that the amount of energy required in their use must be low, taking into account the prevailing conditions and the occupants. This is a joint initiative of the EU Member States and the European Commission. The main objective is to improve the sharing of information and experiences in implementing this important European legislation [15].

Starting by the "R" reduction, as described before, increased efficiency in water use in buildings leads to the reduction of water consumption and wastewater effluents, increasing energy efficiency and contributing to the reduction of GHG emissions [16, 17]. As referred, one way of reducing consumption is the use of efficient products. Below it is described, as an example, a study that assessed the consequences in the water consumption, energy efficiency and CO_2 emissions of some measures of water efficiency in buildings, taking as a case study a single family dwelling, located in Almacave, in the municipality of Lamego, North of Portugal. It is inhabited by three people and consists of three floors and a garden. In the basement there is a garage, two storage rooms, a laundry, a drying area, a bedroom and a bathroom. In the ground floor there are a family room, a bathroom, a kitchen, an entrance hall and an office. The first floor consists of two suites, two bedrooms and a bathroom. The garden surrounds the house and has approximately 170 m^2.

The first step consisted in the estimative of water consumption in the dwelling. Consulting the water bill it was possible to know the consumption of water in the housing and then estimate the water used for each device, with the help of the percentages proposed in Almeida *et al.* [18] (Fig. **2**).

Then it was estimated the time used in each domestic device. After knowing the characteristics of various types of existing devices, in particular brands and models, it was possible to know the flows and make a subsequent estimate of the time of use.

The second step consisted in the estimation of the water consumption obtained with efficient devices. It was made a search for more efficient equipment in order to reduce the consumption as well as the characteristics of these products. In addition, it was proposed a rainwater harvesting system for irrigation and general washes. To know the water saving by the implementation of these measures, a comparison between the existing consumption and the consumption with efficient equipment was made: using more efficient devices a total of 24236 L/month was achieved, which corresponds to a reduction of 3374 L/month in the total consumption (12.22%). With the rainwater harvesting system, considering that it will be used for irrigation and washing, for the months of June, July, August and September (summer - high consumes), the saving accomplish is 3313 L/month, which makes a total of 6687 L/month, corresponding to a water saving of 24%.

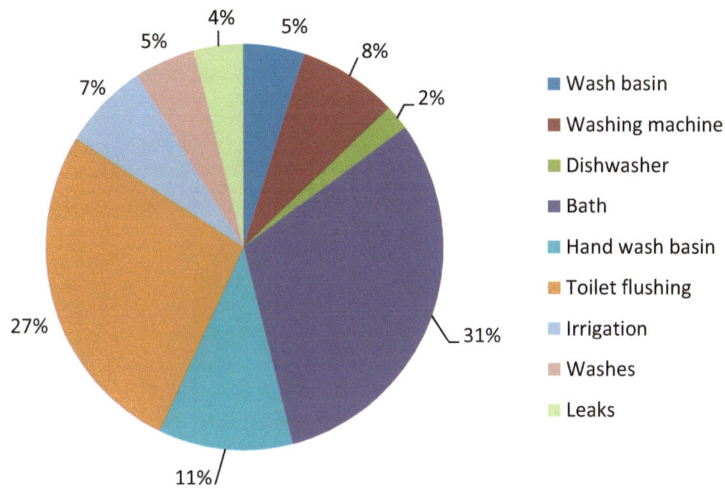

Fig. (2). Water consumption distribution by device.

This is not however the only saving obtained with efficient devices, because the reduction of water consumption in buildings implies a reduction of the volume of water abstracted treated and pumped, and also a reduction in the volume of treated and pumped wastewater. In a study conducted in the city of Aveiro, in Portugal, with 73000 inhabitants (27000 homes), it was concluded that energy consumption in the public water supply system and in the wastewater drainage and treatment system was 1.97 kWh/m^3 [19]. The city of Aveiro presents the typical characteristics of medium-sized cities in Portugal, so this value can be regarded as a normal value in the country.

In order to calculate the energy savings and CO_2 emissions reduction with the application of the efficient devices, it can be considered that the heating of a m^3 of water at 37ºC spends 30 kWh of energy and that 369.23 g of CO_2 are emitted in average per kWh of electricity consumed in Portugal [19] (considering the mix of technologies used to produce electrical energy: hydro, coal, wind, natural gas and others, also involving the import).

These data allow us to conclude that energy savings in the single family dwelling from Almacave, correspond to 45.29 kWh/month and the CO_2 emissions are reduced in 16724.13 g CO_2/month. In this case, about 13% savings in energy consumption and CO_2 emissions have been achieved, only with respect to the devices with hot water but the adoption of efficient products can lead to savings of up to 33% [19]. Considering also the savings in energy in the public network, resulting from the use of efficient devices in the building, we can add:

$$6687.75 \text{ L/month x } 1.97 \text{ x } 10^{-3} \text{ kWh/L} = 13.17 \text{ kWh} \quad \textbf{(1)}$$

In terms of CO_2 reduction:

$$13.17 \text{ kWh x } 369.23 \text{ g } CO_2\text{/kWh} = 4855.37 \text{ g } CO_2 \quad \textbf{(2)}$$

The total savings are:

$$\text{Energy: } 45.29 + 13.15 = 58.44 \text{ kWh/month} \quad \textbf{(3)}$$

$$CO_2 \text{ emissions: } 58.44 \text{ x } 369.23 = 21577.80 \text{ g } CO_2\text{/month} \quad \textbf{(4)}$$

These results show the importance of efficient water use in buildings, not only as a way to reduce water consumption, but also to improve the energy efficiency of buildings and reducing emissions of greenhouse gases.

Water "R"euse has been addressed in the perspective of reusing the wastewater after treatment in a centralized wastewater treatment plant. Various reuse strategies had been developed in terms of treatment and reuse [20]. The main benefit is to reduce the consumption of potable water, helping to reduce water bills and minimize the risk of water shortages or extreme drought. Another benefit is water conservation, helping to preserve the environment, and decrease treatment costs and capacity of the wastewater network downstream. Despite all these benefits, the applicability of water reuse and the degree of risk for the environment and public health several must be taken into account.

The main advantages of implementing a reuse system are: better management of water resources; reduction of water volume to be treated; reducing the flow of potable water; reduction of treatment costs; bill reduction for the consumer; amortization in the medium term of the initial investment; reduced maintenance costs; reduction of pollutant load washed away by surface water; reducing the volume of drinking water reservoirs; possibility of reduction of surplus flood flows; control of minimum water quality for use in question; boost the market in the sector of sanitary facilities allowing the recycling of older facilities; and specialized construction companies. The main disadvantages are the poor water quality, the high initial investment and maintenance costs and the space for the water storage tank.

As an alternative, greywater decentralised reuse seems to be more advantageous than total wastewater reuse, since it is less polluted and so, it is easier and cheaper to treat. To guarantee the success of a greywater reuse strategy it is necessary to know in detail the available flow by appliance and its variation along the day, as well as its quality. There are some investigation carried out on this subject, which has shown how usage varies significantly along the day and between domestic devices [21, 22]. In fact the quantity of wastewater generated inside a dwelling depends on several factors and it has highly variation patterns. In some European capitals the wastewater flow generated per day in a dwelling may reach 586 L. According with NSW Government [23] greywater represents 68% of the total wastewater generated in a house and 49% of this comes from baths, 34% from laundry, 10% from the kitchen and 7% from the washing basin. Friedler [24] refers a capitation of 100-150 L/person/day and in Portugal, it is estimated a range of 100-180 L/person/day, 66% of it consumed in baths [25]. Depending on the type of reuse that is considered, all the studies agree on the fact that greywater generated in a house is enough to supply a reuse strategy. In their study, Imteaz and Shanableh [26] refer that in multi storey buildings, the available greywater can significantly exceed the demand, which suggests that greywater collection from some floors would be enough to meet the demand of all the floors in a building.

A study made by Friedler [27] concluded, through an economic analysis, that on-site greywater reuse can be economically feasible in multi-storey buildings, especially if water prices continually increase. However, once the installation cost is significant (due mainly to the need of a treatment unit), in smaller buildings, with less occupants, payback periods can be unrealistic. For each case, the solution must be carefully planned and studied.

"R"esorting to alternative sources can also bring some environmental and economic advantages. Coombes & Kuczera [28] found that a rainwater harvesting system installed in a building with a 150 m^2 roof area and 1 to 5 m^3 tank in Sydney, can return 10 to 58% of water savings, depending on the number of people using the building. In Sweden, Villarreal and Dixon [29] investigated water savings potential of RWH systems from roof areas and noted that a main water saving of 30% can be achieved using a 40 m3 sized tank (toilet and washing machine end-use only) [30]. In Brazil, it has already been reported that the potential for potable water savings by using rainwater varies greatly depending on the geographic region, varying from 48% in the southeast region to 100% in the north region [31]. In some Brazilian cities, the average potential for potable water savings range from 12 to 79% [32, 33]. However, researchers refer that, although the supply capacity of rainwater is sufficient to meet the total demand in households in this region, this alternative water supply scheme is not suitable for potable water supply due to rainwater quality. Also, Muthukumaran *et al.* [34] found that the use of rainwater inside a home in regional Victoria in Australia can save up to 40% of potable water use. Example like these are registered worldwide.

A proper analysis and design before implementing these systems is important to improve their performance and amplify the benefits and effectiveness [35, 36]. It is necessary to provide criteria for the spreading of RWH systems, taking into consideration that the proper application of economic principles to environmental problems is essential to identify and implement the most cost-effective solutions [37] and move towards economic and environmental sustainable strategies for urban water management [6]. The storage tank size is usually the largest factor of the total installation cost [38] hence its optimization is essential to the feasibility and short payback of the system. Normative and technical documents existing in Germany, UK and Portugal [39 - 41] recommend the use of a "detailed approach" based on daily simulations of the system's operation by using a model of yield and demand that considers a continuous daily rainfall data corresponding to a time series from 3 to 10 years. However, the criteria to calculate the tank size using this approach is not referred, leading to the necessity to simulate several scenarios in order to obtain the storage volume that leads to an efficient and feasible system [42].

Matos *et al.* [43] presented the economic analysis of a RWH tank of a big commercial surface (Dolce Vita Braga), in Braga – Portugal, studying various scenarios of regularization and end-uses, analysing the economic feasibility and potential decrease of reliance on centralised water supply systems, which follow from these scenarios. The use of rainwater in periods in which it is available involves small tanks to the RWH system, which causes a high economic viability for these systems. This type of partial regularization precludes the use of rainwater for purposes where the needs are bigger in dry periods, like gardens irrigation, but can be highly attractive for other uses, since the price charged for drinking water is always the same regardless of the time of year. The results of this research indicate that the RWH scenarios proposed are cost-efficient. This cost–benefit analysis considers the savings of mains water as the only benefit of RWH systems. Maintenance costs and environmental, social, financial and energy saving benefits were not considered. The environmental benefits are so important that in the Guide to cost-benefit analysis of investment projects [44] the discount rate is expected to be very low, or even negative, although in Portuguese reality this may be considered a potential risk.

C. Santos [45] has investigated the application of WSS in different types of Portuguese buildings, from a dwelling where a RWH system was installed, to several commercial buildings with RWH and GWR systems. In terms of economic feasibility, it was concluded that small buildings are not completely feasible, once the payback periods for these systems are still too high (15 to 30 years), especially due to the installation costs. In large buildings such as the commercial ones, the payback periods were under 10 years, making these systems feasible is these cases, once the saving on water bill were major.

Main Obstacles

Besides their important advantages, both economic and environmental, as explained above, water saving systems have several obstacles that have been prevented their large-scale implementation. The most significant ones are the installation costs and the public acceptance.

Observing the 5R principle explained in Fig. (**1**), it is clear that:

- Reduce consumption can be cost free if based in behavioural measures from the users. In fact, many people can reduce water consumption in simple attitudes such as closing the taps while brushing their teeth. However, more powerful measures that include the use of water efficient appliances (A or plus, labels in the ANQIP labelling system) or systems that control water losses and reduces water waste do have an installation cost, especially in refurbishments. These measures are more cost-effective if considered in new constructions.
- Reuse and recycle water, as well as the use of alternatives, are great ways to eliminate a significant part of water bills. However these type of water saving systems include plumbing (and sometimes pumping) systems, once these waters cannot be in contact with potable water and they need infrastructures to collect, store and treat, if necessary, the non-potable water that will be used.

Despite the efforts in promoting worldwide the implementation of water saving systems in buildings, homeowners and stakeholders are still sceptic about their safety.

A study made in Western Sydney [46] reports the outcomes of a detailed survey about people's views on general water issues and their motivation, practices and concerns related to greywater reuse for irrigation. One in five participants (19%) thought the cost associated with the reuse was a very important consideration for them. When asked about health risk to people, plants and soil, one in three participants (38%) considered health risk to be very important for any decision on the reuse of greywater.

In Sant Cugat del Vallès (metropolitan Area of Barcelona, Spain) a similar study was carried out in 2008, considering 13 greywater reuse systems in operation serving 278 households [7]. This study was an exceptional opportunity to assess community perceptions during the socio-technical transition process of a six years' experience on greywater reuse provided opportunity. Results show that the perception of health risks, operation regimes, perceived costs and environmental awareness are, in different degrees, significant determinants of public acceptance.

Also in Israel, the findings of a survey (256 participants) conducted to determine the attitude of the Israeli urban public towards various urban water reuse options was made in 2004 [47]. The survey included 21 reuse options, which were clustered into three reuse categories, namely: low, medium, and high contact levels. Results show that a high proportion of the participants supported medium contact reuse options such as sidewalk landscaping (95%), domestic WC flushing (85%) and firefighting (96%). Higher contact reuse options such as domestic laundry (38%), preserved food (13%), and potable aquifer recharge (11%) found much lesser support, which shows that people relate medium and high contact reuse option to health concerns and reject them.

To reverse this tendency and solve the problem, recent investigation carried out in Australia show that social marketing or persuasion is ineffective in influencing people to use recycled water [48]. One essential factor to reduce people's concerns about reuse is information provision including risk communication in a very honest manner, enhancing the benefits of water reuse and showing successful reuse projects.

WATER QUALITY

Health Issues that can Occur by Using Non-Potable Water in Buildings

The most common WSS using non-potable water have already been study for many years. The technologies for rainwater harvesting or greywater reuse in buildings, for example, are relatively well known. The main issues that still arise in relation to these water efficiency measures are of sanitary nature and problems were registered in several countries resulting from lack of health security in installations of this type. Usually, problems appear when:

- Cross-connections between non-potable and drinking water networks are made, leading to a dangerous scenario where people can drink and use non-potable water for food preparing;
- Non-potable water treatment is inexistent or insufficient, which can be due to both misevaluation of water quality in the source or a change of its quality through time;

- There is lack of maintenance, which leads to the malfunctioning of the treatment system.

Rainwater has, in general, great quality but it should never be considered potable water as it can pick up impurities by rains form the atmosphere and its quality decreases during harvesting and storage. In terms of physical-chemical parameters, collected roof water, rainwater and urban storm water tend to exhibit quality levels that are generally comparable to the World Health Organization (WHO) guideline values for drinking water [49]. However, a number of collected rainwater samples have exceeded the WHO values in terms of total coliform and faecal coliform, being the droppings of birds, rodents, *etc.* indicated as the source of this problem. Also, low pH rainwater can occur as a result of Sulphur dioxide, nitrous oxide and other industrial emissions, hence air quality standards must be reviewed and enforced. In addition, high lead values can sometimes be attributed to the composition of certain roofing materials - thus it is recommended that for roof water collection systems, the type of roofing material should be carefully considered. If rainwater is used for watering the gardens, then chemicals and low pH values can be harmful for the pants. Contamination of human beings can also occur if there is any possibility for formation of aerosols (sprinklers in the garden, for example). To avoid this, areas and pavements used by people should be excluded form catchment areas and some types of roof coverers should be avoided [8]:

- Asbestos-cement roofs can cause filters to block, and collect water may pose a health risk;
- Metal roofs (except stainless steel) can release small amounts of leachates, which can stain water fixtures;
- Bitumen felt coated roofs can lead to discoloration and odour problems;
- Grass (and other vegetation) covered roofs reduce the amount and speed of run-off water, which might need extra treatment as the water is likely to be discoloured by soil.

As for greywater, its quality is very variable and influenced by many factors such as the dynamics of households, number of occupants, their age distribution, lifestyle characteristics, water usage patterns, local cost of water and also the climate [50]. Countries and areas with very low water consumption tend to have particularly low greywater quality, as contaminants are concentrated in the small quantities of water used. GW quality is expected to vary widely due to different uses of sanitary installations. In the public buildings this variation should be more significant since those installations are used by diverse people and, consequently, the amount of dirtiness, soaps, bathing products and other pollutants varies accordingly with every usage [51].

Studies also show that greywater may contain large number of potentially pathogenic microorganisms but it should also be considered that the incidence of diseases is dependent upon more than just the concentration of organisms (other factors include the degree of exposure and the health and age of affected individuals). Research made by Dixon *et al.* [52] has shown that counts of total coliform and faecal coliforms increased from 10^0/100 ml to above 10^5/ 100 ml within 48 h in stored greywater from various sources. The number of viruses found in greywater is dependent upon the health of the population generating the liquid because only infected individuals may excrete a virus, although the individual may not show any symptoms of infection. In those cases, transmission of pathogens can occur by drinking or having direct contact with greywater, eating products exposed to contaminated water or soil and by inhaling contaminated aerosols. Table **1** resumes the scale of risks form greywater reuse.

Table 1. Conceptual analyses of range of risk from greywater reuse [53].

	Lower Risk	Intermediate Risk	Higher Risk
Population	Small population (single family)		Large population (multi-occupancy)
Exposure	No body contact (sub-surface irrigation)	Soma contact (WC flushing, bathing)	Ingestion (drinking)
Dose-response	<1 virus per sample <1 bacteria per sample		>1 virus per sample >105 bacteria per sample
Delay before reuse	Immediate reuse	Reused within hours	Reused within days

Studies of Water Quality in Existing WSS

The use of rainwater has been a subject of study in Portugal including a comparison of results with existing standards and guidelines in several countries (Table **2**).

Table 2. International Comparative Values [55].

Parameters	NBR 15.527/07 Brazil	EPA	Australia	Canada
Total coliforms (NMP/100ml)	Absence	Absence	-	-
Heat-resistant coliforms (NMP/100ml)	Absence	Absence	< 150	<150
Turbidity (NTU)	< 2.0 or < 5,0 for less restrictive uses	< or = 2	-	50
Color (Hz)	< 15 µH	-	-	-
pH	6.0 - 8.0	6 - 9	6.5 – 8.5	6.5 – 8.5

Taking into account the need to consider long-term water storage periods in rainwater harvesting systems built in some regions (such as the Mediterranean basin, for example), in particular when rainwater is used for watering gardens, it was developed by ANQIP, in Portugal, a water quality control monitoring study at a pilot facility, analysing, over several weeks, the essential chemical, physical and micro-biological parameters (including Legionella) in the tank and in the end-uses (sprinkler irrigation) [54].

The major purpose of monitoring control was the evaluation of water quality during all the summer season, trough analysis of essential chemical, physical, and microbiological parameters over the weeks. Sampling was conducted in two parts of the system, particularly in the storage tank and sprinklers, since they are the most critical locations in the facility.

An analytical plan was established with two batteries of tests with ranges of parameters and different frequencies:

- Level 1: Monthly complete analysis (physic-chemical and microbiological), allowing an assessment not only of the characteristics of rainwater but also those arising from precipitation over the coverage areas with eventual contamination, and from the collection;
- Level 2: Weekly analysis of the main physic-chemical and microbiological contaminants control, to allow the characterization of the profile quality changes, as a function of storage time;
- Level 3: If the results determine an eventual need of disinfection procedures, or water treatment, this level will introduce the methodology for monitoring and control to define.

The parameters were chosen according to the applicable legal values for irrigation water and surface water, considering the specified in the European Directives to Maximum Recommended Values for water for irrigation, (MRV1), and Maximum Recommended Values for water for human consumption (MRV2). Some analytical values found are shown in Table **3** for the monthly monitoring and in Table **4** for the weekly control.

During the sampling period short periods of rainfall occurred, that had no meaning in refilling of the tank, resulting in the absence of water for operation of the irrigation system. So the last sample corresponding to the sprinklers was not available.

For the pilot installation used, corresponding to a conventional solution built according the Portuguese technical specifications, without filtration or disinfection, the results demonstrate the possibility of considering a storage period higher than the values adopted in most international standards that does not exceed 30 days in general.

Table 3. Level 1 – Monthly sample analytical results [54].

PARAMETER/(UNIT)	Month 1		Month 3		MRV1	MRV2
	Tank	Sprinkler	Tank	Sprinkler		
Total Coliforms (UFC/100ml)	1	0	0	Insufficient water for operation of the sprinklers		50
Faecal Coliforms (UFC/100ml)	0	0	0		100	20
Temperature (ºC)	23.5	22.8	23.1			22
Turbidity (NTU)	2.77	1.20	4			
pH (E. Sorensen)	7.57	7.81	7.92		6.5–8.4	6.5–8.5
Conductivity (µS/cm)	130	135	153			3
Color (mg PtCo/L)	<5	<5	<5			10
Dissolved oxygen (%)	75.1	99.2	84.9			70
Chemical Oxygen Demand COD (mg O2/L)	<10	<10	<10			-
Biochemical Oxygen Demand – BOD (mg O_2/L)	<3	<3	<3			3
Total hardness (mg $CaCO_3$/L)	35.0	37.0	34.5			
Ammonium (mg NH4/L)	<0.050	<0.050	<0.050			0.05
Nitrate (mg NO_3/L)	2.2	1.2	1.8		50	25
Chloride (mg CL-/L)	9.7	<4	<4		70	200
Nitrite (mg NO_2/L)	<0.020	<0.020	<0.020			
Sulfate (mg SO_4/L)	8.7	6.2	6.7		575	150
Total Suspended Solids (mg/L)	<3	<3	<3		60	25
Total Solids (mg/L)	74	75	94			
Zinc (µg Zn/L)	1.6x102	<10	<10		2	0.5
Iron (µg Fe/L)	4.6x102	27	<10			
Cadmium (µg Cd/L)	<1.0	<1.0	<1.0		0.01	0.001
Lead (µg Pb/L)	<5	<5	<5		5	-
Legionella Pneumophila (UFC/100ml)	Not detected	Not detected	Not detected			

Table 4. Level 2 – weekly sample analytical results [54].

<table>
<tr><th rowspan="2">PARAMETER (Unity)</th><th colspan="2">Week 1</th><th colspan="2">Week 2</th><th colspan="2">Week 3</th><th colspan="2">Week 6</th><th colspan="2">Week 12</th></tr>
<tr><th>Tank</th><th>Sprink</th><th>Tank</th><th>Sprink</th><th>Tank</th><th>Sprink</th><th>Tank</th><th>Sprink</th><th>Tank</th><th>Sprink</th></tr>
<tr><td>Total Coliforms (UFC/100ml)</td><td>1</td><td>0</td><td>0</td><td>0</td><td>0</td><td>0</td><td>0</td><td>0</td><td>10</td><td rowspan="11">Insufficient water for operation of the sprinklers</td></tr>
<tr><td>Faecal Coliforms (UFC/100ml)</td><td>0</td><td>0</td><td>0</td><td>0</td><td>0</td><td>0</td><td>0</td><td>0</td><td>0</td></tr>
<tr><td>Temperature (ºC)</td><td>23.8</td><td>22.2</td><td>22.8</td><td>23.4</td><td>24.4</td><td>24.4</td><td>23.7</td><td>23.7</td><td>25.7</td></tr>
<tr><td>Turbidity (NTU)</td><td>0.74</td><td>0.55</td><td>1.95</td><td>1.65</td><td>1.98</td><td>1.98</td><td>12.5</td><td>3.72</td><td>1.22</td></tr>
<tr><td>pH (E. Sorensen)</td><td>8.16</td><td>8.55</td><td>8.73</td><td>8.66</td><td>8.64</td><td>8.64</td><td>7.85</td><td>7.8</td><td>8.03</td></tr>
<tr><td>Conductivity (μS/cm)</td><td>140</td><td>139</td><td>139</td><td>139</td><td>140</td><td>140</td><td>139</td><td>146</td><td>163</td></tr>
<tr><td>Color (mg PtCo/L)</td><td><5</td><td><5</td><td><5</td><td><5</td><td>-</td><td>-</td><td>5</td><td><5</td><td><5</td></tr>
<tr><td>Dissolved oxygen (%)</td><td>75.9</td><td>78.2</td><td>59</td><td>75.2</td><td>-</td><td>-</td><td>-</td><td>-</td><td>77.7</td></tr>
<tr><td>Total Suspended Solids (mg/L)</td><td><3</td><td><3</td><td><10</td><td><10</td><td>-</td><td>-</td><td>-</td><td>-</td><td><3</td></tr>
<tr><td>Total Solids (mg/L)</td><td>68</td><td>69</td><td>90</td><td>87</td><td>-</td><td>-</td><td>87</td><td>100</td><td>114.5</td></tr>
</table>

About existing greywater reuse systems, a study made by Domènech and Saurí [3] in Spain reveals the most compromising parameters of treated greywater in three different types of systems: type A with filtration (nylon filter), sedimentation and disinfection (hypochlorite); type B comprising sedimentation, filtration (sliex anthracite filter and cartridge filter), sedimentation again and disinfection (hypochlorite); type C with filtration (cylindrical sieve), oxygenation (OM degradation) and disinfection (UV light).

Treated greywater presented the quality parameters described in Table **5**.

Table 5. Quality of treated greywater [3].

Parameters	Unit	System A	System B	System C	MAC
Escherichia coli	UCF/100 mL	2400	0	0	0*
Temperature	◦C	25	25.9	19.5	-
Free chlorine	ppm	<0.10	4	0.3	0.5-2**
Legionella spp.	UCF/1 L	Not detected	<50	<50	100*
BOD5	mg O2/L	120	34	<3	
QOD	mg O2/L	276	101	<301	
Total phosphorous	mg/L	<3	Not detected	Not detected	
Suspended solids	mg/L	57	<10	<10	<10*

(Table 5) contd.....

Parameters	Unit	System A	System B	System C	MAC
Nitrates	mg/L	Not detected	Not detected	11	
Organic and ammonia nitrogen	mg/L	11	<3	36	2*
Turbidity	NTU	22	19	2	1*
Nematode eggs	Eggs/L	0	-	-	
Taenia eggs	Eggs/L	0	-	-	
Comments		Sodium hypochlorite was used up when the sample was taken		Drinking water entered the system when the sample was taken	

MAC: Maximum accepted concentration.
* Royal Decree 1620/2007 (Spain) about reclaimed water reuse.
** Decree 95/2000 (Spain) about water quality in swimming pools.

Another study, developed by Friedler and Gilboa [4] examines the microbial quality of treated greywater along a continuous pilot-scale reuse system for toilet flushing. The research was carried out in a pilot plant which treated light greywater (originating from the baths, showers and washbasins) from 14 flats. The treatment system consisted of two sub-systems, functioning in parallel: a Rotating Biological Contactor (RBC) followed by a sedimentation basin and a Membrane Bioreactor (MBR). Effluents from the two sub-systems were disinfected by two independent UV irradiation units.

Samples were taken from toilet bowls which were flushed with RBC treated greywater without disinfection, from toilet bowls flushed with UV-disinfected RBC treated greywater, and from toilet bowls flushed with potable water, at different days of the week and varying times from homes of six families. Results are presented in Table **6** and Table **7**.

Table 6. Raw greywater and RBC and MBR effluent quality in the pilot-scale systems [4].

Parameter		Raw Greywater	RBC Effluent	MBR Effluent
CODt [mg O_2/L]	Avg	148	47	-[a]
	Std	49	60	
	n	61	32	
CODd [mg O_2/L]	Avg	86	45	42
	Std	27	36	32
	n	60	60	28

(Table 6) contd.....

Parameter		Raw Greywater	RBC Effluent	MBR Effluent
BODt [mg O_2/L]	Avg	95	3.7	1.1
	Std	29	5.5	1.7
	n	8	18	9
Turbidity [NTU]	Avg	33	1.5	0.21
	Std	24	0.95	0.11
	n	65	65	26
Transmittance $_{254\ nm}$ [%]	Avg	52	93	94
	Std	20	3.7	4.3
	n	30	30	30
Absorbance $_{254\ nm}$ [cm^{-1}]	Avg	0.282	0.032	0.026
	Std	0.076	0.014	0.018
	n	30	30	30

[a] COD in the MBR effluent was considered as COD_d.

Table 7. Required volumes to cause infection by P. Aeruginosa and S. Aureus [4].

Parameter	Exposure	Infection Dose (number)	T. Greywater (L)	T. Greywater after UV (L)	Potable Water (L)
Pseudomonas aeruginosa	Oral	Traumatised: 10^6	238	83	∞
		Healthy: 10^8	$238 \cdot 10^2$	$83 \cdot 10^2$	∞
Staphylococcus aureus	Skin	Traumatised: 10^2	1.8	∞	0.19
		Healthy: 10^6	$1.8 \cdot 10^4$	∞	$1.9 \cdot 10^3$

It is clear that the added health risk associated with reusing UV-disinfected greywater for toilet flushing (regarding P. aeruginosa and S. aureus), was found to be insignificant. Considering the infection doses of these pathogens, it was assessed that, in order to be infected, people should be in contact with very large volumes of water, large enough to be actually quite impossible. Further, the microbial quality of the UV-disinfected greywater was found to be about the same or even better than the microbial quality of "clean" water in toilet bowls flushed with potable water (which were used for excretion).

Investigators concluded that, since the microbial quality of the disinfected greywater was found to be equal or even better than the microbial quality of "clean" water in toilet bowls used by people for defecation and flushed with potable water, the added health risk due to reusing properly treated greywater for toilet flushing should be insignificant.

SAFETY ASSURANCE OF WATER SAVING SYSTEMS

Implications on the Infrastructures

It should be noted that health problems can occur not only by reusing or recycling waste water or using alternative sources, but also by installing efficient products or devices. Indeed, very low flow rates or much reduced volumes can have important consequences on the health of users. The reduction of the volume in flushing cisterns, for example, can have some constraints at the level of the drainage safety (non-occurrence of blockages) and cleaning performance [56]. In the case of showers, low flow rates can increase the risk of burns [57].

Unfortunately, this information is not generally known by the public. Even most of the recognized water efficiency labelling systems doesn't take into account health concerns, not being more than an indication of water consumption.

An exception is the labelling system from ANQIP (a Portuguese non-governmental association, which aims to promote water efficiency in buildings), used in Portugal and accepted in other countries [58], with categories form E to A++. In this scheme, the label "A" signifies the ideal efficiency and the devices with very low flow rates or volumes, rated with "A+" or "A++", should bear indications on the conditions of application or use, taking into account aspects of health, comfort or proper functioning of the building networks (Fig. **3** and Fig. **4**). In the case of WaterSense label (US Environmental Protection Agency), their specifications are also drawn up considering aspects of safety and health, but this concern is not explicit in the label.

In public infrastructures, there is still a lack of investigation on the real impacts that a large-scale implementation of WSS might have. It is essential to assess the impacts of a large-scale implementation of wastewater reuse systems, in the existing sewerage networks, since it is expected that these systems modify the quality and quantity of the wastewater that flows into the public network and treatment plants.

In a first approach, a reuse system will change the amount of water that will enter the network and this may threaten the operation of wastewater drainage networks. The general criteria for the calculation of domestic wastewater flows, which are mainly resulting from domestic uses of drinking water, are based on a demographic study, a forecast of the evolution of capitation and the use of appropriate peak factors. The hydraulic-sanitary design of wastewater drainage systems is fundamentally based on the assessment of the flow rate conditions and water depth. The network is dimensioned for a certain flow, which will impose a fixed diameter to collectors. With wastewater reuse the network will be working with a different flow, which may cause constraints such as blockages and bad cleaning performances. Thus it is important to study what changes the network will suffer from the introduction of wastewater reuse systems.

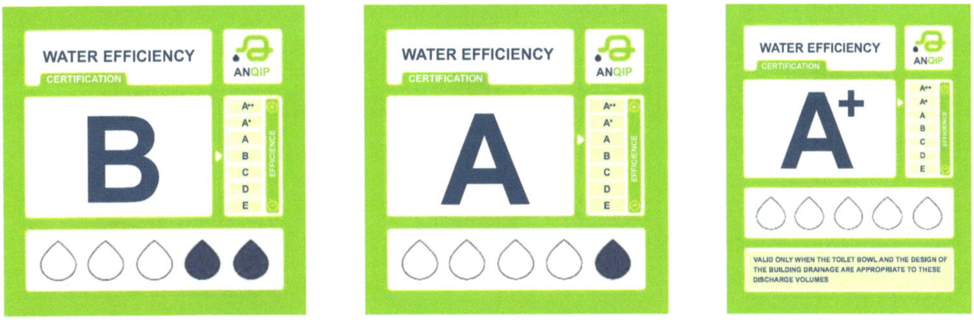

Fig. (3). Examples of ANQIP general labels of water efficiency [59].

Fig. (4). ANQIP special label A++ for low volume flushing cisterns [60].

According to a study conducted at the University of Trás-os-Montes and Alto Douro [61], the reuse of greywater causes a reduction of the influent flow to the network even though, the height of the water depth required by the Portuguese Regulation [62] is accomplished. Velocities experienced a minor decrease: the values obtained in the studied scenarios meet both the minimum and the maximum speed imposed on the Regulation. In this case study, it was possible to verify that, regarding the operation of the network, there is no major change, with the reuse of light or total greywater. The network would operate according to the criteria imposed by the Regulation. As for pollution and volumetric loads, in the case study, the pollutant loads that would stream to the wastewater drainage network would decrease when wastewater reuse is applied, mainly due to the decrease of the influent flow. For the volumetric loads, as expected, it would be lower.

In this study the impacts on the existing wastewater treatment plants were not analysed, although it is an important aspect that has to be addressed.

Water Quality Requirements for Non-Potable Uses

Nowadays, the water security has been consolidating itself as a global concern to ensure the health of populations [63]. In this sense, the International Water Association (IWA) published the Bonn Charter for the supply, which describes the institutional and operational conditions to manage water supply from the source to the consumer, in accordance with World Health Organization (WHO) guidelines.

Within this perspective, the management and control of the supply systems that use non potable water should be based on a Security Plan, and this guidance should also be understood in relation to building systems and not only in relation to public systems.

There are legislation or recommendations about quality parameters in some countries for the regenerated water intended to reuse in buildings, as it is the case, for example, of the Spanish Royal Decree 1620/2007 [64], the Information Sheet H201 from the German Association FBR and the Portuguese Technical Specifications ETA 0905 (reuse of greywater) [65] and ETA 0701 (rainwater harvesting) [39] from ANQIP. However, the widespread lack of standards and regulations led to some instances where the risks to public health due to deficiencies in design, operation, or maintenance became apparent.

To reduce potential risks, a direct use of rainwater or greywater is not recommended. Taking as an example the Technical Specification ANQIP ETA 0905 [65], it is considered that regenerated water can be used in flushing of toilets, washing clothes, and watering gardens, after appropriate treatment. The infiltration in the soil or direct discharge into waterways can also be considered in relation to the disposal of excessive water in the circuit.

According to the European Environment Agency [8], rainwater harvesting systems should include at least a simple treatment with a filter (that automatically prevent leaves and other solid debris from entering the system) which is enough if the harvested water is used in WC flushing, water gardening or uses that don't include drinking or washing with this water. BSI [40] goes further and recommends that the filter should have an efficiency of at least 90% and allow the passage of a maximum particle size of 1.25 mm.

A frequent cleaning of the gutter to prevent dirt to enter the system is also recommended, as well as the labelling of any taps fed by rainwater (possibly they should also have their handles removed to prevent children from drinking it). It's not necessary to test harvested rainwater frequently because it is used for purposes that do not present significant risk. However, water quality observations should be carried out during maintenance visits to identify any problems early [8].

To flush toilets, the requirements in Table **8** should be considered, where the values for total coliforms and faecal coliforms are defined as in the quality standards for interior bathing waters, under the terms of the European directives (Directive No. 2006/7/EC of the European Parliament and the Council). In this table, MAV means Maximum Acceptable Value and MRV means Maximum Recommended Value, as usual.

Table 8. Quality requirements for the flushing of toilets [65].

Parameter	MAV	MRV
Total coliforms	-	10^4 UFC /100 mL
Faecal streptococci (Enterococci)	400 UFC/100 mL	-
Faecal coliforms (Escherichia coli)	10^3 UFC /100 mL	0 UFC/100 mL
Pseudomonas aeruginosa	1 UFC/mL	
Enteric parasites	1 egg/ 10 L	
Suspended solids	10 mg/L	
Turbidity	2 NTU	

Additional safety measures may be considered, such as the placement of signs requiring the closing of the toilet lid in the case of the flushing of toilets, or coloration the non-potable water used. The requirements indicated in Table **8** are also applicable to the washing of clothes, which should not be carried out at temperatures less than 55° C.

For the watering of private gardens, the requirements given in Table **9** must be complied, without the need for adding chemical products. In the case of irrigation of products that can be eaten raw, the MAV indicated in the Portuguese Standard NP 4434 (2005) must be considered, and is not recommend the use of greywater from kitchens.

Regarding the rainwater harvesting in buildings, the Technical Specification ANQIP ETA 0701 considers the following possible uses [39]:

- Toilet storage tank flushing
- Washing clothes
- Washing floors, cars and so on.
- Watering gardens, lawns, parks *etc.*
- Industrial uses (cooling towers, firefighting systems, HVAC, *etc.*)

Table 9. Quality requirements for the watering of private gardens [65].

Parameter	MAV	MRV
Legionella spp.(*)	100 UFC/100 mL	-
Total coliforms	-	104 UFC /100 mL
Faecal streptococci (Enterococci)	100 UFC/100 mL	-
Faecal coliforms (Escherichia coli)	200 UFC/100 mL	0 UFC/100 mL
Salmonellae	Not detectable	-
Enteric parasites	1 egg/ 10 L	Not detectable
Suspended solids	10 mg/L	-
Turbidity	2 UNT	–

It is felt, moreover, that the use of untreated rainwater for toilet flushing should only be acceptable if the water quality is at least up to that of bathing water, defined on the European directives (Directive 76/160/EEC, of the Council) [66]. It may be disinfected with chlorine or a similar process, if necessary. Clothes should only be washed with rainwater that has had no specific treatment if the washing water temperature reaches at least 55°C. A micro filter with a minimum mesh of 100 μm should be fitted if the water is to be used for this purpose.

However, in some countries, such as Germany (Table **10**) less conservative values are allowed for restricted uses such as toilet flushing and laundry. It is quite logical that countries in a very high hydrological stress should decrease quality requirements for non-potable water uses, if a risk assessment is made for each particular case, in order to promote the installation of greywater reuse and rainwater harvesting systems.

Table 10. Quality requirements for service water in Germany and the EU Directive for Bathing Water [67].

Parameter	Service Water Reused for Toilet Flushing	EU Directive for Bathing Water (2006/7/EC)	
		Excellent Quality (Guide)	**Good Quality (MSandatory)**
Sensory test	colorless, clear, almost odorless	n/a	no abnormal change in color
Suspended solids	minor	n/a	-
Oxygen Saturation (%)	> 50	80-120	n/a
Transmission $_{254nm}$ (%)	> 60	n/a	n/a
BOD_7 (mg/L)	< 5 mg/l	n/a	n/a

(Table 10) contd.....

Parameter	**Service Water Reused for Toilet Flushing**	**EU Directive for Bathing Water (2006/7/EC)**	
		Excellent Quality (Guide)	**Good Quality (MSandatory)**
Total coliforms (cfu/mL)	< 100/ml	n/a	n/a
Faecal coliforms (cfu/100 mL)	< 1.000	250	500
Faecal streptococci (cfu/100 mL)	< 1.000	100	200
P. aeruginosa (cfu/mL)	< 1/mL	n/a	n/a

About greywater in particularly, a quick comparison between Table **6** and Table **10** allow us to see that raw greywater does not meet the quality parameters needed for a safe reuse. A treatment unit is mandatory with a secondary level of treatment to reduce the organic material and with a disinfection unit to remove harmful bacteria and viruses.

Another important matter that should be kept in mind is that all forms of wastewater when stored will turn septic unless an adequate treatment in made upstream. Similarly, when greywater is stored, it will also turn septic, giving rise to offensive odours and providing suitable conditions for microorganisms to multiply. Thermotolerant coliforms have been found to be multiplied by between 10 and 100 times during the first 24 to 48 hours of storage. Therefore, greywater must only be stored temporarily in a surge tank, unless it is adequately treated [48]. According to Dixon *et al.* [52], four major processes occur in storage units: settlement of suspended solids, aerobic microbial growth and anaerobic release of soluble COD from settled organic matter and atmospheric reaeration. Storing greywater for 24 hours may significantly improve water quality through rapid settlement of organic particles, however, storage beyond 48 hours leads to depleted DO levels and potential aesthetic problems.

Recommendations for a Correct and Safe Reuse

To minimize the risk to human health and to prevent the malfunction of rainwater harvesting and greywater reuse systems, the following considerations are important [50, 67]:

- Cisterns and tanks should be tight, properly dimensioned (according to local regulations and recommendations) vented and leakage free. An anti-backflow device should be installed if there is an overflow connected to a drain or sewer. They should also be easily accessible for emptying or cleaning the systems.
- Systems must dispose the water below ground surface to prevent the entry of settled materials.
- The system must be designed and operated to prevent direct human contact with

non-potable water

- Cross-connection between the drinking water and service water networks must be completely excluded. This is usually achieved by installing colour designated pipes for the service water supply and testing for any cross-connection, for example, by using food dyes at the beginning of operation.
- Regular inspection and maintenance of the system according to the instructions of the manufacturer contributes to operational safety in addition to increasing service life and energy efficiency.
- Greywater must not be allowed to enter any storm water drainage system.
- Non-potable water must not be allowed to pond on the surface or to run off the property.
- Taps and other devices supplied with non-potable water must be clearly labelled to advise that contact with the water must be avoided.
- Sprinkler irrigation must be avoided.

In general, the definition of standards and guidelines for ensuring the safe use of water is the first step to the promotion of sustainable practices to abstraction, reuse, and storage of water.

In Portugal, the use of rainwater in buildings was object of the development of voluntary Technical Specifications from ANQIP (ETA 0701 and ETA 0702) [39] [68]. The certification scheme implies intervention from ANQIP in the prior analysis of the project, the execution of inspection of the works, on the requirement for a maintenance plan and also in the certification of installers.

One aspect to which special attention was given in the Specification ANQIP ETA 0701 was the need to divert the first flush, since the prolonged dry periods typical of the Portuguese climate (Mediterranean) can aggravate the problem of pollution of these first waters.

On the sizing of the tank, and due to the long dry periods, it is generally important to prolong the storage period, in particular when rainwater is used for watering gardens. To establish the maximum storage period allowed, ANQIP developed a water quality control monitoring system at a pilot facility (described in item 2.2) analysing, over several weeks, the essential chemical, physical and micro-biological parameters (including Legionella) in the tank and in the uses (watering gardens). As a result of this study, ANQIP established in ETA 0701, a maximum storage period of up to three months, significantly higher than that set in other foreign standards, but which was found to be feasible in the context of the Mediterranean climate and may contribute in this region to the increase in rainwater harvesting installations.

With regard to installations for reuse of greywater in buildings, ANQIP developed the Technical Specifications ANQIP ETA 0905 [65], which requires prior analysis of the project by ANQIP, inspections during construction, the certification of installers, a maintenance plan and a safety plan. The certification scheme was established in the Technical Specification ANQIP ETA 0906 [69].

Under the Specification ETA 0905, a safety plan must be prepared for all systems of reuse and recycling of greywater in buildings, with an initial version the installer's responsibility, but periodically updated by the user.

This safety plan must include, at least, the following chapters:

- Description of the installation;
- Risk analysis;
- Criteria for the evaluation of the conformity of the quality of reclaimed water (minimum frequency of analysis for the various installation operation periods, *etc.*);
- Procedures in the event of a fault or serious problem (Action Plan).

The establishment of certification schemes for these facilities is a solution mentioned in some documents of the European Union [70], procedure which was adopted, for example, in Portugal on a voluntary basis, as described above.

The Importance of Maintenance

Experience has shown that controlling the construction, maintenance, and operation of water saving systems can be more important that the existence of rules about the quality of regenerated water. This is clear from Domènech and Saurí [3] study, where high levels of turbidity (Table **5**) were indicated as the main reason for the existence of *E. Coli* in regenerated water. However, the control system must not be so excessive that becomes an obstacle to the spread of these water efficiency measures (as the case of the referred Spanish document, where many parameters need to be analysed, requiring an excessive frequency of tests) [71 - 73].

Regular maintenance contributes considerably for RHW and GWR systems safe operation and increases their service life and energy efficiency. Monitoring plans must take into account the manufacturer's indications and usually define the following instructions [39, 40, 65, 74, 75]:

- Annual maintenance to: filters, pumps and pump controls, storage tanks, back up water supplies, control units, water level gauge, wiring, pipework, markings, supports and fixings; in RWH systems, gutters and downpipes; in GWR systems, membranes, biological support, media and strainers;

- Maintenance every 6 months to UV lamps;
- Monthly control of biocides, disinfectants or other consumable chemical in GWR systems;
- Every 10 years: drain down and clean the tanks.

Before any maintenance operation, systems must be drained and flushed with clean water to reduce the risk of contaminations.

CONCLUSION

The 5Rs principle covers all aspects for an efficient water use in buildings. Amongst them, the use of efficient devices (taps, showers, WCs), rainwater harvesting and greywater reuse systems are the most significant.

The use of efficient devices is, nowadays, a reality in many buildings and they can be found easily in the market. Labelling systems like the one presented by ANQIP in Portugal help people to easily identify the most efficient devices. However, rainwater harvesting and water reuse systems are not being installed as generally as expected. Besides many studies proving their environmental and economic advantages, two main obstacles are still slowing their large-scale implementation: high installation costs and public acceptance.

Studies made in several countries reveal that health issues are still a major concern for people and the best way to revert this situation is to clearly inform house owners, stakeholders and the general population, how to install and maintain a safe and feasible system. For that propose, this chapter focuses norms and reference guidelines with all the aspects to consider. For example, it is clear that a rainwater and greywater should not be used without a treatment and usually, only WC flushing and garden irrigation should be considered as end uses. If rainwater treatment, for the described purposes, only requires a filter as treatment, greywater also needs at least a secondary treatment and disinfection in order to prevent human contamination. However, studies reveal that treated greywater and even rainwater can still have harmful microorganisms but many times the amount is smaller than in WC supplied by potable water and the infection dose is so that people had to be in contact or ingest such a high amount of treated water to be infected that it becomes unrealistic. That is why it can be concluded that rainwater harvesting and greywater reuse systems, designed, installed and maintained in accordance with legal and normative documents, with certification schemes, are safe and health issues can be insignificant.

Other aspect focused on this study is the problems that can occur in infrastructures. The use of water saving systems can bring consequences such as possible occurrence of blockages and cleaning performance in buildings and public networks. These aspects need further investigations to prevent future problems.

Finally, if health issues are still a concern and an obstacle to the efficient water use in buildings, the best way to promote them is by installing and operate rainwater harvesting and greywater reuse systems in public buildings, for example, giving people the opportunity to see that they are safe and that health risks are identified and minimized. Nevertheless, water management authorities must be the main drives of this process once public perception of water quality is mainly related to public authorities, more that in water quality studies.

CONFLICT OF INTEREST

The authors confirm that they have no conflict of interest to declare for this publication.

ACKNOWLEDGEMENTS

The authors of this chapter are thankful to ANQIP for the authorization to include images of their labels of water efficiency.

REFERENCES

[1] A. Silva-Afonso, and C. Pimentel Rodrigues, "Water Policy for Buildings: A Portuguese Perspective", In: *Water Efficiency in Buildings: Theory and Practice.* John Wiley & Sons, Ltd.: UK, 2014, pp. 42-55.

[2] E. Shaki, Z. Zahraa, and A. Al-Obaidy, "Environmental and health risks associated with reuse of wastewater for irrigation", *Egyptian Journal of Petroleum,* Feb 2016. Available online, ISSN 1110-0621

[3] L. Domènech, and D. Saurí, "Socio-technical transitions in water scarcity contexts: Public acceptance of greywater reuse technologies in the Metropolitan Area of Barcelona", *Resour. Conserv. Recycling,* vol. 55, no. 1, pp. 53-62, 2010.
[http://dx.doi.org/10.1016/j.resconrec.2010.07.001]

[4] E. Friedler, and Y. Gilboa, "Performance of UV disinfection and the microbial quality of greywater effluent along a reuse system for toilet flushing", *Sci. Total Environ.,* vol. 408, no. 9, pp. 2109-2117, 2010.
[http://dx.doi.org/10.1016/j.scitotenv.2010.01.051] [PMID: 20172592]

[5] E. Friedler, "Quality of individual domestic greywater streams and its implication for on-site treatment and reuse possibilities", *Environ. Technol.,* vol. 25, no. 9, pp. 997-1008, 2004.
[http://dx.doi.org/10.1080/09593330.2004.9619393] [PMID: 15515266]

[6] R. Farrency, X. Gabarrell, and J. Rieradevall, "Cost-efficiency of rainwater harvesting strategies in dense Mediterranean neighbourhoods", *Resour. Conserv. Recycling,* vol. 55, pp. 686-694, 2011.
[http://dx.doi.org/10.1016/j.resconrec.2011.01.008]

[7] L. Domènech, and D. Saurí, "A comparative appraisal of the use of rainwater harvesting in single and multifamily buildings of the Metropolitan Area of Barcelona (Spain): Social experience, drinking water savings and economic cost", *J. Clean. Prod.,* vol. 19, no. 6-7, pp. 598-608, 2011.

[http://dx.doi.org/10.1016/j.jclepro.2010.11.010]

[8] Environmental Agency UK, *Harvesting Rain water for Domestic Uses: An Information Guide,* 2010.[Accessed 2016.09.09] Available: http://publications.environment-agency.gov.uk/pdf/GEHO0108BNPN-E-E.pdf

[9] M.A. Zavala, *Development of Sustainable Systems for Water Supply and Sanitation in Rural and Urban Zones.* Technical Report GTO-2005-C02-66; Tecnológico de Monterrey: Monterrey, Mexico, 2008. (In Spanish)

[10] E. Eriksson, H.R. Andersen, T.S. Madsen, and A. Ledin, "Greywater pollution variability and loadings", *Ecol. Eng.,* vol. 35, no. 5, pp. 661-669, 2009. [http://dx.doi.org/10.1016/j.ecoleng.2008.10.015]

[11] G.P. Winward, L.M. Avery, R. Frazer-Williams, M. Pidou, P. Jeffrey, T. Stephenson, and B. Jefferson, "A study of the microbial quality of grey water and an evaluation of treatment technologies for reuse", *Ecol. Eng.,* vol. 32, no. 2, pp. 187-197, 2008. [http://dx.doi.org/10.1016/j.ecoleng.2007.11.001]

[12] A. Gross, D. Kaplan, and K. Baker, "Removal of chemical and microbiological contaminants from domestic greywater using a recycled vertical flow bioreactor (RVFB)", *Ecol. Eng.,* vol. 31, no. 2, pp. 107-114, 2007. [http://dx.doi.org/10.1016/j.ecoleng.2007.06.006]

[13] A. Duarte, D. Coelho, and N. Tomás, "Photovoltaic Integration in Buildings: A Case Study in Portugal", *Proceedings of the 4th WSEAS International Conference on Energy Planning, Energy Saving, Environmental Education (EPESE'10) and The 4th WSEAS International Conference on Renewable Energy Sources (RES '10),* pp. 119-123 Kantaoui, Sousse, Tunisia

[14] ADENE- Agência para a Energia, *Sistema Nacional de Certificação Energética e da Qualidade do Ar Interior nos Edifícios,* Portuguese, November 2011.[Accessed 2016.08.20] Available: http://www.adene.pt/pt-pt/NavegacaoDeTopo/ EnergiaEmCasa/CertificacaoEnergetica/Paginas/SCE0912-9793.aspx

[15] Energy Performance of Buildings (EPBD), *Available at: http://www.epbd-ca.eu,* 2011. [Accessed Nov. 2011]

[16] A. Silva-Afonso, and C. Pimentel-Rodrigues, "The importance of water efficiency in buildings in Mediterranean countries; The Portuguese experience – Recent Advances in Urban Planning, Cultural Sustainability and Green Development", In: *Proceedings of the IEEEAM International Conference on Urban Sustainability, Cultural Sustainability, Green Development, Green structures and Clean Cars – USCUDAR 201.* WSEAS Press: Malta, 2010, pp. 217-132.

[17] M.S. Kyvelou, "Present and future of building performance assessment tools", *Manag. Environ. Qual.,* vol. 17, no. 5, pp. 570-586, 2006. [http://dx.doi.org/10.1108/14777830610684530]

[18] M.C. Almeida, J.M. Baptista, P. Vieira, A. Moura e Silva, and R. Ribeiro, "O uso eficiente de água em Portugal no sector urbano: Que medidas e que estratégia de implementação?", *Proceedings of the Jornadas de Engenharia Civil, Universidade do Minho,* 2001

[19] A. Silva-Afonso, F. Rodrigues, and C. Pimentel-Rodrigues, "Water efficiency in buildings: Assessment of its impact on energy efficiency and reducing GHG emissions", *6th IASME/WSEAS International Conference on Energy & Environment – EE'11,* , 2011pp. 191-195

[20] M.H. Marecos do Monte, *Contributo para a utilização de águas residuais tratadas para a irrigação em Portugal.* LNEC: Lisboa, 1996.

[21] D. Butler, "A small-scale study of wastewater discharges from domestic appliances", *J. IWEM,* vol. 5, pp. 178-185, 1991.

[22] D. Butler, E. Friedler, and K. Gatt, "Characterising the quantity and quality of domestic wastewater", *Water Sci. Technol.,* vol. 31, no. 7, pp. 13-24, 1995.

[http://dx.doi.org/10.1016/0273-1223(95)00318-H]

[23] NSW, *Guideline for sewered Residential Premises (Single households).* Greywater Reuse, 2006.

[24] E. Friedler, R. Kovalio, and N.I. Galil, "On-site greywater treatment and reuse in multi-storey buildings", *Water Sci. Technol.,* vol. 51, no. 10, pp. 187-194, 2005.
[PMID: 16104421]

[25] M.A. Portugal, *Programa Nacional para o Uso Eficiente da Água.* PNUEA: Lisbon, 2001. (In Portuguese)

[26] M. Imteaz, and A. Shanableh, "Feasibility of Recycling Grey-water in Multi-Storey Buildings in Melbourne", *2nd World Sustainability Forum,* November 2012

[27] E. Friedler, "The water saving potential and the socio□economic feasibility of greywater reuse within the urban sector – Israel as a case study", *Int. J. Environ. Stud.,* vol. 65, no. 1, pp. 57-69, 2008.
[http://dx.doi.org/10.1080/00207230701846697]

[28] P. Coombes, and G. Kuczera, "Analysis of the performance of rainwater tanks in Australian capital cities", *28th International Hydrology and Water Resources Symposium,* 2003 Wollongong, NSW, Australia

[29] E.L. Villarreal, and A. Dixon, "Analysis of a rainwater collection system for domestic water supply in Ringdansen, Norrkoping, Sweden", *Build. Environ.,* vol. 40, pp. 1174-1184, 2005.
[http://dx.doi.org/10.1016/j.buildenv.2004.10.018]

[30] M.A. Imteaz, O. Adeboye, S. Rayburg, and A. Shanableh, "Rainwater harvesting potential for southwest Nigeria using daily water balance model", *Resour. Conserv. Recycling,* vol. 62, pp. 51-55, 2012.
[http://dx.doi.org/10.1016/j.resconrec.2012.02.007]

[31] E. Ghisi, A. Montibeller, and R.W. Schmidt, "Potential for potable water savings by using rainwater: an analysis over 62 cities in southern Brazil", *Build. Environ.,* vol. 41, no. 2, pp. 204-210, 2006.
[http://dx.doi.org/10.1016/j.buildenv.2005.01.014]

[32] E. Ghisi, D.L. Bressan, and M. Martini, "Rainwater tank capacity and potential for potable water savings by using rainwater in the residential sector of southeastern Brazil", *Build. Environ.,* vol. 42, no. 4, pp. 1654-1666, 2007.
[http://dx.doi.org/10.1016/j.buildenv.2006.02.007]

[33] E. Ghisi, D.F. Tavares, and V.L. Rocha, "Rainwater harvesting in petrol stations in Brasilia: potential for potable water savings and investment feasibility analysis", *Resour. Conserv. Recycling,* vol. 54, pp. 79-85, 2009.
[http://dx.doi.org/10.1016/j.resconrec.2009.06.010]

[34] S. Muthukumaran, K. Baskaran, and N. Sexton, "Quantification of potable water savings by residential water conservation and reuse –a case study", *Resour. Conserv. Recycling,* vol. 55, pp. 945-952, 2011.
[http://dx.doi.org/10.1016/j.resconrec.2011.04.013]

[35] M.A. Imteaz, A. Shanableh, A. Rahman, and A. Ahsan, "Optimization of rainwater tank design from large roofs: A case study in Melbourne, Australia", *Resour. Conserv. Recycling,* vol. 55, no. 11, pp. 1022-1029, 2011.
[http://dx.doi.org/10.1016/j.resconrec.2011.05.013]

[36] J.S. Mun, and M.Y. Han, "Design and operational parameters of a rooftop rainwater harvesting system: definition, sensitivity and verification", *J. Environ. Manage.,* vol. 93, no. 1, pp. 147-153, 2012.
[http://dx.doi.org/10.1016/j.jenvman.2011.08.024] [PMID: 22054581]

[37] R.A. Corbitt, *Standard handbook of environmental engineering.* 2nd ed McGraw-Hill: New York, 1998.

[38] J.C. Chilton, G.G. Maidment, D. Marriott, A. Francis, and G. Tobias, "Case study of a Rainwater harvesting system in a commercial building with a large roof", *Urban Water,* vol. 1, no. 4, pp. 345-354, 2000. [http://dx.doi.org/10.1016/S1462-0758(00)00032-7]

[39] ANQIP, *ETA 0701 - Sistemas de Abastecimento de Águas Pluviais em Edifícios (SAAP).* Portuguese National Association for Quality in Building Services - ANQIP, Version 7.: portuguese, 2009.

[40] BSI, *BS 8515:2009 Rainwater harvesting systems - Code of practice.* British Standard Institution: UK, 2009.

[41] fbr, "Translation of DIN 1989-1:2001-10. Rainwater harvesting systems – Part 1: Planning, installation, operation and maintenance", In: *Fachvereinigung Betriebs - und Regenwassernutzung e.V.* Darmstadt, Germany, 2002, p. 34.

[42] C. Santos, and F. Taveira-Pinto, "Analysis of different criteria to size rainwater storage tanks using detailed methods", *Resour. Conserv. Recycling,* vol. 71, no. 1-6, 2013.

[43] C. Matos, C. Santos, S. Pereira, I. Bentes, and M. Imteaz, "Rainwater storage tank sizing: case study of a comercial building", *International journal of sustainable built environment,* vol. 2, no. 2, 2013. [http://dx.doi.org/10.1016/j.ijsbe.2014.04.004]

[44] European Comission, "Guide to COST-BENEFIT ANALYSIS of investment projects Structural Funds, Cohesion Fund and Instrument for Pre-Accession", *Directorate General Regional Policy,* 2008. http://ec.europa.eu/regional_policy/sources/docgener/ guides/cost/guide2008_en.pdf

[45] C. Santos, *"Otimização Ambiental do Uso de Água em Edifícios"*, (Environamental improvement of water use in buildings), Portuguese, Faculdade de Engenharia da Universidade do Porto, 2012.

[46] U. Pinto, and B. Maheshwari, *Issues and Challenges of Greywater Reuse for Irrigation in Australia — A Case Study of Western Sydney Region.* University of Western Sydney: Sydney, Australia, 2007.

[47] E. Friedler, O. Lahav, H. Jizhaki, and T. Lahav, "Study of urban population attitudes towards various wastewater reuse options: Israel as a case study", *J. Environ. Manage.,* vol. 81, no. 4, pp. 360-370, 2006. [http://dx.doi.org/10.1016/j.jenvman.2005.11.013] [PMID: 16574307]

[48] M. Po, J.D. Kaercher, and B.E. Nancarrow, "Literature Review Of Factors Influencing Public Perceptions Of Water Reuse", In: *CSIRO Land and Water.* Technical Report 54/03, December 2003.

[49] United Nations Environment Programme (UNEP), "An Environmentally Sound Approach for Sustainable Urban Water Management: An Introductory Guide for Decision-Makers", *Division of Technology, Industry and Economics,* 2016.(Last access: 2016.09.09) Available at: http://www.unep.or.jp/ietc/publications/urban/urbanenv-2/5.asp

[50] World Health Organization (WHO), *Overview of greywater management Health considerations.* Regional Office for the Eastern Mediterranean Centre for Environmental Health activities: Amman, Jordan, 2006.

[51] C. Santos, C. Matos, and F. Taveira-Pinto, "A comparative study of greywater from domestic and public buildings", *Water Science and Technology: Water Supply,* vol. 14, no. 1, pp. 135-141, 2014.

[52] A.M. Dixon, D. Butler, A. Fewkes, and M. Robinson, "Measurement and Modelling of Quality Changes in Stored Untreated Grey Water", *Urban Water,* vol. 1, no. 4, pp. 293-306, 2000. [http://dx.doi.org/10.1016/S1462-0758(00)00031-5]

[53] A.M. Dixon, D. Butler, and A. Fewkes, "Guidelines for Greywater Re□Use: Health Issues", *Water Environ. J.,* vol. 13, no. 5, pp. 322-326, 1999. [http://dx.doi.org/10.1111/j.1747-6593.1999.tb01056.x]

[54] I. Lança, and A. Silva-Afonso, "Rainwater storage and reuse. A safety case study in a groundwater storage installation", In: *Anais do CIB W062 2011.* Water Supply and Drainage for Buildings: Aveiro, Portugal, 2011, pp. 293-300.

[55] Adapted from ABNT, *EPA (2004), Australia (2000), Canada (1992),* 2007. [Accessed: Aug 2016] Available at: http://www.crea-pr.org.br/crea3/blog/cadernos/usoreuso.pdf

[56] A. Silva-Afonso, J. Bernardo, and C. Pimentel-Rodrigues, "Implications of reduced flow volumes in building drainage: An experimental study", In: *Anais do CIB W062 2013 – Water Supply and Drainage for Buildings.* Nagano, 2013.

[57] A. Silva-Afonso, and C. Pimentel Rodrigues, "Measures of water efficiency in buildings. Comfort and performance issues", *in XXXIX IAHS – World Congress on Housing Science (Changing needs, adaptive buildings, smart cities),* 2013

[58] A. Silva-Afonso, and C. Pimentel Rodrigues, "Water efficiency of products. The Portuguese system of certification and labeling", *Journal AWWA, American Water Works Association, Denver, USA,* vol. 102, no. 2, pp. 52-56, 2010.

[59] ANQIP, "ETA 0803 – Rótulos de eficiência hídrica de produtos. Características e condições de utilização", *Portuguese National Association for Quality in Building Services – ANQIP,* vol. Version 4, portuguese, . Available at: http://www.anqip.com/images/stories/ETA_0803_v.4_novo.1.pdf

[60] ANQIP, "ETA 0804 – Especificações para a Atribuição de Rótulos de Eficiência Hídrica ANQIP a Autoclismos de Bacias de Retrete", *Portuguese National Association for Quality in Building Services – ANQIP,* vol. Version 4, portuguese, . Available at: http://www.anqip.com/images/stories/ ETA_0804_v.4_novo.1.pdf

[61] A.M. Monteiro, *"Estudo do impacte provocado pela reutilização de água cinzenta in situ nas Redes de Drenagem de Águas Residuais Domésticas"*, Msc Thesis in Portuguese, Universidade de Trás-o-Montes e Alto Douro, 2012.

[62] *Decreto-Lei nº 23/95 de 23 de janeiro, Ministério das Obras Públicas, Transportes e Comunicações, Diário da República-I Série-B nº 194.* Portuguese, 1995.

[63] A. Silva-Afonso, and I. Lança, "Safe Water. The importance of risk prevention. The situation in Portugal", In: *World Plumbing Revue.* Notting Hill, Australia, 2008.

[64] *Real Decreto 1620/2007, de 7 de diciembre, por el que se establece el régimen jurídico de la reutilización de las aguas depuradas.* Gobierno de España, Ministerio de la Presidencia: Madrid, Spain, 2007.

[65] ANQIP, "*ETA 0905 - Sistemas Prediais de Reutilização e Reciclagem de Águas Cinzentas (Sprac)*", *Portuguese National Association for Quality in Building Services – ANQIP,* vol. Version 1, portuguese, 2007. Available at: http://www.anqip.com/images/stories/ETA_0905_v.1_novo.pdf

[66] *Council Directive 76/160/EEC of 8 December 1975 concerning the quality of bathing water as amended by Council Directive 91/692/EEC (further amended by Council Regulation 1882/2003/EC), and Council Regulation 807/2003/EC, European Council,* 1975. Available at: http://rod.eionet.europa.eu/instruments/204.

[67] E. Nolde, "Regulatory framework and standards for rainwater harvesting and greywater recycling", In: *MEDA Water Programm (Zero-M) and Fachvereinigung Betriebs - und Regenwassernutzung e.V.*

[68] ANQIP, "*ETA 0702* – Certificação de Sistemas de Aproveitamento de Águas Pluviais", *Portuguese National Association for Quality in Building Services – ANQIP,* vol. Version 5, . Available at: http://www.anqip.com/images/stories/ETA_0702_v.5_novo.pdf

[69] ANQIP, "*ETA 0906* – Certificação de Sistemas Prediais de Reutilização e Reciclagem de Águas Cinzentas", *Portuguese National Association for Quality in Building Services – ANQIP,* vol. Version 1, . Available at: http://www.anqip.com/images/stories/ETA_0906_v.2_novo.pdf

[70] European Union, *Blueprint to safeguard Europe's water resources,* 2011. Available at: http://ec.europa.eu/environment/ water/pdf/blueprint_leaflet.pdf.

[71] A. Silva-Afonso, C. Pimentel Rodrigues, and I. Lança, "Reuse and recycling of grey water in buildings. The Portuguese approach", In: *Anais do CIB W062 2011.* Water Supply and Drainage for Buildings: Aveiro, Portugal, 2011, pp. 56-61.

[72] A. Silva-Afonso, "Building rainwater harvesting systems. Doubts and certainties", In: *Anais do CIB W062 2009.* Water Supply and Drainage for Buildings: Dusseldorf, Germany, 2009, pp. 238-246.

[73] A. Silva-Afonso, C. Pimentel Rodrigues, and I. Lança, "Reuse of grey water in buildings. Technical, Environmental and health concerns", *in XXXVIII IAHS – World Congress on Housing Science (Visions for the future of housing – Mega cities),* 2012

[74] BSI, *BS 8525-1:2010 Greywater systems - Code of practice.* British Standard Institution: UK, 2010.

[75] fbr, "Greywater recycling: Planning fundamentals and operation information", In: *Fachvereinigung Betriebs - und Regenwassernutzung e.V.* Darmstadt, Germany, 2005.

SUBJECT INDEX

E

F

G

H

I

S

T

U

V

W

Y

www.ingramcontent.com/pod-product-compliance
Lightning Source LLC
LaVergne TN
LVHW070933210926
848482LV00003B/234
* 9 7 8 1 6 8 1 0 8 4 8 4 8 *